Yingkui Li and Michael A. Urban (Eds.)

Water Resource Variability and Climate Change

MDPI

This book is a reprint of the Special Issue that appeared in the online, open access journal, *Water* (ISSN 2073-4441) from 2015–2016, available at:

http://www.mdpi.com/journal/water/special_issues/water-resource-variability

Guest Editors
Yingkui Li
Department of Geography University of Tennessee Knoxville
USA

Michael A. Urban
Department of Geography University of Missouri Columbia
USA

Editorial Office
MDPI AG
St. Alban-Anlage 66
Basel, Switzerland

Publisher
Shu-Kun Lin

Managing Editor
Cherry Gong

1. Edition 2016

MDPI • Basel • Beijing • Wuhan • Barcelona • Belgrade

ISBN 978-3-03842-244-0 (Hbk)
ISBN 978-3-03842-230-3 (PDF)

Table of Contents

Vivek Shandas, Rosa Lehman, Kelli L. Larson, Jeremy Bunn and Heejun Chang
Stressors and Strategies for Managing Urban Water Scarcity: Perspectives from the Field

List of Contributors

Mukand S. Babel Water Engineering and Management, Asian Institute of Technology, Pathumthani 12120, Thailand.

Thomas Bosshard Swedish Meteorological and Hydrological Institute, Norrköping 60176, Sweden.

Willibroad Gabila Buma Department of Civil and Environmental Engineering, Dongguk University, Seoul 04620, Korea.

Jeremy Bunn Herrera Inc., Seattle, WA 98121, USA.

Wen-Ting Chan Department of Civil Disaster Prevention Engineering, National United University, Miaoli 36063, Taiwan.

Heejun Chang Department of Geography, Portland State University, Portland, OR 97201-0751, USA.

Yung-Ming Chen National Science and Technology Center for Disaster Reduction, New Taipei City 23143, Taiwan.

Chao-Tzuen Cheng National Science and Technology Center for Disaster Reduction, New Taipei City 23143, Taiwan.

Hiroyuki Cho Faculty of Agriculture, Saga University, 1 Honjo-machi, Saga 840-8502, Japan.

Kyung Hwa Cho School of Urban and Environmental Engineering, Ulsan National Institute of Science and Technology, Ulsan 689-798, Korea.

Yuan Gao Key Laboratory of Water and Sediment Sciences, Ministry of Education, Beijing 100871, China; College of Environmental Sciences and Engineering, Peking University, Beijing 100871, China.

Huadong Guo Key Laboratory of Digital Earth Science, Institute of Remote Sensing and Digital Earth, Chinese Academy of Sciences, No. 9 Dengzhuang South Road, Haidian District, Beijing 100094, China.

Cornelia Hesse Potsdam-Institute for Climate Impact Research, Post Box 601203, Potsdam 14412, Germany.

Baodeng Hou State Key Laboratory of Simulation and Regulation of Water Cycle in River Basin, China Institute of Water Resources and Hydropower Research, 1-A Fuxing Road, Haidian District, Beijing 100038, China.

Yicheng Wang Ya Huang College of Civil Engineering and Architecture, Guangxi University, Nanning 530004, China; State Key Laboratory of Simulation and Regulation of Water Cycle in River Basin, China Institute of Water Resources and Hydropower Research, 1-A Fuxing Road, Haidian District, Beijing 100038, China.

Yutaka Ichikawa Department of Civil and Earth Resources Engineering, Kyoto University, C1, Kyoto-Daigaku-Katsura, Nishikyo-ku, Kyoto-shi, Kyoto 615-8540, Japan.

Hiroshi Ishidaira International Research Center for River Basin Environment (ICRE), University of Yamanashi, Takeda 4-3-11, Kofu, Yamanashi 400-8511, Japan.

Shaofeng Jia Institute of Geographic Science and Natural Resources Research, and Key Laboratory of Water Cycle and Related Land Surface Processes, Chinese Academy of Sciences, Beijing 100101, China.

Chong Jiang College of Global Change and Earth System Science, Beijing Normal University, Beijing 100875, China; Joint Center for Global Change Studies, Beijing 100875, China; Key Laboratory of Regional Eco-Process and Function Assessment and State Environment Protection; State Key Laboratory of Environmental Criteria and Risk Assessment, Chinese Research Academy of Environmental Sciences, Beijing 100012, China.

Gilho Kim Department of Hydro Science and Engineering, Korea Institute of Civil Engineering and Building Technology, Goyang-si, Gyeonggi-do 10223, Korea.

Hanna Kim Water Institute, 1689beon-gil 125, Yuseong-Daero, Yuseong-gu, Daejeon 305-730, Korea.

Hung Soo Kim Department of Civil Engineering, Inha University, Incheon 22212, Korea.

Joon Ha Kim Department of Environmental Science and Engineering, Gwangju Institute of Science and Technology, Gwangju 500-712, Korea.

Soojun Kim Columbia Water Center, Columbia University, New York, NY 10027, USA.

Valentina Krysanova Potsdam-Institute for Climate Impact Research, Post Box 601203, Potsdam 14412, Germany.

Jaewon Kwak Forecast and Control Division, Nakdong River Flood Control Office, Busan 49300, Korea.

Kelli L. Larson Schools of Geographical Sciences and Urban Planning and Sustainability, Arizona State University, AZ 85287-5302, USA.

Sang-Il Lee Department of Civil and Environmental Engineering, Dongguk University, Seoul 04620, Korea.

Seungwon Lee Environmental and Plant Engineering Research Institute, Korea Institute of Civil Engineering and Building Technology (KICT), 283, Goyangdae-Ro Ilsanseo-Gu, Goyang-Si, Gyeonggi-Do 10223, Korea.

Rosa Lehman Toulan School of Urban Studies and Planning, Portland State University, Portland, OR 97201-0751, USA.

Daiqing Li Key Laboratory of Regional Eco-Process and Function Assessment and State Environment Protection, Chinese Academy of Environmental Sciences, Beijing 100012, China; State Key Laboratory of Environmental Criteria and Risk Assessment, Chinese Research Academy of Environmental Sciences, Beijing 100012, China.

Fen Li Key Laboratory of Regional Eco-Process and Function Assessment and State Environment Protection, Chinese Academy of Environmental Sciences, Beijing 100012, China; State Key Laboratory of Environmental Criteria and Risk Assessment, Chinese Research Academy of Environmental Sciences, Beijing 100012, China.

Qingting Li Key Laboratory of Digital Earth Science, Institute of Remote Sensing and Digital Earth, Chinese Academy of Sciences, No. 9 Dengzhuang South Road, Haidian District, Beijing 100094, China.

Tianhong Li Key Laboratory of Water and Sediment Sciences, Ministry of Education, Beijing 100871, China; College of Environmental Sciences and Engineering, Peking University, Beijing 100871, China.

Yingkui Li Department of Geography, University of Tennessee, Knoxville, TN 37996, USA.

Mayzonee Ligaray School of Urban and Environmental Engineering, Ulsan National Institute of Science and Technology, Ulsan 689-798, Korea.

Jun-Jih Liou National Science and Technology Center for Disaster Reduction, New Taipei City 23143, Taiwan.

Wen-Cheng Liu Taiwan Typhoon and Flood Research Institute, National Applied Research Laboratories, Taipei 10093, Taiwan; Department of Civil Disaster Prevention Engineering, National United University, Miaoli 36063, Taiwan.

Chuiyu Lu State Key Laboratory of Simulation and Regulation of Water Cycle in River Basin, China Institute of Water Resources and Hydropower Research, 1-A Fuxing Road, Haidian District, Beijing 100038, China.

Linlin Lu State Key Laboratory of Remote Sensing Science, Institute of Remote Sensing and Digital Earth, Chinese Academy of Sciences, Beijing 100101, China; Key Laboratory of Digital Earth Science, Institute of Remote Sensing and Digital Earth, Chinese Academy of Sciences, No. 9 Dengzhuang South Road, Haidian District, Beijing 100094, China.

Rashid Mahmood Institute of Geographic Science and Natural Resources Research, and Key Laboratory of Water Cycle and Related Land Surface Processes, Chinese Academy of Sciences, Beijing 100101, China.

Shinya Nakamura Faculty of Agriculture, University of the Ryukyus, 1 Senbaru, Nishihara-cho, Okinawa 903-0213, Japan.

Tamotsu Nakandakari Faculty of Agriculture, University of the Ryukyus, 1 Senbaru, Nishihara-cho, Okinawa 903-0213, Japan.

Francis Ndamani Graduate School of Engineering, Kochi University of Technology, 2-22 Eikokuji, Kochi City, Kochi 780-8515, Japan.

Shaowei Ning International Research Center for River Basin Environment (ICRE), University of Yamanashi, Takeda 4-3-11, Kofu, Yamanashi 400-8511, Japan.

Ken Okamoto United Graduate School of Agricultural Sciences, Kagoshima University, 1-21-24 Korimoto, Kagoshima-shi, Kagoshima 890-0065, Japan.

Jonas Olsson Swedish Meteorological and Hydrological Institute, Norrköping 60176, Sweden.

Shota Ootani Eight-Japan Engineering Consultants Inc.; 33-11 Honcho 5 Chome, Nakano-ku, Tokyo 164-0012, Japan.

Jungsool Park Forecast and Control Division, Nakdong River Flood Control Office, Busan 49300, Korea.

Ilias G. Pechlivanidis Swedish Meteorological and Hydrological Institute, Norrköping 60176, Sweden.

Kazuhito Sakai Faculty of Agriculture, University of the Ryukyus, 1 Senbaru, Nishihara-cho, Okinawa 903-0213, Japan.

Jae Young Seo Department of Civil and Environmental Engineering, Dongguk University, Seoul 04620, Korea.

Vivek Shandas Toulan School of Urban Studies and Planning, Portland State University, Portland, OR 97201-0751, USA.

Devesh Sharma Department of Environmental Science, Central University of Rajasthan, Kishangarh, Dist-Ajmer, Rajasthan 305817, India.

K.C. Sharma Department of Environmental Science, Central University of Rajasthan, Kishangarh, Dist-Ajmer, Rajasthan 305817, India.

Vijay P. Singh Department of Biological and Agricultural Engineering and Zachry Department of Civil Environmental Engineering, Texas A & M University, College Station, TX 77843-2117, USA.

Suthipong Sthiannopkao Department of Environmental Engineering, Dong-A University, Busan 604-714, Korea.

Yue Sui Key Laboratory of Digital Earth Science, Institute of Remote Sensing and Digital Earth, Chinese Academy of Sciences, No. 9 Dengzhuang South Road, Haidian District, Beijing 100094, China.

Parmeshwar Udmale International Research Center for River Basin Environment (ICRE), University of Yamanashi, Takeda 4-3-11, Kofu, Yamanashi 400-8511, Japan.

Cuizhen Wang Department of Geography, University of South Carolina, Columbia, SC 29208, USA.

Hejia Wang State Key Laboratory of Simulation and Regulation of Water Cycle in River Basin, China Institute of Water Resources and Hydropower Research, 1-A Fuxing Road, Haidian District, Beijing 100038, China; Department of Hydraulic Engineering, Tsinghua University, Haidian District, Beijing 100084, China.

Jianhua Wang State Key Laboratory of Simulation and Regulation of Water Cycle in River Basin, China Institute of Water Resources and Hydropower Research, 1-A Fuxing Road, Haidian District, Beijing 100038, China.

Yicheng Wang State Key Laboratory of Simulation and Regulation of Water Cycle in River Basin, China Institute of Water Resources and Hydropower Research, 1-A Fuxing Road, Haidian District, Beijing 100038, China.

Tsunemi Watanabe School of Economics and Management, Kochi University of Technology, 2-22 Eikokuji, Kochi City, Kochi 780-8515, Japan.

Hsiao-Ping Wei National Science and Technology Center for Disaster Reduction, New Taipei City 23143, Taiwan.

Weihua Xiao State Key Laboratory of Simulation and Regulation of Water Cycle in River Basin, China Institute of Water Resources and Hydropower Research, 1-A Fuxing Road, Haidian District, Beijing 100038, China.

Keh-Chia Yeh Department of Civil Engineering in National Chiao Tung University, Hsinchu 300, Taiwan.

Linbo Zhang Key Laboratory of Regional Eco-Process and Function Assessment and State Environment Protection, Chinese Academy of Environmental Sciences, Beijing 100012, China; State Key Laboratory of Environmental Criteria and Risk Assessment, Chinese Research Academy of Environmental Sciences, Beijing 100012, China.

About the Guest Editors

Yingkui Li, PhD, is Associate Professor in Geography at the University of Tennessee. He has broad research interests in reconstructing paleo-climate and environments, assessing climate and human impacts on the environment, and applying LiDAR (Light Detection and Ranging) technique to quantify Earth surface processes. He has investigated the pattern and potential drivers of dramatic lake changes on the Tibetan Plateau since the 1970s and assessed the impact of land use change/urbanization on surface runoff and water quality for different watersheds in the United States and China. He has also applied cosmogenic nuclides and optically stimulated luminescence dating techniques in paleo-climate and environmental reconstruction on the Tibetan Plateau, Tian Shan in Central Asia, and highlands in Costa Rica. More recently, he is developing and applying terrestrial LiDAR technique to quantify Earth surface processes, such as hillslope erosion, gully erosion, glacier retreat, and terrain modeling. His research has been supported by the National Science Foundation, the National Natural Science Foundation of China, Chinese Academy of Sciences, and the University of Tennessee.

Michael Urban, Ph.D. is Associate Professor and Chair of Geography at the University of Missouri-Columbia whose research specialty is geomorphology, water resources and environmental management. The focus of much of his work has been on how river systems and water resources have changed in response to patterns of climate, human behavior, and human impacts on the environment over the past century. Recent work involves modeling 20th and 21st century climatic water budgets and examining the nature of human–environment interaction on changing landscapes. He has attended the United Nations Climate Change Conference as a representative of the Association of American Geographers and is interested in how changing climates will affect water resources and environmental management over local and regional scales. He has also published on a number of other different but related topics including how perception of natural environments influences public policy, the ethics and philosophy behind environmental restoration, and the effect of agriculture on the physical integrity of river channels and stream networks. He was a Fulbright scholar from 2006–2007 at Northeastern University in Shenyang, China and has lectured in Universities throughout China and South Korea.

Preface to "Water Resource Variability and Climate Change"

Climate itself represents one of the most significant external forcing mechanisms with respect to surficial environmental processes. Climates that are systematically changing in response to anthropogenic warming necessitate a closer examination of global and regional water cycling and surficial and subsurface water availability. These changes to surficial water are already increasing vulnerabilities of ecosystems and of human society. Understanding the ways in which climate change affects water resource variability is essential to the broader policy goal of sustainable development in different parts of the world. The sixteen papers included in this Special Issue of "Water Resource Variability and Climate Change" address three broad perspectives: (1) the quantification of water resource variability altered by changing climates using remote sensing assessment, meteorological station-based observational datasets, and tree-ring record reconstruction; (2) modeling climate and hydrology to simulate the various impacts on water resource variability; and (3) an evaluation of social perceptions and adaptation strategies in the face of unstable water resource variability. The findings and methods presented in this collection of papers provide important contributions to the increased study and awareness of climate change on water resources.

Yingkui Li and Michael Urban
Guest Editors

Water Resource Variability and Climate Change

Yingkui Li and Michael A. Urban

Abstract: A significant challenge posed by changing climates is how water cycling and surficial and subsurface water availability will be affected at global and regional scales. Such alterations are critical as they often lead to increased vulnerability in ecosystems and human society. Understanding specifically how climate change affects water resource variability in different locations is of critical importance to sustainable development in different parts of the world. The papers included in this special issue focus on three broad perspectives associated with water resource variability and climate change. Six papers employ remote sensing, meteorological station-based observational data, and tree-ring records to empirically determine how water resources have been changing over historical time periods. Eight of the contributions focus on modeling approaches to determine how known processes are likely to manifest themselves as climate shifts over time. Two others focus on human perceptions and adaptation strategies in the midst of unstable or unsettled water availability. The findings and methods presented in this collection of papers provide important contributions to the increased study and awareness of climate change on water resources.

Reprinted from *Water*. Cite as: Li, Y.; Urban, M.A. Water Resource Variability and Climate Change. *Water* **2016**, *8*, 384.

1. Introduction

Climate change and increased anthropogenic pressure on earth–atmosphere interactions affect water quantity, quality, and water-related processes, such as sediment yield, on local, regional, and global scales [1–3]. Recent decades have seen continuously increasing temperatures in most parts of the world, and changes in precipitation patterns have increased the frequency of extreme climate events such as drought and flooding [4]. The impact of changing baseline conditions coupled with increased variability can be especially complicated in regions with rapid changes in population, land development (especially urbanization), and economic disruptions. While public discussions often focus more on temperature than water availability, ecosystems and human society are highly vulnerable to water stress [5–8]. Understanding the mechanisms and geographic patterns by which anthropogenic climate change is impacting water resource variability is of critical importance to sustainable development, environmental management, and human health.

A variety of approaches have been used to examine the relationships between atmospheric variability and surficial water resources. Instrumental data collected from meteorological and hydrological gauging stations can be used to investigate altered hydrologic regimes over the timespan of several decades to a few centuries in certain areas. Relatively long-term (hundreds to thousands of years) climate and environmental records can be reconstructed using various proxies such as tree rings, sediment cores, ice cores, and landform features [9,10]. More recently, the availability of various remote sensing datasets, such as Landsat/MODIS (Moderate Resolution Imaging Spectroradiometer) imagery, ICESAT (Ice, Clouds, and Land Elevation Satellite) altimetry, GRACE (Gravity Recovery and Climate Experiment) gravity, and LiDAR (Light Detection And Ranging) measurements have facilitated remote sensing-based approaches to quantifying water resource changes [11–13]. Computational modeling approaches, ranging from global circulation models (GCMs) to regional or watershed hydrological models, are serving to simulate and forecast the projected nature of climate variability on water resources [14,15]. Social scientists have also been investigating how groups or local communities perceive the impacts of climate change and climate vulnerability in order to implement better adaptation practices and sustainable development in coping with changing water resources of different regions [16,17].

The papers included in this special issue address three broad perspectives associated with water resource variability and climate change: (1) the *quantification* of water resource variability altered by changing climates using remote sensing assessment, meteorological station-based observational datasets, and tree-ring record reconstruction; (2) the *simulation* of such impacts on water resource variability using modeling approaches; and (3) evaluating social *perceptions and adaptation strategies* in the face of unstable water resource variability. The following section summarizes the individual contributions within each perspective.

2. Contributions

Six of the papers assess the various impacts of climate change on water resources using a variety of datasets, empirical observations, and proxies. Li et al. [18] examine surface area fluctuations occurring in 10 major lakes in the arid province of Xinjiang, China, from 2000 to 2014 using MODIS time series imagery. The authors develop a classification method to accommodate varied spectral characteristics of water pixels and derived water bodies for April, July, and September in each year for 10 major lakes (>100 km^2) in the study area. Lakes in the lowland (close to urban and agriculture areas) showed a shrinking trend, while mountain lakes have diverse changing patterns (some shrinking, some expanding), and lakes on the Tibetan Plateau exhibited significant expanding trends. By observing varied patterns of lake surface changes across the region, the authors conclude that observed lake expansion is likely driven by rising temperature, leading to accelerated melting of snow and glaciers in high mountains and on the Tibetan Plateau, and increased precipitation in this region (especially in 2010), whereas the shrinking of some lakes is likely related to anthropogenic utilization based on agricultural and industrial needs.

Ning et al. [19] analyze recent changes in water resources and grassland in the Hulun Lake region, a semi-arid region in northeastern China, using monthly GRACE and Tropical Rainfall Measuring Mission (TRMM) data. Results indicate decreasing trends in overall water storage and precipitation between 2002 and 2007, followed by increasing trends in the period from 2007 to 2012. Water storage trends are mainly correlated to precipitation and temperature patterns. As a result, a large proportion of grassland recovered to its normal state in 2008–2012, and only a small proportion of grassland (16.5% of the study area) is classified as degraded. The authors conclude that degraded grassland areas in the region are more vulnerable to climate variability and require protective strategies to prevent further degradation.

Buma et al. [20] assess observed changes in hydrological conditions of Lake Chad basin based on the total water storage (TWS) derived from GRACE, lake levels taken from satellite altimetry, and water fluxes and soil moisture obtained from the Global Land Data Assimilation System (GLDAS). The authors observe a similar pattern between TWS and lake level changes and subsurface water volume changes. The derived values for subsurface water volume changes are found to be consistent with groundwater outputs calculated from the WaterGAP Global Hydrology Model (WGHM). By utilizing recently developed remote sensing datasets, this study provides an alternative means of generating information for the management of water resources in the Lake Chad basin.

Jiang et al. [21] summarize the changing patterns, causes, and implications of surface water discharge and sediment load in Chinese rivers from 1956 to 2012 based on monthly hydrological and daily meteorological data obtained from 725 rain gauge stations across the country. Numerous patterns can be observed during this period. Streamflow discharges manifest a decreasing trend, a relatively stable state, and an increasing trend within northern, southern, and western China, respectively. Excepting the Lancang River and Yarlung Zangbo River basins, sediment loads in most Chinese river basins show gradually decreasing trends, especially after 2000. Although patterns of streamflow and sediment load are affected by the interaction of varied meso-scale climate systems—including East and South Asian monsoons and westerlies—the authors determine that water consumption for industrial and residential purposes, soil, and water conservation engineering, hydraulic engineering, and land surface changes induced by other factors are likely the main causes of observed patterns of streamflow and sediment reduction.

Wang et al. [22] investigate the impact that climate change has had on the duration of flood seasons in the Fenhe River, China, from 1957 to 2014, based on daily precipitation data from 14 meteorological stations in the basin and an analysis of the variations in the onset and retreat dates of yearly flood seasons. The results show that the observed duration of the flood season has been extended since 1975. In particular, the onset of floods has advanced 15 days, although the retreat date is relatively stable. Based on these results, the authors recommend corresponding measures to adapt to the flood season variations.

Kwak et al. [23] conduct a drought analysis using a long-term streamflow record reconstructed using tree ring indices within the Sacramento Basin, California, USA. By first identifying annual streamflow patterns of the Sacramento River from 1560 to 1871 and then analyzing the hydrological drought return period in this river basin, the authors argue that drought with a 20-year return period can be considered a critical indicator of drought for water shortages in the Sacramento River basin.

Eight of the papers aim to simulate the impact of climate change on water resource variability using various climatological and hydrological models. Pechlivanidis et al. [24] investigate the impact of changing climates have on the hydro-climatology of the Indian subcontinent by comparing current and projected future water fluxes from three RCP (Representative Concentration Pathway) scenarios (RCP2.6, RCP4.5 and RCP8.5). These results are used to depict expected changes in the annual flow cycles of three major rivers from different hydro-climatic regions, while acknowledging that conclusions can be significantly influenced by statistical uncertainty embedded in the RCP scenarios. Based on this study, the models project a gradual increase in temperature and uneven changes (ranging from -20% to $+50\%$) in long-term average precipitation and evapotranspiration. Potential surface runoff is also expected to change anywhere from -100% to $+100\%$. The analysis of annual cycles for the three selected regions show that the impact of climate change on discharge and evapotranspiration varies between seasons, and the magnitude of change is primarily dependent on the hydro-climatic gradient in different regions.

Li and Gao [25] simulate the impact of various precipitation change scenarios on runoff and sediment yield in a hilly-gullied watershed typical of the Loess Plateau in China using the Soil and Water Assessment Tool (SWAT). This study indicates that runoff and sediment yield both increase with increasing precipitation, while the variation in sediment yield is more sensitive to smaller rainfall events. The authors determine that under these conditions, annual runoff and sediment yield fluctuate greatly and the magnitude of the variations was especially amplified when precipitation increased by 20%. Overall escalation in runoff and sediment caused by increased precipitation is greater than corresponding decreases coincident with reduced precipitation, and runoff is the more sensitive variable compared to sediment yield.

Ligaray et al. [26] assess the hydrological response of climate change in the Chao Phraya River Basin, Thailand. Streamflow variations were simulated using a combination of SWAT and meteorological data from 2003 to 2011 for various climate sensitivity and greenhouse gas emission scenarios. Simulation results reveal that streamflow variations correspond to the changes in rainfall totals and intensity, while increased air temperature likely leads to future water shortages. The simulation also suggests that high CO_2 concentration drives plant responses that may lead to a dramatic increase in streamflow. Specifically, increased streamflow variations to 6.8%, 41.9%, and 38.4% were simulated for the three greenhouse gas emission scenarios (A1B, A2, B1) in the reference period of 2003–2011.

Mahmood et al. [27] investigate the potential impacts of climate change on the water resources of the Kunhar River basin, Pakistan under A2 and B2 climate scenarios. Using the HEC-HMS (Hydrologic Engineering Center's Hydraulic Modeling System) hydrological model, the authors simulate streamflow for the periods: 2011–2040, 2041–2070, and 2071–2099, and compare them with the baseline period (1961–1990) to explore changes in different streamflow variables. The results indicate an overall increase in mean annual flow projected under both A2 and B2 scenarios, but with a high degree of variability. Stream discharge increases mainly in summer and autumn, but decreases throughout the spring and winter months. High and median flows are predicted to increase, with peak discharges shifting from June to July, while low flow conditions are projected to decrease. The Kunhar basin will face a higher degree of variability—both more floods and droughts—by the end of the 21st century, due to the projected increase in high flow, the decrease in low flow, and greater variations in peak discharges. This study highlights key impacts of climate change on water resources to help develop suitable policies for water resource use and management in this river basin.

Hesse and Krysanova [28] simulate the impacts of climatic shifts and changing management practices on water quality and in-stream processes in the Elbe River Basin using a semi-distributed watershed model (SWIM) with implemented in-stream nutrient (N+P) turnover and algal growth processes. The set of modeled climate scenarios show a projected increase in temperature (+3 °C) and precipitation (+57 mm) on average until the end of the century, leading to varied changes in discharge (+20%), nutrient loads (NO_3-N: 5%; NH_4-N: 24%; PO_4-P: +5%), phytoplankton biomass (4%), and dissolved oxygen concentration (5%) in the Elbe River Basin. The authors utilize the model to examine the ways in which changes in climatic variables fundamentally impact the ways by which land use and nutrients are managed to reduce nutrient emissions to the river.

Liu and Chan [29] assess impacts on water quality in the Danshuei River estuarine system in northern Taiwan using a coupled three-dimensional hydrodynamic and water quality model driven by changes in climatic variables. The model is calibrated and validated using observed data and then applied to simulate water quality projections under various climate change scenarios. Results indicate that dissolved oxygen concentrations are likely to significantly decrease in the Danshuei, whereas nutrients will increase in response to expected climate changes. In particular, dissolved oxygen concentrations will be reduced to less than 2 mg/L in the main stream, failing to meet accepted water quality standards. This study suggests an appropriate strategy for effective water quality management in estuarine systems such as the Danshuei is needed to adapt to the water quality changes likely to accompany anthropogenic climate change.

Wei et al. [30] estimate flood risk that is likely to occur under the heightened hydrologic variability driven by climate change in the Tsengwen River Basin, Taiwan, using a SOBEK model (Deltares, The Netherlands). Simulated results indicate that the discharge of the Tsengwen is at increasing risk of exceeding the designed maximum streamflow at three stations from different areas of the watershed for three projected periods of 1979–2003, 2015–2039, and 2075–2099. Model results indicate that the exceedance frequency for the designed flood is 2 in 88 events in the base period (1979–2003), 6 in 82 events in the near future (2015–2039), and 10 in 81 events at the end of the century (2075–2099).

Okamoto et al. [31] turn our attention from streamflow to water fluxes driving hillslope processes. They investigate the optimal soil hydraulic parameters for simulating unsaturated flow based on a case study from the island of Miyakojima, Japan. The authors optimize the parameters for root water uptake and then examine the influence of soil hydraulic parameters on simulations of evapotranspiration. From there, they compare volumetric water content between the simulation results and those using pedotransfer estimates obtained from ROSETTA software. The resulting comparison highlights the importance of using soil hydraulic parameters based on measured data to simulate evapotranspiration and unsaturated water flow processes.

The last two papers in this special issue examine the ways by which different perceptions of climate change and adaptation strategies impact management and water resource variability. Ndamani and Watanabe [32] analyze farmer perceptions of adaptation practices using semi-structured questionnaires and focus group discussions of 100 farmer-households from four communities in the Lawra district of Ghana. The results show that adaptation is largely driven by response to dry spells and droughts (93.2%) rather than floods. Farmers in the region ranked improved crop varieties and irrigation as the most important adaptation measures, but largely lacked the capacity to implement these adaptation practices. The study also revealed that unpredictable weather, high cost of farm inputs, limited access to reliable weather information, and lack of water resources were the most critical barriers to successful adaptation. This study highlights the critical linkage between climate, hydrology, perception, and environmental management.

Shandas et al. [33] present a study of differing perspectives from the field on stressors and strategies for managing urban water scarcity in two urbanizing regions of the western US: Portland, Oregon and Phoenix, Arizona. The results show that long-term drought, population growth, and outdoor water use are the most important stressors to urban water systems, and indicate more agreement across cities than across professions in terms of effective strategies, suggesting that land-use planners and water managers remain divided in their conception of the solutions to urban water management. The authors also recommend potential pathways for coordinating the fields of land and water management to streamline strategies for urban sustainability.

3. Conclusions

This collection of papers focuses on a range of research topics influenced by the overriding hydrologic mechanisms associated with anthropogenic climate change and associated water resource variability. This includes a wide range of problems ranging from changes in surficial water levels, streamflow, sediment yields, and water quality in lakes, rivers, watersheds, and estuarine systems. The authors have brought a number of methodological tools to bear on these problems by examining various datasets and techniques, such as remote sensing, meteorological station-based observational data, tree-ring records, climate forecasts, and hydrological models used to simulate climatic impacts on streamflow, sediment yield, and water quality. Because consequent environmental problems and strategies for coping and mitigating deleterious effects must be defined in a social context, it is also important to include research examining perception, vulnerability, and adaptation. This collection of 16 papers emphasizes the importance of understanding the various interrelated facets that changing climates have on water resource variability and how focused investigations will help ground suitable strategies for mitigating and adapting to anthropogenic climate change.

Acknowledgments: The authors of this paper, who served as the guest-editors of this special issue, wish to thank the journal editors, all authors submitting papers to this special issue, and the many referees who contributed to paper revision and improvement of the 16 published papers.

Conflicts of Interest: The authors declare no conflict of interest.

References

1.	Gulick, S.P.S.; Jaeger, J.M.; Mix, A.C.; Asahi, H.; Bahlburg, H.; Belanger, C.L.; Berbel, G.B.B.; Childress, L.; Cowan, E.; Drab, L.; et al. Mid-Pleistocene climate transition drives net mass loss from rapidly uplifting St. Elias Mountains, Alaska. *Proc. Natl. Acad. Sci. USA* **2015**, *112*, 15042–15047.

2.   Kouchak, A.A.; Feldman, D.; Hoerling, M.; Huxman, T.; Lund, J. Water and climate: Recognize anthropogenic drought. *Nature* **2015**, *524*, 409–411.

3.   Castle, S.L.; Thomas, B.F.; Reager, J.T.; Rodell, M.; Swenson, S.C.; Famiglietti, J.S. Groundwater depletion during drought threatens future water security of the Colorado River Basin. *Geophys. Res. Lett.* **2014**, *41*, 5904–5911.

4.   National Academies of Sciences, Engineering and Medicine. *Attribution of Extreme Weather Events in the Context of Climate Change*; The National Academies Press: Washington, DC, USA, 2016.

5.   Vörösmarty, C.J.; Green, P.; Salisbury, J.; Lammers, R.B. Global water resources: Vulnerability from climate change and population growth. *Science* **2000**, *289*, 284–288.

6.   Phillips, O.L.; Aragão, L.E.O.C.; Lewis, S.L.; Fisher, J.B.; Lloyd, J.; López-González, G.; Malhi, Y.; Monteagudo, A.; Peacock, J.; Quesada, C.A.; et al. Drought sensitivity of the amazon rainforest. *Science* **2009**, *323*, 1344–1347.

7.   Piao, S.; Ciais, P.; Huang, Y.; Shen, Z.; Peng, S.; Li, J.; Zhou, L.; Liu, H.; Ma, Y.; Ding, Y.; et al. The impacts of climate change on water resources and agriculture in China. *Nature* **2010**, *467*, 43–51.

8.   Cowell, C.M.; Urban, M.A. The changing geography of the U.S. water budget: Twentieth-century patterns and twenty-first-century projections. *Ann. Assoc. Am. Geogr.* **2010**, *100*, 740–754.

9.   Jones, P.D.; Mann, M.E. Climate over past millennia. *Rev. Geophys.* **2004**, *42*, RG2002.

10.   Zhang, Q.B.; Cheng, G.; Yao, T.; Kang, X.; Huang, J. A 2326-year tree-ring record of climate variability on the northeastern Qinghai-Tibetan Plateau. *Geophys. Res. Lett.* **2003**, *30*.

11.   Trenberth, K.E.; Dai, A.; Rasmussen, R.M.; Parsons, D.B. The changing character of precipitation. *Bull. Am. Meteorol. Soc.* **2003**, *84*, 1205–1217.

12.   Song, C.; Huang, B.; Ke, L. Modeling and analysis of lake water storage changes on the Tibetan Plateau using multi-mission satellite data. *Remote Sens. Environ.* **2013**, *135*, 25–35.

13.   Li, Y.K.; Liao, J.J.; Guo, H.D.; Liu, Z.W.; Shen, G.Z. Patterns and potential drivers of dramatic changes in Tibetan lakes, 1972–2010. *PLoS ONE* **2014**, *9*, e111890.

14.   Seager, R.; Ting, M.; Held, I.; Kushnir, Y.; Lu, J.; Vecchi, G.; Huang, H.P.; Harnik, N.; Leetmaa, A.; Lau, N.C.; et al. Model projections of an imminent transition to a more arid climate in southwestern North America. *Science* **2007**, *316*, 1181–1184.

15.   Xu, C.Y. From GCMs to river flow: A review of downscaling methods and hydrologic modelling approaches. *Prog. Phys. Geogr.* **1999**, *23*, 229–249.

16.   Sheppard, S.R.J. Landscape visualisation and climate change: The potential for influencing perceptions and behavior. *Environ. Sci. Policy* **2005**, *8*, 637–654.

17.   Preston, B.L.; Yuen, E.J.; Westaway, R.M. Putting vulnerability to climate change on the map: A review of approaches, benefits, and risks. *Sustain. Sci.* **2011**, *6*, 177–202.

18.   Li, Q.T.; Lu, L.L.; Wang, C.Z.; Li, Y.K.; Sui, Y.; Guo, H.D. MODIS-Derived Spatiotemporal Changes of Major Lake Surface Areas in Arid Xinjiang, China, 2000–2014. *Water* **2015**, *7*, 5731–5751.

19. Ning, S.; Ishidaira, H.; Udmale, P.; Ichikawa, Y. Remote Sensing Based Analysis of Recent Variations in Water Resources and Vegetation of a Semi-Arid Region. *Water* **2015**, *7*, 6039–6055.

20. Buma, W.G.; Lee, S.I.; Seo, J.Y. Hydrological Evaluation of Lake Chad Basin using Space Borne and Hydrological Model Observation. *Water* **2016**, *8*, 205.

21. Jiang, C.; Zhang, L.B.; Li, D.Q.; Li, F. Water Discharge and Sediment Load Changes in China: Change Patterns, Causes, and Implications. *Water* **2015**, *7*, 5849–5875.

22. Wang, H.J.; Xiao, W.H.; Wang, J.H.; Wang, Y.C.; Huang, Y.; Hou, B.D.; Lu, C.Y. The Impact of Climate Change on the Duration and Division of Flood Season in the Fenhe River Basin, China. *Water* **2016**, *8*, 105.

23. Kwak, J.; Kim, S.; Kim, G.; Singh, V.P.; Park, J.; Kim, H.S. Bivariate Drought Analysis Using Streamflow Reconstruction with Tree Ring Indices in the Sacramento Basin, California, USA. *Water* **2016**, *8*, 122.

24. Pechlivanidis, I.G.; Olsson, J.; Bosshard, T.; Sharma, D.; Sharma, K.C. Multi-Basin Modelling of Future Hydrological Fluxes in the Indian Subcontinent. *Water* **2016**, *8*, 177.

25. Li, T.H.; Gao, Y. Runoff and Sediment Yield Variations in Response to Precipitation Changes: A Case Study of Xichuan Watershed in the Loess Plateau, China. *Water* **2015**, *7*, 5638–5656.

26. Ligaray, M.; Kim, H.; Sthiannopkao, S.; Lee, S.; Cho, K.H.; Kim, J.H. Assessment on Hydrologic Response by Climate Change in the Chao Phraya River Basin, Thailand. *Water* **2015**, *7*, 6892–6909.

27. Mahmood, R.; Jia, S.F.; Babel, M.S. Potential Impacts of Climate Change on Water Resources in the Kunhar River Basin, Pakistan. *Water* **2016**, *8*, 23.

28. Hesse, C.; Krysanova, V. Modeling Climate and Management Change Impacts on Water Quality and In-Stream Processes in the Elbe River Basin. *Water* **2016**, *8*, 40.

29. Liu, W.C.; Chan, W.T. Assessment of Climate Change Impacts on Water Quality in a Tidal Estuarine System Using a Three-Dimensional Model. *Water* **2016**, *8*, 60.

30. Wei, H.P.; Yeh, K.C.; Liou, J.J.; Chen, Y.M.; Cheng, C.T. Estimating the Risk of River Flow under Climate Change in the Tsengwen River Basin. *Water* **2016**, *8*, 81.

31. Okamoto, K.; Sakai, K.; Nakamura, S.; Cho, H.; Nakandakari, T.; Ootani, S. Optimal Choice of Soil Hydraulic Parameters for Simulating the Unsaturated Flow: A Case Study on the Island of Miyakojima, Japan. *Water* **2015**, *7*, 5676–5688.

32. Ndamani, F.; Watanabe, T. Farmers' Perceptions about Adaptation Practices to Climate Change and Barriers to Adaptation: A Micro-Level Study in Ghana. *Water* **2015**, *7*, 4592–4604.

33. Shandas, V.; Lehman, R.; Larson, K.L.; Bunn, J.; Chang, H. Stressors and Strategies for Managing Urban Water Scarcity: Perspectives from the Field. *Water* **2015**, *7*, 6775–6787.

MODIS-Derived Spatiotemporal Changes of Major Lake Surface Areas in Arid Xinjiang, China, 2000–2014

Qingting Li, Linlin Lu, Cuizhen Wang, Yingkui Li, Yue Sui and Huadong Guo

Abstract: Inland water bodies, which are critical freshwater resources for arid and semi-arid areas, are very sensitive to climate change and human disturbance. In this paper, we derived a time series of major lake surface areas across Xinjiang Uygur Autonomous Region (XUAR), China, based on an eight-day MODIS time series in 500 m resolution from 2000 to 2014. A classification approach based on water index and dynamic threshold selection was first developed to accommodate varied spectral features of water pixels at different temporal steps. The overall classification accuracy for a MODIS-derived water body is 97% compared to a water body derived using Landsat imagery. Then, monthly composites of water bodies were derived for the months of April, July, and September to identify seasonal patterns and inter-annual dynamics of 10 major lakes (>100 km^2) in XUAR. Our results indicate that the changing trends of surface area of major lakes varied across the region. The surface areas of the Ebinur and Bosten Lakes showed a significant shrinking trend. The Ulungur-Jili Lake remained relatively stable during the entire period. For mountain lakes, the Barkol Lake showed a decreasing trend in April and July, but the Sayram Lake showed a significant expanding trend in September. The four plateau lakes exhibited significant expanding trends in all three seasons except for Arkatag Lake in July. The shrinking of major lakes reflects severe anthropogenic impacts due to agricultural and industrial needs, in addition to the impact of climate change. The pattern of lake changes across the XUAR can provide insight into the impact of climate change and human activities on regional water resources in this arid and semi-arid region.

Reprinted from *Water*. Cite as: Li, Q.; Lu, L.; Wang, C.; Li, Y.; Sui, Y.; Guo, H. MODIS-Derived Spatiotemporal Changes of Major Lake Surface Areas in Arid Xinjiang, China, 2000–2014. *Water* **2015**, *7*, 5731–5751.

1. Introduction

Inland water bodies are important parts of the hydrosphere, serving as an essential source of freshwater for human consumption, agriculture, industry, and other uses. Due to climate change, uneven distribution of precipitation, and human activities, water resources show tremendous temporal variability worldwide [1]. Lakes and rivers are primary freshwater sources available to the local population

10

and their livestock in arid and semi-arid areas [2,3]. Spatial dynamics and up-to-date information on surface water resources are essential for understanding water resource-related issues in these areas. Temporal water bodies that provide habitats for plant and animal communities in these areas have rarely been included in global datasets, such as the Global Lakes and Wetlands Database (GLWD) [4,5] and Vector Map Level 0 (VMAP0) [6].

With the capability of synoptic view and repeated coverage of the earth's surface, satellite remote sensing is an effective means of extracting water bodies across a variety of spatial and temporal scales. Due to their strong absorption in the near-infrared (NIR) spectrum, optical remote sensing platforms, such as Landsat [7,8], Advanced Spaceborne Thermal Emission and Reflection Radiometer (ASTER) [9,10], and Satellite Pour l'Observation de la Terre (SPOT) [11], have been used to map the area of the water body at various spatial resolutions. However, the high costs, narrow swath, and long revisit intervals of the medium- and high-resolution images limit their applications on monitoring the dynamics of lake systems across large spatial scales. Remote sensing data with high temporal resolution have the advantage of documenting detailed water area variation. Time series data from satellite sensors, such as the SPOT VEGETATION, Moderate Resolution Imaging Spectroradiometer (MODIS), and Advanced Very High Resolution Radiometer (AVHRR), have been applied to seasonal and inter-annual change detection of water bodies over large areas [12–16].

One of the common image processing methods for extracting water extent is based on a threshold of a water detection index. Spectral indexes, such as Normalized Difference Water Index (NDWI) [17], Modified Normalized Difference Water Index (MNDWI) [18], and Normalized Difference Pond Index (NDPI) [11], have been developed for water detection using remote sensing imagery. NDWI uses green band and near-infrared band to distinguish water from vegetation and soil [17]. In order to enhance the ability of water detection, especially for areas with built-up land in the background, the middle infrared band was integrated into MNDWI and NDPI instead of NIR band in NDWI [11,18]. Built-up areas and water bodies show discriminating spectral responses at the MODIS short-wavelength infrared (SWIR) band. A Combined Water Index (CWI) combining SWIR's and NDVI's ability to represent vegetation information was proposed for water body identification using MODIS data [19]. In addition, spectral reflectance of water shows spatiotemporal variability across different scenes and acquisition dates. Thus, delineating water bodies using a standard threshold may become problematic in large-area applications [20]. A strategy for threshold computation of different satellite images is needed.

The Xinjiang Uygur Autonomous Region (XUAR), with a widespread area of 1,660,000 km^2, is the largest autonomous region in China. There are 113 lakes with an area of >1 km^2 in the XUAR. Many lakes are important wetlands for threatened species, five of which are designated as the National Nature Reserves of China [21]. The water levels of inland lakes in XUAR are influenced by the runoff of their inflowing rivers, and are sensitive to climate change and human activities altering the rivers' inflows [22]. It has been reported in past studies that the water extent has been shrinking in past decades, causing severe environmental problems, such as land desertification, salinization, vegetation degradation, water shortage, and biodiversity loss in this arid/semi-arid area [22].

Several studies have examined changes of lakes in the XUAR using satellite imagery, such as Landsat [22–24], SPOT VEGETATION [25], and MODIS [12,26] data. Landsat TM imagery has been used to interpret changes in area of inland lakes in Xinjiang over the period 1975–2007, but only selected lakes in spring and autumn seasons were analyzed [22]. Comparing with nationwide lake surveys undertaken in the 1960s–1980s, Ma *et al.* (2010) reported that 62 lakes vanished in the XUAR from the 1960s to 2000s; one of the completely dried lakes is the Lop Nur Lake with an original lake area of 5500 km^2 [23]. Water extents were found to have decreased significantly due to anthropogenic impacts, such as agricultural water consumption and damming in this region [24]. The Ebinur Lake, the largest salt lake in the XUAR, exhibited a significant inter-annual and inter-seasonal variation based on SPOT VEGETATION data [25]. However, due to the large area of XUAR, previous studies have mainly focused on selected water bodies and annual or seasonal temporal intervals. The temporal fluctuations in surface area of major lakes were monitored with MODIS data in XUAR, but only inter-annual variations were analyzed [12,26]. The temporal and spatial dynamics of lake surface areas across XUAR are rarely documented in detail.

The purpose of this study is to examine the spatiotemporal variation of water bodies from 2000 to 2014 based on MODIS time series data. Detailed objectives include: (1) to develop an automatic approach to extract water bodies from MODIS data; (2) to generate a 15-year water body mask and document the changes of water extent; and (3) to examine the driving factors of the changes in major lakes in the XUAR.

2. Study Area

The XUAR is located in north-western China and encompasses the Altay Mountains, Junngar Basin, Tianshan Mountain, Tarim Basin, and Kunlun Mountains from north to south (Figure 1). It is an arid and semi-arid area with mean annual precipitation ranging between 100 and 200 mm [27]. Vast areas of the XUAR are covered by grassland and desert [28]. Forests are sparsely scattered within high mountains and along rivers. Oasis landscapes characterized by human settlements and agriculture lands are distributed within inland river deltas, alluvial-diluvial plains, and along the edges of diluvial-alluvial fans.

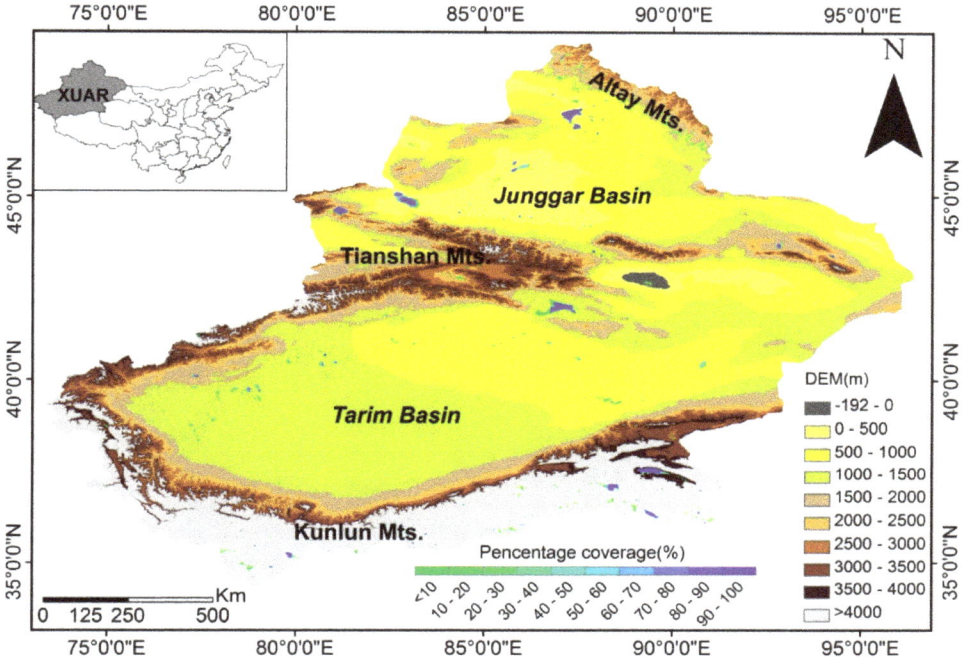

Figure 1. Geographic location and topographic map of study area. The percentage coverage of water bodies was combined for April, July, and September from 2000 to 2014.

According to the topographic characteristics, lakes in the XUAR can be categorized as four types [26]: (1) plateau lakes (>3500 m), with snow and glacier ice melt and surface runoffs as their main charge; (2) mountain lakes (1000–3500 m), with snow and glacier ice melt, underground runoff as their main influx; (3) plain lakes (<1000 m), heavily influenced by human activities; and (4) transition lakes, located at the transition area between mountains and plains.

Among all the lakes in the XUAR, 10 lakes have areas larger than 100 km². Their total area accounts for more than 80% of the lake surface area (Table 1) [26].

Table 1. Ten major lakes in the study area (with surface area >100 km^2).

Name	Area (km^2)	Altitude (m)	Mean Water Depth (m)	Region	Type
Ebinur	673.46	194	1.2	Junggar	Plain
Manas	259.81	244	6	Junggar	Plain
Ulungur-Jili	1041.60	478	10.4	Junggar	Plain
Bosten	1004.33	1050	9	Tarim	Transition
Sayram	462.63	2072	46	Junggar	Mountain
Barkol	118.57	1577	0.6	Junggar	Mountain
Ayakkum	200.46	3876	10	Kumukuli	Plateau
Aqqikkol	168.93	4251	8	Kumukuli	Plateau
Arkatag	110.33	4713	8	Kumukuli	Plateau
Aksayquin	88.54	4844	8	Northern Tibet	Plateau

3. Materials and Methods

3.1. Remote Sensing Datasets

Terra MODIS images were selected as the main data source for monitoring water body variation in XUAR. The MODIS Surface Reflectance (MOD09A1) dataset was used to build an image time series from 2000 to 2014 [29]. It provides surface reflectance at bands 1–7 with 500 m spatial resolution and eight-day temporal resolution. The XUAR region is entirely covered by six tiles (h23v04, h23v05, h24v04, h24v05, h25v04, and h25v05). Lakes in XUAR were relatively stable during the spring and autumn seasons. In summer, some were influenced by extensive agricultural irrigation and high evaporation. In winter, some of the water bodies were frozen and may have caused high uncertainty in lake extent extraction. Therefore, we downloaded MODIS data for the months of April, July, and September to analyze the spatiotemporal dynamics of lake surface area. The Shuttle Radar Topography Mission (SRTM) digital elevation model with a spatial resolution of 90 m was used to correct the water body extraction affected by shadow and snow in mountainous regions [14,30].

Landsat data of 30 m resolution was used to assess the accuracy of remote sensing products with 250 m [31] and 1 km resolution [32]. In our study, Landsat TM, ETM+, and OLI images were used to validate the results of water extraction from MODIS data. A total of 71 Landsat image scenes (including Landsat 5 TM, Landsat 7 ETM+, and Landsat8 OLI_TIRS) were processed for the validation (Table 2).

The MODIS 500-m land-cover product (MCD12Q1) was used to identify primary land covers in the study area [33]. The MCD12Q1 is produced using an ensemble supervised classification algorithm with MODIS band 1–7 surface reflectance, an enhanced vegetation index, and land surface temperature as the main input. Post-processing refinements with ancillary datasets were also conducted. The MCD12Q1 product of 2012 covering the study area was downloaded. With the

International Geosphere-Biosphere Program (IGBP) classification scheme, the land cover types were mapped and used to assess the impacts of human activities on lake changes in this study.

Table 2. Landsat data used in our study.

Lake	Ulungur	Manas	Bosten	Ayakekumu
Sensors	Landsat5 TM Landsat7 ETM+ Landsat8 OLI_TIRS	Landsat5 TM Landsat7 ETM+ Landsat8 OLI_TIRS	Landsat5 TM Landsat7 ETM+ Landsat8 OLI_TIRS	Landsat5 TM Landsat7 ETM+ Landsat8 OLI_TIRS
Path/Row	143/27, 144/27	144/28	143/31	140/34
Date	7 July 2000 24 September 2000 13 April 2001 27 July 2002 14 September 2002 10 April 2003 24 July 2006 17 September 2006 6 April 2007 23 July 2009 18 September 2009 10 May 2011 13 July 2011 15 September 2011 13 April 2013 11 July 2013 29 September 2013 25 April 2014	16 March 2000 6 July 2000 24 September 2000 19 March 2001 10 April 2003 31 July 2006 17 September 2006 4 September 2007 23 July 2009 9 September 2009 10 May 2011 13 July 2011 1 October 2011 13 April 2013 2 July 2013 4 September 2013 2 May 2014 21 July 2014	25 March 2000 31 July 2000 17 September 2000 1 August2006 10 September 2006 14 April 2007 27 July 2007 13 September 2007 2 May 2008 16 July 2009 3 April 2009 6 July 2011 24 September 2011 22 April 2013 27 July 2013 29 September 2013 31 August2014	4 September 2000 29 July 2001 1 October 2001 11 April 2002 1 August2002 4 October 2002 14 April 2003 19 April 2005 27 July 2006 13 September 2006 6 July 2007 25 April 2010 30 July 2010 31 August2010 8 May 2012 7 August2013 24 September 2013 6 May 2014

3.2. MODIS-Based Water Body Extraction

In XUAR, large areas of deserts and bare rocks have high spectral responses in short-wavelength infrared (SWIR) band [34]. Water usually has low reflectance along the spectrum. In this study we use CWI to detect water bodies from MODIS data. The computation formula of CWI is as follows [19]:

$$CWI = (NDVI + SWIR + A) \times C \tag{1}$$

$$NDVI = \frac{b2 - b1}{b2 + b1} \tag{2}$$

$$SWIR = \frac{b7}{b7,} \tag{3}$$

where b1, b2, and b7 represent the reflectance of band 1 (Red band, 620–670 nm), band 2 (NIR band, 871–876 nm), and band 7 (SWIR band, 2105–2155 nm) of the MOD09 data, respectively. A and C are correction factors to adjust the data ranges of *CWI* values. They are empirically determined by comparing *CWI* values between water pixels and background in the study area. We set A as 0.4 and C as 100 in our study [19]. Figure 2 illustrates the procedure for extracting water bodies based on MODIS data.

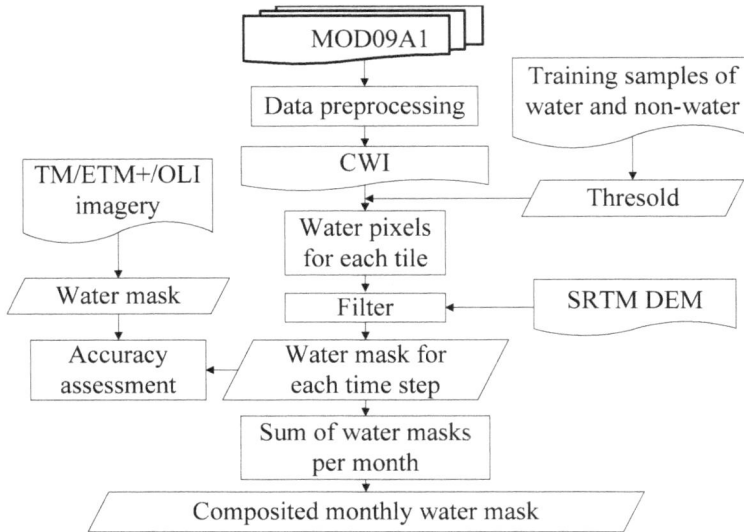

Figure 2. Workflow for water body extraction in our study.

The MODIS tiled data in Sinusoidal projection were mosaicked, re-projected using the nearest neighbor re-sampling method, and saved as GeoTIFF format using the MODIS Reprojection Tool. The DEM data is resampled to the pixel size of MODIS to refine the water detection results in the following steps.

The quality assessment information of MODIS data was used to exclude pixels labeled as cloudy or snow/ice. Then CWI was calculated for each MODIS tile. The atmosphere condition, water depth, and chlorophyll content all have influence on the spectral features of water on remote sensing images. A single threshold value derived for one image might not be suitable for another. Since there is no standard threshold for the whole study period, an optimized threshold must be identified for each scene or each month. In this study, we set a different threshold for each time step and extract water pixels. For the threshold selection, the training datasets that as pixels were covered by water for all time steps were collected manually from MODIS data in July. The statistics of CWI values were calculated based on training samples. We choose two standard deviations of the mean CWI value as the threshold value and classify pixels within it as water and *vice versa*.

Shadows in mountainous areas can lead to confusion with water bodies. To eliminate the snow and shadow effect, a slope map was used to refine the water extent. All pixels with a slope >1° were removed from the classification result because water bodies usually have flat surfaces. In this way, misclassification of shadow in mountainous areas can be corrected. In addition, to reduce the noise caused by small and temporary water bodies in mountainous areas, we removed water bodies smaller than 4 km^2 in the detection results. After these steps, a binary mask with water and non-water pixels for each time step was derived.

A binary water mask was calculated for each eight-day interval. For each month, we summed up the binary masks at four time steps. For all pixels, we obtained the number of times they were classified as water. Only pixels classified as water three out of four times in a month were marked as water [14]. In this way, composited monthly water masks were generated for April, July, and September, 2000–2014. The seasonality of each lake was calculated using the maximum/minimum ratio of area extent of one representative year for each lake [35]. Finally, combining the monthly masks for April, July, and September from 2000 to 2014, a percentage coverage layer was derived. In this layer, pixel values show the percentage of times a pixel was classified as water from 2000 to 2014. A 100% percentage means the pixel was identified as water at all the 45 monthly masks from 2000 to 2014.

3.3. Accuracy Assessment

Water extents classified from Landsat images served as our validation sources in this study. An integrated water body mapping method combining the NDVI, NDWI, NIR, and slope layers was applied [36]. The commonly used threshold method was used to calculate threshold values and segment water bodies [36]. After that, the water bodies detected from Landsat images were resampled to MODIS pixel size to perform a pixel-to-pixel comparison. Confusion matrices were calculated to represent the accuracy of the classification results [37]. Three measurements, namely user's accuracy, producer's accuracy, and overall accuracy were calculated to assess the accuracies of MODIS detection. The user's accuracy is defined as the number of correctly classified water pixels divided by the total number of classified water pixels in the MODIS detection results. The producer's accuracy is defined as the total number of correctly classified water pixels divided by the total number of water pixels in the Landsat detection results. Overall accuracy is defined as the sum of all correctly classified water/non-water pixels divided by the total number of validation samples.

We selected four water bodies to conduct accuracy assessment for MODIS water detection results. For each lake, Landsat data acquired in April, July, and September or the nearest month were collected for each year. The MODIS results on the nearest neighboring date were selected and compared with the Landsat interpreted results to ensure the images were consistent in acquisition time.

It is worth noting that classification error exists in the water extent mapped from Landsat data and may influence the accuracy assessment results. For the 30-m resolution Landsat images, mixed pixels of lake fringes and small water bodies contain significant spectral response from backgrounds such as grasslands and croplands. In addition, the turbidity of water can increase its response at the near-infrared band and lead to a lower NDWI value [38]. A more accurate validation exercise could be conducted based on comparison with manually collected validation samples, which is very time-consuming and was not conducted in this study.

3.4. Data Sources of Climate and Human Activities

We also analyzed possible drivers of lake area variations in XUAR. Specifically, annual mean temperature and annual precipitation were used as indicators of regional climate. Cropland and built-up areas were used as indicators of human activities. The monthly air temperatures and precipitations for XUAR were obtained from 55 meteorological stations (National Meteorological Information Center of China Meteorological Administration; http://cdc.nmic.cn/home.do). Monthly values were averaged or summed (for temperature and precipitation, respectively) to acquire annual values. For each variable, annual time series graphs were plotted for the 15-year study period, 2000–2014. The linear regression method and t-test are used to estimate the changing trend and test statistical significance of the changing trends of climate data. If the P value derived from the statistical analysis is less than the significance level, a significant changing trend is observed. Areas of irrigated croplands and built-up areas from 2000 to 2013 were collected from the XUAR Statistical Year Book [39].

4. Results and Discussion

4.1. Intra- and Inter-Annual Dynamics of Water Bodies

The water body mapping results of the Ebinur Lake using NDWI, MNDWI, and CWI from MODIS data were compared in Figure 3. The spectral response of water is very similar to bare lands around the lake for NDWI (Figure 3b). For MNDWI, the spectral signatures of dried-up lake and snow at the lower right of the image are identical with water pixels (Figure 3d). We applied a threshold value of zero to extract water pixels from the NDWI and MNDWI images [18]. The mapping result from NDWI shows an obvious underestimation of the lake surface area (Figure 3c). An overestimation of lake surface area is observed from the MNDWI results (Figure 3e). Comparing with water bodies extracted from NDWI and MNDWI visually, the water pixels identified from CWI have higher accuracy (Figure 3g).

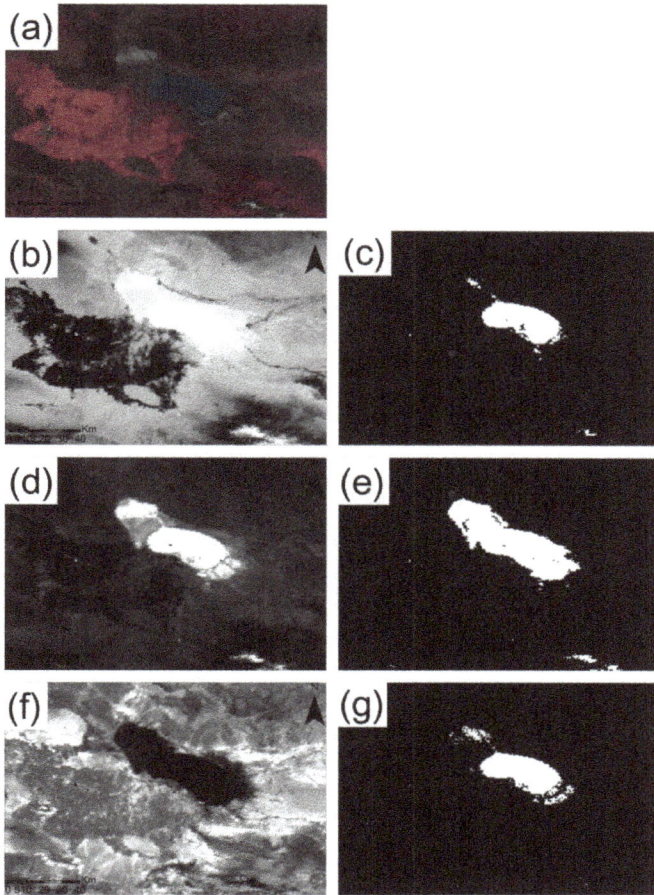

Figure 3. Water body detection results from MODIS data. (**a**) Landsat image; (**b**) NDWI [17]; (**c**) extracted water bodies from NDWI; (**d**) MNDWI [18]; (**e**) extracted water bodies from MNDWI; (**f**) CWI; and (**g**) extracted water bodies from CWI. Water bodies are shown in white in (**c**), (**e**), and (**g**).

In our study, the water body detection results from 2000 to 2014 in April, July, and September were combined and illustrated in Figure 1. In addition to the permanent water bodies that were detected at each time step, many temporal pools that were rarely mapped in global datasets were revealed. For lakes located in high elevation areas, they were frozen and cannot be detected in winter and spring months. For lakes located in arid and semi-arid areas, some of them dried up in the summer months.

We calculated the total area of 10 large lakes and analyzed their intra- and inter-annual dynamics in XUAR. Their total surface area varied from 4217 km^2

19

(September 2010) to 5014.75 km^2 (April 2003), with high seasonal variability. The largest surface area is usually in April and the smallest value in September. A decreasing trend of total lake surface area can also be observed for all three months from 2000 to 2014, even though the trend is insignificant with a 0.1 significance level (Figure 4). Significant shrinking trends were found for April (34.65 km^2/year), July (22.91 km^2/year), and September (31.02 km^2/year) from 2000 to 2010 with a 0.1 significance level.

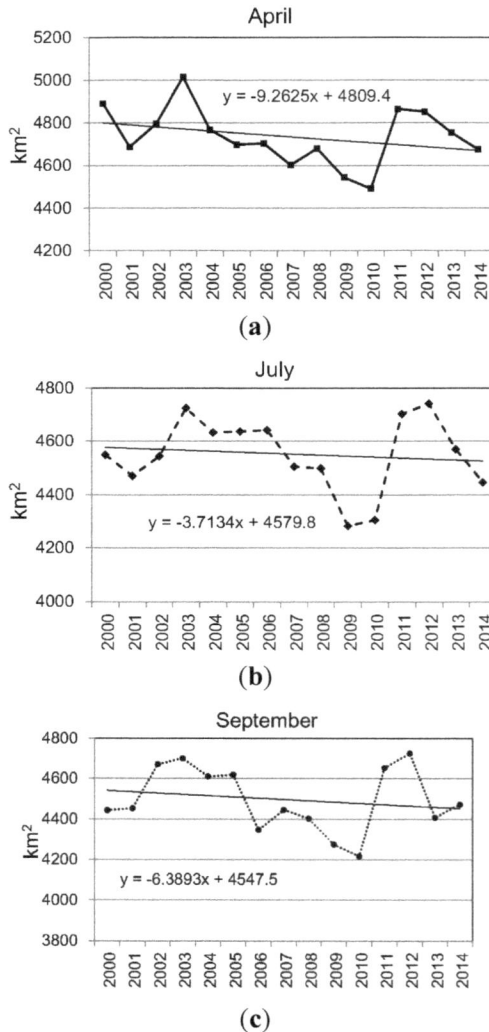

Figure 4. Inter-annual variation of the total area of major lakes for the months of (a) April; (b) July; and (c) September in XUAR from 2000 to 2014.

4.2. Temporal Variations of Major Water Bodies

The inter-annual variation in the surface area of each major lake was illustrated for April, July, and September (Figure 5). The changing trend in the surface area of each lake in July was calculated (Figure 6). The statistics of inter-annual variations of surface area for each lake were listed in Table 3. These results revealed that the lakes along the Tianshan Mountains (central Xinjiang) are shrinking, while the lakes in the northern and southern parts are expanding. This may indicate that human activities cause the lake to shrink, since most human settlements and agricultural lands are distributed along the Tianshan Mountains (Figure 6). Lakes where there is less human influence exhibited expanding trends, mainly due to changes in climate variables.

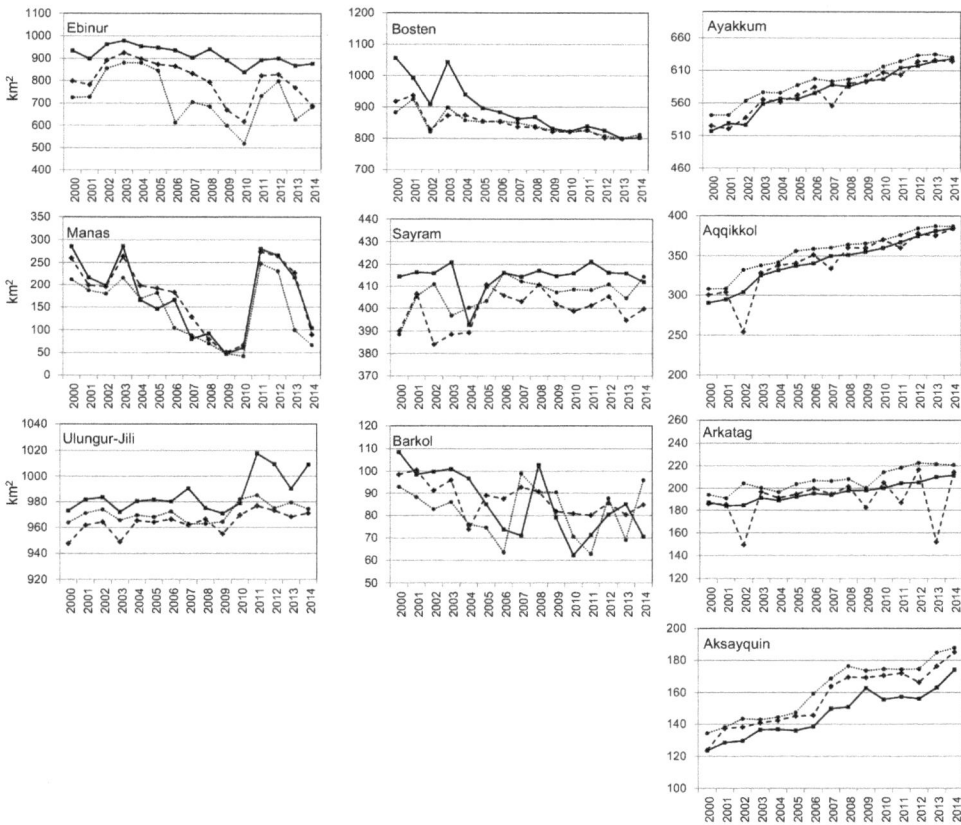

Figure 5. Inter-annual variation in the surface area of major lakes for the months of April, July, and September in XUAR from 2000 to 2014. Solid, dashed, and dotted lines represent April, July, and September, respectively.

Table 3. Changing trend in the surface area of major lakes in XUAR from 2000 to 2014.

Name	April	July	September	Seasonality (max/min ratio)
Ebinur	−6.09	−10.36	−11.74	1.29
Manas	−5.53	−5.08	−6.59	1.52
Ulungur-Jili	2.08	1.25	0.81	1.03
Bosten	−16.71	−7.78	−6.36	1.06
Sayram	0.22	0.61	0.90	1.03
Barkol	−2.32	−1.07	−0.43	1.38
Ayakkum	7.80	7.48	6.72	1.05
Aqqikkol	6.55	6.82	5.52	1.06
Arkatag	1.9	1.13	2.08	1.04
Aksayquin	3.21	3.88	3.92	1.15

Note: The values in bold indicate significant trends (p value <0.1).

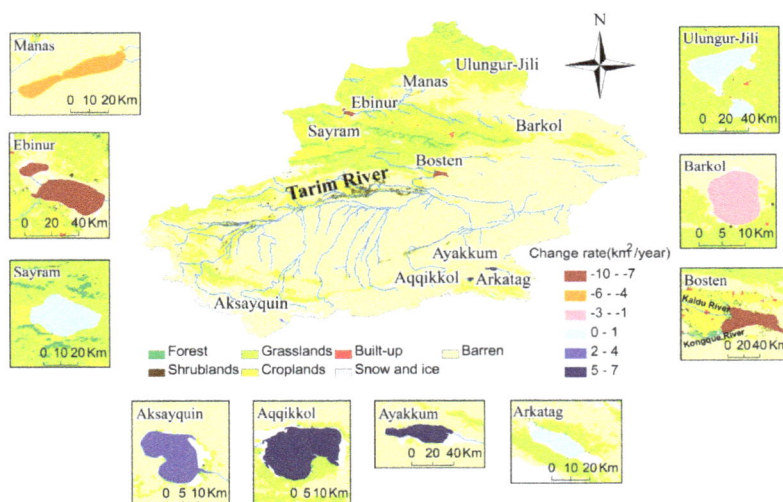

Figure 6. Changing trends and rates in the surface area of major lakes in XUAR in July from 2001 to 2014. Spatial distribution of land cover/use types was extracted from MODIS land cover type product in 2012 [33].

The highest seasonality is at Manas Lake, Barkol Lake, and Ebinur Lake. The lowest seasonality exists at Ulungur-Jili Lake, Sayram Lake, and Arkatag Lake. The seasonal dynamics of Ebinur Lake, Manas Lake, Barkol Lake, and Aksayquin Lake, which have the highest seasonality, are illustrated in Figure 7. For each pixel, we summed the number of times it was classified as water for April, July, and September over the entire study period. The monthly composite was classified into four intervals to indicate areas that stayed stable over the entire time series, and areas that were covered by water for several years during the 2000–2014 period. Different seasonal behaviors were observed for the four lakes. Ebinur Lake showed the largest area in April and the smallest in September. Manas Lake was largest in July and smallest in September.

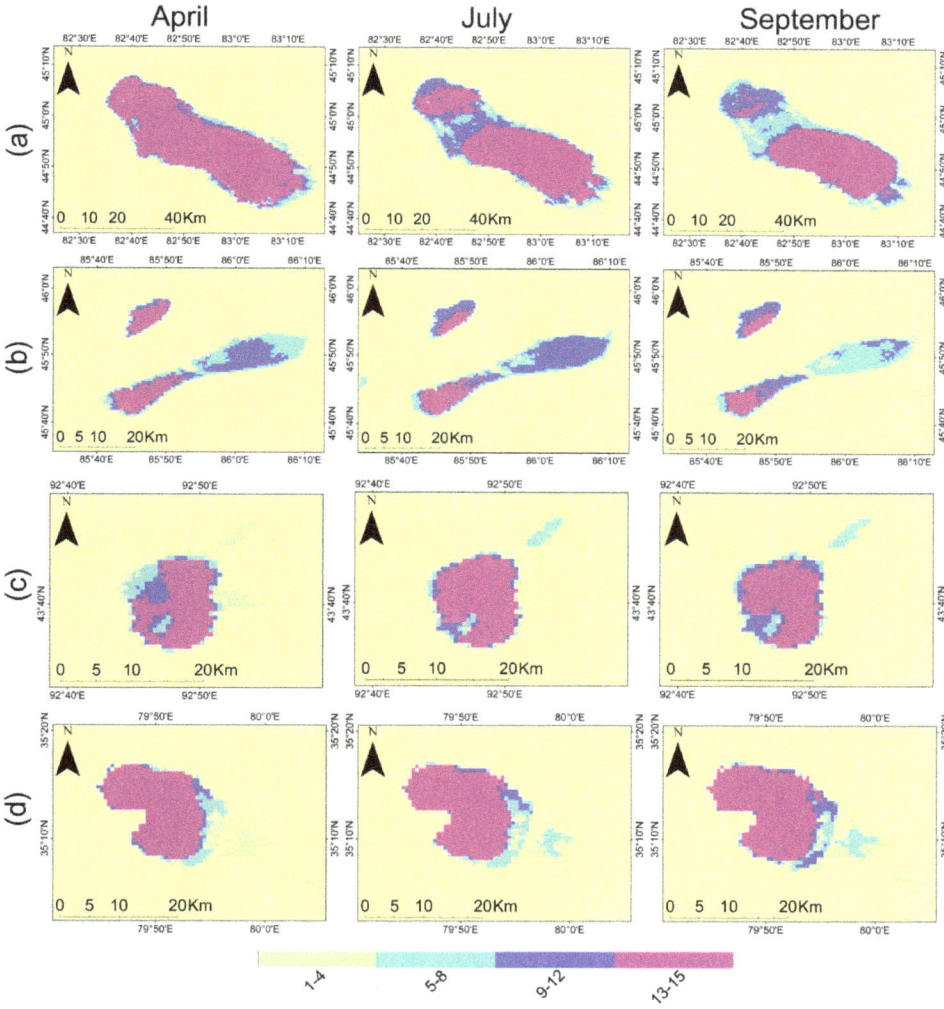

Figure 7. Typical seasonal water bodies of XUAR and their temporal dynamics for the months of April, July, and September from 2000 to 2014. (**a**) Ebinur; (**b**) Manas; (**c**) Barkol and (**d**) Aksayquin Lakes. The numbers at each pixel indicate the number of times it was detected as a water body during the 2000–2014 period.

4.3. Accuracy Assessment

Figure 8 illustrates the comparison of a monthly water mask derived from MODIS data with a water mask derived from Landsat images of the same month. The small water bodies that were detected from Landsat data cannot be extracted from MODIS imagery at 500 m resolution. This leads to a partial disagreement of detection results between Landsat and MODIS data. Based on our assessment, an overall accuracy of 0.97 is obtained (Table 4), which is adequate for the dynamic analysis of lake surface area. The detection results for Manas Lake showed lower user accuracy than the other lakes. Due to the high concentration of salts and other dissolved minerals, the spectral feature of water in Manas Lake is different from other water bodies. Therefore, the identification of water pixels using a general threshold for all of the water bodies in the entire study area may lead to misclassifications. For Bosten Lake, mixed pixels of wetland and small water bodies can be misclassified as non-water bodies due to the coarse resolution of MODIS data. The lake ice and snow can also lead to lower accuracies in April for plateau lakes like Ayakkum.

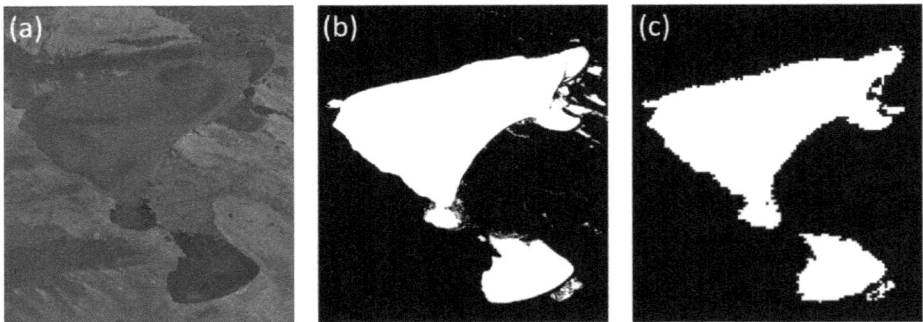

Figure 8. Comparison of a water body derived from MODIS data and Landsat for April 2014, Ulungur Lake. (**a**) Landsat image; (**b**) water mask derived from Landsat data; and (**c**) water mask derived from MODIS data.

Table 4. The error matrix, overall, producer's, and user's accuracies of water bodies and non-water bodies resulting from MODIS time series data, 2000–2014.

Month	Ulungur		Manas		Bosten		Ayakkum	
	User	Prod	User	Prod	User	Prod	User	Prod
April	0.92	0.93	0.72	0.97	0.95	0.89	0.97	0.95
July	0.96	0.92	0.66	0.95	0.99	0.86	0.95	0.94
September	0.97	0.93	0.63	0.93	0.99	0.89	0.97	0.97
Overall Accuracy	0.97							

4.4. Effects of Regional Climate and Human Activities on Lake Changes

The linear trend of temperature and precipitation are not statistically significant at the 0.1 significance level for 2000–2014. In our study period, an increase in precipitation occurred in 2010, which corresponds well to the extending lake area in 2011 and 2012 (Figure 9). The area of cropland increased from 3.39 million hectares in 2000 to 5.21 million hectares in 2013. Built-up areas increased from 473 km^2 in 2000 to 1065 km^2 in 2013. As we can see in Figure 5, the oases comprised of agricultural lands and built-up areas are distributed in river plains near major lakes. The water demands of oases can influence aerial changes in lakes that have river runoff as their main inflows, such as Ebinur, Bosten, and Manas Lakes. For a regional comparison, the changing trends of temperature and precipitation at meteorological stations near each major lake were analyzed (Table 5). Their relationships with lake area changes were also analyzed. Due to the lack of observed climate data, the analysis was not performed for the four plateau lakes.

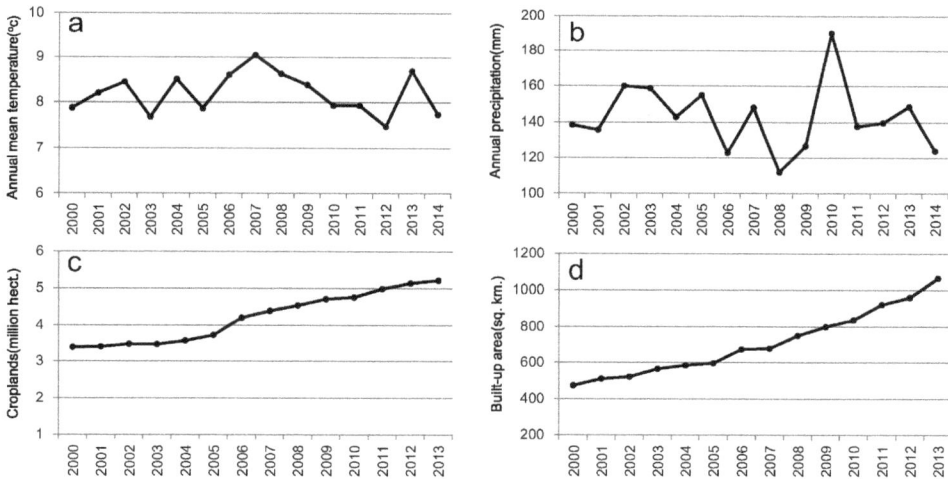

Figure 9. Variations in (**a**) annual mean temperature; (**b**) annual precipitation; (**c**) area of croplands; and (**d**) built-up areas in XUAR over the study period.

Table 5. Changing trends in temperature and precipitation at meteorological stations near the major lakes from 2000 to 2014.

Climate Variable	Ebinur	Manas	Ulungur-Jili	Bosten	Sayram	Barkol
Annual Mean Temperature (°C)	−0.20	−0.06	0.31	−0.26	−0.27	0.51
Annual Precipitation (mm)	−0.28	−1.70	−2.28	−2.72	−3.21	2.82

Among the major lakes, three plain lake or lake systems, Ebinur, Manas, and Ulungur-Jili, were analyzed. Located in low elevation plain areas, their main supply is river runoff, which can be influenced tremendously by human activities. Ebinur is the largest saline lake in XUAR. With a maximum depth of 3.5 m and a mean depth of 1.2 m, it is a closed lake without outlet. Its hydrologic input is mainly from the Bo and Jing rivers, which originate from precipitation in mountainous areas. In their analysis of multi-temporal VGT-S10 data from 1 April 1998 to 31 December 2005, Ma *et al.* (2007) revealed that this lake had a peak area of 903 km^2 in 2003 and subsequently decreased to an area of 847 km^2 in 2004 [23]. Our study showed a decreasing trend of surface area from 2000 to 2014 in all three seasons, with a short increase in 2002–2004 (Figure 5). The significant decreasing trend of the Ebinur Lake is 6.09 km^2/year for April, 10.36 km^2/year for July, and 11.74 km^2/year for September, respectively. A slight decreasing trend of precipitation and temperature was revealed (Table 5). According to Zhang *et al.* (2015), human factors, such as population growth and agricultural development, led to the increase in desertification area in the Ebinur Lake region between 1990 and 2010 [40].

Manas Lake is the terminal lake of Manas River. This lake is about 55 km long and 15–20 km wide, with an area of about 550 km^2 and an average depth 6 m [41]. Our results show a decreasing trend of Manas Lake from 2000 to 2014. The smallest lake size occurred in 2009 and 2010 (Figure 5). A decreasing trend of precipitation and temperature was observed at the meteorological station near Manas Lake (Table 5). Manas River Valley is a representative area for oasis exploitation in Xinjiang as a primary agricultural production region and core region of economic development of the northern slope of the Tianshan [41]. The expansion of oasis in Manas River Valley was characterized by the spread of settlements and agricultural lands. In addition to reclaiming agricultural lands, many hydrological constructions, including reservoirs, wells, and canals, were built along the Manas River for irrigation of the croplands.

Ulungur-Jili lake system is a closed inland lake, with its water supplied by the Ulungur River, groundwater, precipitation, and snow melting. As the second largest lake in XUAR, the Ulungur-Jili lake system can be divided into two sections, the Ulungur Lake and the smaller Jili Lake, connected by a narrow channel. In 2011, Ulungur Lake had an average depth of 10.4 m and a surface area of 859 km^2, and Jili Lake had an average depth of 8.8 m and an area of 169 km^2 [42]. Based on an analysis of the lake area with Landsat images acquired in August, a stable or slightly increasing trend of lake surface area is revealed from 2000 to 2011 [43]. A slight extending trend of Ulungur-Jili Lake was also observed in our study from 2000 to 2014. Increasing temperature and decreasing precipitation were observed at the nearby meteorological stations. Since snow and ice melting off the Altay Mountain is the main runoff supply of Irtysh River, snow and glacier changes caused by the increasing temperature contributed to the variation of lake area of the Ulungur-Jili lake system.

Bosten Lake, as a transition lake located between the mountains and plains, is the largest inland freshwater lake in China. As an open catchment lake with outlets, Bosten Lake lies at the end of Kaidu River and the beginning of Kongque River. The lake inflow mainly comes from Kaidu River, which contributes about 95% of the total water inflow [44,45]. The main outflow is Kongque River and evaporation [46]. With a large area in 2001 and 2003 (Figure 5), our study revealed an obvious decreasing trend in the surface area of Bosten Lake from 2000 to 2014. The nearby meteorological station observed the decreasing trend of precipitation and temperature. According to the monthly mean lake level data, the lake level decreased dramatically from 2002 to 2010 [46]. The sharply decreasing lake level during 2003 to 2010 was reported to be caused by the emergency project of transferring water to Tarim River and increasing lake outflow in addition to the reduced precipitation [46].

Sayram Lake is the largest and highest alpine lake in XUAR. The lake is located in a mountain basin in the western part of the Tian Shan with an average water depth of 46.1 m. It has a frozen period extending for six months from October to May [47]. Though a decreasing trend of precipitation and temperature was observed at the nearby meteorological station, a slight increasing trend was observed from 2000 to 2014 in our study. Located in a natural environment, Sayram Lake was less impacted by anthropogenic disturbances and showed a stable surface area from 1975 to 2007 [22].

Barkol Lake is a closed saline lake with an elevation of about 1580 m. Located in the Barkol Basin, Barkol Lake is bordered by the Barkol Range, the eastern Tianshan to the south, and the Moqinwula Range to the north. The average water depth is only 0.6 m, with a maximum water depth of about 1 m [48]. In our study, a shrinking trend of water surface area was observed from 2000 to 2014, while an increasing trend of precipitation was observed (Table 5). The primary cause for lake size decrease may be attributed to human interference. Due to the mirabilite production and reducing precipitation near the lake, the lake area shrunk from 233 km^2 in the 1950s to 60 km^2 in 2011 [49]. The severe mineralization of lake water led to large area degradation of the surrounding wetlands [49].

Ayakkum, Aqqikkol, and Arkatag Lakes are located in the Kumkol Basin between the Altyn Mountains and the Kunlun Mountains. Aksayquin Lake is located in the western Kunlun Mountains of the northern Tibetan Plateau. The main supply of the four lakes is river runoff from the melting of glaciers and snow. In our study, significant expanding trends were observed for these four lakes from 2000 to 2014 except Arkatag Lake in July. Based on Landsat image analysis over the entire Tibetan Plateau, Aqqikkol had undergone surface extent increases in excess of 10% from 2000 to 2011 [8]. The expansion of lake areas and increasing trend of lake level on northern Tibetan Plateau was observed and documented by several studies [8,50,51]. Due to the increasing temperature, the water recharge increased a lot from accelerated melting of glaciers and perennial snow cover [52], and permafrost degradation [53,54]. However, for areas where surface water resources are generated mainly in the mountain glaciers of XUAR, the increasing flows caused by melting of mountain glaciers cannot be sustained in the long term if the glaciers disappear due to increasing temperature [55].

5. Conclusions

This study presented a time series analysis of lake water surfaces across XUAR using MODIS data from 2000 to 2014. A classification approach based on water index calculation and dynamic threshold selection was developed. Compositing water detection results of each time step, water masks were derived for the months of April, July, and September. The major lakes with an area of >100 km^2 were categorized into four classes based on their topographic locations. The seasonal and inter-annual surface area variation of the 10 major lakes was revealed in detail.

For plain lakes, the surface area of Ebinur Lake showed a significant shrinking trend and Manas lake an insignificant shrinking trend. They are both influenced by the expanding oasis and increasing water consumption. The Ulungur-Jili Lake had a stable area throughout the entire time period. The decreasing area of Bosten Lake may have been caused by the construction of hydrological projects in addition to the reduced precipitation. For mountain lakes, overexploitation has caused the shrinking of the Barkol Lake and the degradation of surrounding wetlands. As the largest and highest alpine lake, Sayram Lake showed a significant expanding trend in September. The four plateau lakes exhibited significant expanding trends for all three seasons except Arkatag Lake in July.

The lake dynamics revealed by MODIS time series are useful for ecological assessment of XUAR. Further studies are needed to use satellite imagery with different spatiotemporal resolutions, such as AVHRR and Landsat data since the 1970–80s, to capture the long-term dynamics of lakes in XUAR. It is also important to integrate the analysis of satellite data and climatic datasets for better understanding of the impact of climate change on water bodies in this arid and semi-arid region.

Acknowledgments: This research was supported by the National Natural Science Foundation of China under Grant No. 41471369, the Open Research Fund of State Key Laboratory of Remote Sensing Science under Grant No. OFSLRSS201413, and the Major International Cooperation and Exchange Project "Comparative study on global environmental change using remote sensing technology" under Grant No. 4112011400. The authors appreciate the financial assistance provided by the China Scholarship Council.

Author Contributions: Yingkui Li and Cuizhen Wang designed the research. Qingting Li and Linlin Lu conducted the water body detection and dynamic analysis. Yue Sui collected the

MODIS data and performed quality assessment. All authors interpreted and reviewed the data. Qingting Li, Linlin Lu, Cuizhen Wang, and Yingkui Li wrote the final manuscript.

Conflicts of Interest: The authors declare no conflict of interest.

References

1. Oki, T.; Shinjiro, K. Global hydrological cycles and world water resources. *Science* **2006**, *313*, 1068–1072.
2. Soti, V.; Tran, A.; Bailly, J.S.; Puech, C.; Seen, D.L.; Bégué, A. Assessing optical earth observation systems for mapping and monitoring temporary ponds in arid areas. *Int. J. Appl. Earth Obs.* **2009**, *11*, 344–351.
3. Haas, E.M.; Bartholomé, E.; Lambin, E.F.; Vanacker, V. Remotely sensed surface water extent as an indicator of short-term changes in ecohydrological processes in sub-Saharan Western Africa. *Remote Sens. Environ.* **2011**, *115*, 3436–3445.
4. Haas, E.M.; Bartholomé, E.; Combal, B. Time series analysis of optical remote sensing data for the mapping of temporary surface water bodies in sub-Saharan western Africa. *J. Hydrol.* **2009**, *370*, 52–63.
5. Lehner, B.; Doll, P. Development and validation of a global database of lakes, reservoirs and wetlands. *J. Hydrol.* **2004**, *296*, 1–22.
6. Vector Map Level 0 (VMap0). Available online: http://earth-info.nga.mil/publications/vmap0.html (accessed on 16 February 2015).
7. Li, W.; Du, Z.; Ling, F.; Zhou, D.; Wang, H.; Gui, Y.; Sun, B.; Zhang, X. A comparison of land surface water mapping using the normalized difference water index from TM, ETM+ and ALI. *Remote Sens.* **2013**, *5*, 5530–5549.
8. Song, C.; Huang, B.; Ke, L. Modeling and analysis of lake water storage changes on the Tibetan Plateau using multi-mission satellite data. *Remote Sens. Environ.* **2013**, *135*, 25–35.
9. Fujita, K.; Sakai, A.; Nuimura, T.; Ymaguchi, S.; Sharma, R.R. Recent changes in Imja Glacial Lake and its damming moraine in the Nepal Himalaya revealed by in situ surveys and multi-temporal ASTER imagery. *Environ. Res. Lett.* **2009**, *4*.
10. Lira, J. Segmentation and morphology of open water bodies from multispectral images. *Int. J. Remote Sens.* **2006**, *27*, 4015–4038.
11. Lacaux, J.P.; Tourre, Y.M.; Vignolles, C.; Ndione, J.A.; Lafaye, M. Classification of ponds from high-spatial resolution remote sensing: Application to rift valley fever epidemics in Senegal. *Remote Sens. Environ.* **2007**, *106*, 66–74.
12. Chipman, J.W.; Lillesand, T.M. Satellite-based assessment of the dynamics of new lakes in southern Egypt. *Int. J. Remote Sens.* **2007**, *28*, 4365–4379.
13. Sun, F.; Zhao, Y.; Gong, P.; Ma, R.; Dai, Y. Monitoring dynamic changes of global land cover types: Fluctuations of major lakes in China every 8 days during 2000–2010. *Chin. Sci. Bull.* **2014**, *59*, 171–189.
14. Klein, I.; Dietz, A.J.; Gessner, U.; Galayeva, A.; Myrzakhmetov, A.; Kuenzer, C. Evaluation of seasonal water body extents in Central Asia over the past 27 years derived from medium-resolution remote sensing data. *Int. J. Appl. Earth Obs.* **2014**, *26*, 335–349.

15. Deus, D.; Gloaguen, R. Remote sensing analysis of lake dynamics in semi-arid regions: Implication for water resource management. Lake Manyara, East African Rift, Northern Tanzania. *Water* **2013**, *5*, 698–727.

16. Wang, J.; Sheng, Y.; Tong, T.S.D. Monitoring decadal lake dynamics across the Yangtze Basin downstream of Three Gorges Dam. *Remote Sens. Environ.* **2014**, *152*, 251–269.

17. Mcfeeters, S.K. The use of normalized difference water index (NDWI) in the delineation of open water features. *Int. J. Remote Sens.* **1996**, *17*, 1425–1432.

18. Xu, H. Modification of normalized difference water index (NDWI) to enhance open water features in remotely sensed imagery. *Int. J. Remote Sens.* **2006**, *27*, 3025–3033.

19. Mo, W.; Sun, H.; Zhong, S.; Huang, Y.; He, L. Research on the CIWI model and its application. *Remote Sens. Inf.* **2007**, *5*, 16–21.

20. Song, C.; Huang, B.; Ke, L.; Richards, K.S. Remote sensing of alpine lake water environment changes on the Tibetan Plateau and surroundings: A review. *ISPRS J. Photogramm. Remote Sens.* **2014**, *92*, 26–37.

21. Chinese Academy of Forestry. The List of National Nature Reserve in China. 2013. Available online: http://www.papc.cn/ (accessed on 16 August 2015).

22. Bai, J.; Chen, X.; Li, J.; Yang, L.; Fang, H. Changes in the area of inland lakes in arid regions of central Asia during the past 30 years. *Environ. Monit. Assess.* **2011**, *178*, 247–256.

23. Ma, M. Change in area of Ebinur Lake during the 1998–2005 period. *Int. J. Remote Sens.* **2007**, *28*, 5523–5533.

24. Ma, R.; Duan, H.; Hu, C.; Feng, X.; Li, A.; Ju, W. A half-century of changes in China's lakes: Global warming or human influence? *Geophys. Res. Lett.* **2010**, *37*.

25. Yang, X.K.; Lu, X.X. Drastic change in China's lakes and reservoirs over the past decades. *Sci. Rep.* **2014**, *4*.

26. Bai, R.; He, L.; Wu, J. Analysis on recent change of water area of the main lakes in Xinjiang based on MODIS data. *Arid Zone Res.* **2012**, *29*, 561–566.

27. Wang, T.; Yan, C.Z.; Song, X.; Xie, J.L. Monitoring recent trends in the area of Aeolian desertified land using Landsat images in China's Xinjiang region. *ISPRS J. Photogramm. Remote Sens.* **2012**, *68*, 184–190.

28. Lu, L.; Kuenzer, C.; Wang, C.; Guo, H.; Li, Q. Evaluation of three MODIS-derived vegetation index time series for dryland vegetation dynamics monitoring. *Remote Sens.* **2015**, *7*, 7597–7614.

29. Vermote, E.F.; Kotchenova, S.Y.; Ray, J.P. MODIS surface reflectance user's guide. Available online: http://modis-sr.ltdri.org/guide/MOD09_UserGuide_v1_3.pdf (accessed on 16 August 2015).

30. Huang, S.; Li, J.; Xu, M. Water surface variations monitoring and flood hazard analysis in Dongting Lake area using long-term Terra/MODIS data time series. *Nat. Hazards* **2012**, *62*, 93–100.

31. Caroll, M.L.; Townshend, J.R.; DiMiceli, C.W.; Noojipady, P.; Sohlberg, R.A. A new global raster water mask at 250 m resolution. *Int. J. Digit. Earth* **2009**, *2*, 291–308.

32. Huesler, F.; Jonas, T.; Wunderle, S.; Albrecht, S. Validation of a modified snow cover retrieval algorithm from historical 1-km AVHRR data over the European Alps. *Remote Sens. Environ.* **2012**, *121*, 497–515.

33. Friedl, M.A.; Sulla-Menashe, D.; Tan, B.; Schneider, A.; Ramankutty, N.; Sibley, A.; Huang, X. MODIS Collection 5 global land cover: Algorithm refinements and characterization of new datasets. *Remote Sens. Environ.* **2010**, *114*, 168–182.

34. Lu, L.; Kuenzer, C.; Guo, H.; Li, Q.; Long, T.; Li, X. A novel land cover classification map based on a MODIS time-series in Xinjiang, China. *Remote Sens.* **2014**, *6*, 3387–3408.

35. Feng, L.; Hu, C.; Chen, X.; Cai, X.; Tian, L.; Gan, W. Assessment of inundation changes of Poyang Lake using MODIS observations between 2000 and 2010. *Remote Sens. Environ.* **2012**, *121*, 80–92.

36. Lu, S.; Wu, B.; Yan, N.; Wang, H. Water body mapping method with HJ-1A/B satellite imagery. *Int. J. Appl. Earth Obs.* **2011**, *13*, 428–434.

37. Congalton, R.G.; Green, K. *Assessing the Accuracy of Remotely Sensed Data: Principles and Practices*, 2nd ed.; CRC Press: Boca Raton, FL, USA, 2009.

38. Frazier, P.S.; Page, K.J. Water body detection and delineation with Landsat TM data. *Photogramm. Remote Sens.* **2000**, *66*, 1461–1467.

39. Xinjiang Uygur Autonomous Region Bureau of Statistics. *Xinjiang Uygur Autonomous Region Statistical Yearbook*; China Statistics Press: Beijing, China; pp. 2001–2004.

40. Zhang, F.; Tiyip, T.; Johnson, V.C.; Kung, H.; Ding, J.; Zhou, M.; Fan, Y.; Kelimu, A.; Nurmuhammat, I. Evaluation of land desertification from 1990 to 2010 and its causes in Ebinur Lake region, Xinjiang China. *Environ. Earth Sci.* **2015**, *73*, 5731–5745.

41. Cheng, W.; Zhou, C.; Liu, H.; Zhang, Y.; Jiang, Y.; Zhang, Y.; Yao, Y. The oasis expansion and eco-environment change over the last 50 years in Manas River Valley, Xinjiang. *Sci. China Ser. D* **2006**, *49*, 163–175.

42. Wu, J.; Zeng, H.; Ma, L.; Bai, R. Recent changes of selected lake water resources in arid Xinjiang, northwestern China. *Quat. Sci.* **2012**, *32*, 142–150.

43. Maitiniyazi, A.; Kasimu, A. Change trend of surface water resources in Altai, Xinjiang China. *J. Desert Res.* **2014**, *34*, 1393–1401.

44. Gao, H.; Yao, Y. Quantitative effect of human activities on water level change of Bosten Lake in recent 50 years. *Sci. Geogr. Sin.* **2005**, *25*, 305–309.

45. Liu, L.; Zhao, J.; Zhang, J.; Peng, W.; Fan, J.; Zhang, T. Water balance of Lake Bosten using annual water-budget method for the past 50 years. *Arid Land Geogr.* **2013**, *36*, 33–40.

46. Guo, M.; Wu, W.; Zhou, X.; Chen, Y.; Li, J. Investigation of the dramatic changes in lake level of the Bosten Lake in northwestern China. *Theor. Appl. Climatol.* **2015**, *119*, 341–351.

47. Ma, D.; Zhang, L.; Wang, Q.; Zeng, Q.; Jiang, F.; Wang, Y.; Hu, R. Influence of warm-wet climate on Sailimu Lake. *J. Glaciol. Geocryol.* **2003**, *25*, 219–223.

48. Xue, J.; Zhong, W. Holocene climate variation denoted by Barkol lake sediments in northeastern Xinjiang and its possible linkage to the high and low latitude climates. *Sci. China Earth Sci.* **2011**, *54*, 603–614.

49. Zheng, S.; Luo, L. Variation of water quality in Balikun Lake in the last 18 years and water resource protection strategies. *Environ. Sci. Tech.* **2011**, *34*, 85–88.

50. Zhang, G.; Xie, H.; Kang, S.; Yi, D.; Ackley, S.F. Monitoring lake level changes on the Tibetan Plateau using ICESat altimetry data (2003–2009). *Remote Sens. Environ.* **2011**, *115*, 1733–1742.

51. Liao, J.; Shen, G.; Li, Y. Lake variations in response to climate change in the Tibetan Plateau in the past 40 years. *Int. J. Digit. Earth* **2013**, *6*, 534–549.

52. Yao, T.; Pu, J.; Lu, A.; Wang, Y.; Yu, W. Recent glacial retreat and its impact on hydrological processes on the Tibetan Plateau, China, and surrounding regions. *Arct. Antarct. Alp. Res.* **2007**, *39*, 642–650.

53. Cheng, G.; Wu, T. Responses of permafrost to climate change and their environmental significance, Qinghai-Tibet Plateau. *J. Geophys. Res.* **2007**, *112*.

54. Li, Y.; Liao, J.; Guo, H.; Liu, Z.; Shen, G. Patterns and potential drivers of dramatic changes in Tibetan lakes, 1972–2010. *PLoS ONE* **2014**, *9*.

55. Lioubimtseva, E.; Henebry, G.M. Climate and environmental change in arid Central Asia: Impacts, vulnerability, and adaptations. *J. Arid Environ.* **2009**, *73*, 963–977.

Remote Sensing Based Analysis of Recent Variations in Water Resources and Vegetation of a Semi-Arid Region

Shaowei Ning, Hiroshi Ishidaira, Parmeshwar Udmale and Yutaka Ichikawa

Abstract: This study is designed to demonstrate use of free remote sensing data to analyze response of water resources and grassland vegetation to a climate change induced prolonged drought in a sparsely gauged semi-arid region. Water resource changes over Hulun Lake region derived from monthly Gravity Recovery and Climate Experiment (GRACE) and Tropical Rainfall Measuring Mission (TRMM) products were analyzed. The Empirical Orthogonal Functions (EOF) analysis results from both GRACE and TRMM showed decreasing trends in water storage changes and precipitation over 2002 to 2007 and increasing trends after 2007 to 2012. Water storage and precipitation changes on the spatial and temporal scale showed a very consistent pattern. Further analysis proved that water storage changes were mainly caused by precipitation and temperature changes in this region. It is found that a large proportion of grassland vegetation recovered to its normal state after above average rainfall in the following years (2008–2012) and only a small proportion of grassland vegetation (16.5% of the study area) is degraded and failed to recover. These degraded grassland vegetation areas are categorized as ecologically vulnerable to climate change and protective strategies should be designed to prevent its further degradation.

Reprinted from *Water*. Cite as: Ning, S.; Ishidaira, H.; Udmale, P.; Ichikawa, Y. Remote Sensing Based Analysis of Recent Variations in Water Resources and Vegetation of a Semi-Arid Region. *Water* **2015**, *7*, 6039–6055.

1. Introduction

Freshwater resources are the lifeblood of our planet. It is fundamental to the biochemistry of all living organisms. The Earth's ecosystems are linked and maintained by water; it drives plant growth and provides a permanent habitat for many species, including ourselves. However, freshwater is a resource under considerable pressure. Its stored potential (surface water, ground water, soil moisture, ice, *etc.*) is increasingly facing challenges from climate changes as well as anthropogenic activities. That current and future climate change is expected to significantly impact freshwater systems including rivers, streams and lakes, in terms of flow and direction, timing, availability, temperature, and its inhabitants. So understanding the information about water resource change, its driving force and potential impact in the past and future is very important for water resource management and eco-environmental protection.

In recent years, the response of water resource and vegetation to the changing climate and anthropogenic effects has been discussed extensively at regional or global scales. With the rapid development of remote sensing techniques, the reliability of satellite products relevant to water resource monitoring has greatly improved. For example, changes in terrestrial water storage are measurable through satellite gravity based approximations of equivalent water thickness to a precision of 0.5 cm per year. [1]. Precipitation is monitored by multiple post-processing phases of currently available satellite data (*i.e.*, Tropical Rainfall Measuring Mission (TRMM)) to a resolution of millimeter per day [2]. Water level change in rivers and lakes is derived from altimetry satellites (*i.e.*, Jason-1/2, ENVISAT) to a sub-meter precision [3]. Hence, satellite observations have been increasingly used in such research, exploiting their potential of providing spatially continuous and temporally recurrent estimates over regional to global scales [4].

Zhang *et al.* [5] used monthly precipitation observations over global land areas to analyze precipitation trends in two twentieth century periods (1925–1999 and 1950–1999), and showed that anthropogenic forcing has had a detectable influence on observed changes in average precipitation within latitudinal bands, and that these changes cannot be explained by internal climate variability or natural forcing. Syed *et al.* [6] characterized terrestrial water storage variations using Gravity Recovery and Climate Experiment (GRACE) and Global Land Data Assimilation System (GLDAS) at global scale, the results illustrated spatial-temporal variability of water storage change over land, with implications for a better understanding of how terrestrial water storage responds to climate change and variability. Apart from global scale studies, Fensholt and Du [7,8] assessed the regional/continental precipitation trends and showed their influence on stream flow, water level, soil moisture and vegetation changes. Moiwo *et al.* [9] analyzed water storage dynamics in the North China Region (an important grain-production base) using GRACE, GLDAS products in conjunction with *in situ* hydro-climate data, the results showed a sharp water storage depletion from April 2002 through December 2009 in that area and water loss which was more a human than a natural cause had already negatively influenced millions of people in the region and beyond in terms of water supply crop production, eco-environmental system and social stability.

Besides that, much research also indicates that a remote sensing approach is a cost-efficient and accurate method to monitor inland water surface and water level (case of lakes and reservoirs) dynamics which are also affected directly by climate change and human activity. Dorothea *et al.* [10] used a Moderate Resolution Imaging Spectro-radiometer (MODIS) surface reflectance dataset and a Modified Normalized Difference Water Index (*MNDWI*) to map the variability of Lake Manyara's water surface area over 2000–2011. Their results implied that recent fluctuations of Lake Manyara's surface water area are a direct consequence of global and regional climate fluctuations. Duan *et al.* [11] proposed and evaluated a method that combined operational satellite altimetry databases with satellite imagery data to estimate water volume variations in Lake Tana. Results showed that satellite altimetry products were in good agreement with *in situ* water levels for Lake Tana ($R^2 = 0.97$). Estimated water volume variations derived from satellite altimetry products and LANDSAT TM/ETM+ agreed well with *in situ* water volume for Lake Tana, with R^2 higher than 0.95 and Root Mean Square Error (RMSE) 9.41% of corresponding mean value of *in situ* measurements.

With respect to vegetation, it plays a notably important role in soil conservation, atmosphere adjustment and maintenance of climatic and whole ecosystem stability because of its natural tie connecting atmosphere, water, and soil. Surface vegetation conditions are known for their sensitivity to natural changes and anthropogenic effects, thus serving as important proxies for regional eco-environmental and global climate fluctuations. Satellite based vegetation indexes such as normalized difference vegetation index (*NDVI*) as an efficient tool are widely used to examine the dynamic of vegetation health, density and degradation due to climate changes and anthropogenic effects [12,13].

As mentioned above, satellite remote sensing has shown promising results in the estimation of water resources and vegetation. However, in this study, we focus on the analysis of a combination of available satellite data including GRACE terrestrial water storage (TWS), TRMM, MODIS/LANDSAT, satellite altimetry data (Topex/Poseidon, Jason-1/2) coupled with *in situ* climate data to assess the water resource variation within a sparsely gauged area—the Hulun lake region and its impact on the eco-environment to provide useful information for future water resource management and eco-environmental protection. More specifically, this study aims (1) to provide a framework for a remote sensing based integrated assessment of water resource trends; (2) to detect trends in consistently established time series (from 2002 to 2012) of terrestrial water storage change and precipitation in a spatial distributed manner and (3) to infer the probable causes of water resource variations and its impacts on vegetation in order to contribute towards sustainable eco-environmental management.

2. Study Area

For this study, a representative case of the Daurian Steppe Eco-region (a most intact example of Eurasian Steppe) is selected (Figure 1). It is straddled over borders of three countries, namely, China, Mongolia and Russia (111° E–119° E, 47° N–50° N). The total study area is about 290,400 km². It covers a part of an ecologically important region—The Daurian International Protected Areas (DIPA), namely, The Hulun Lake

Nature Reserve grassland. The Hulun Lake Nature Reserve is a reserved grassland and least influenced by human activities [14,15]. This draws attention to identify the consequences of water resource changes (consecutive years of precipitation deficit and decline in TWS) on representative natural grassland-vegetation with minimum anthropogenic disturbances.

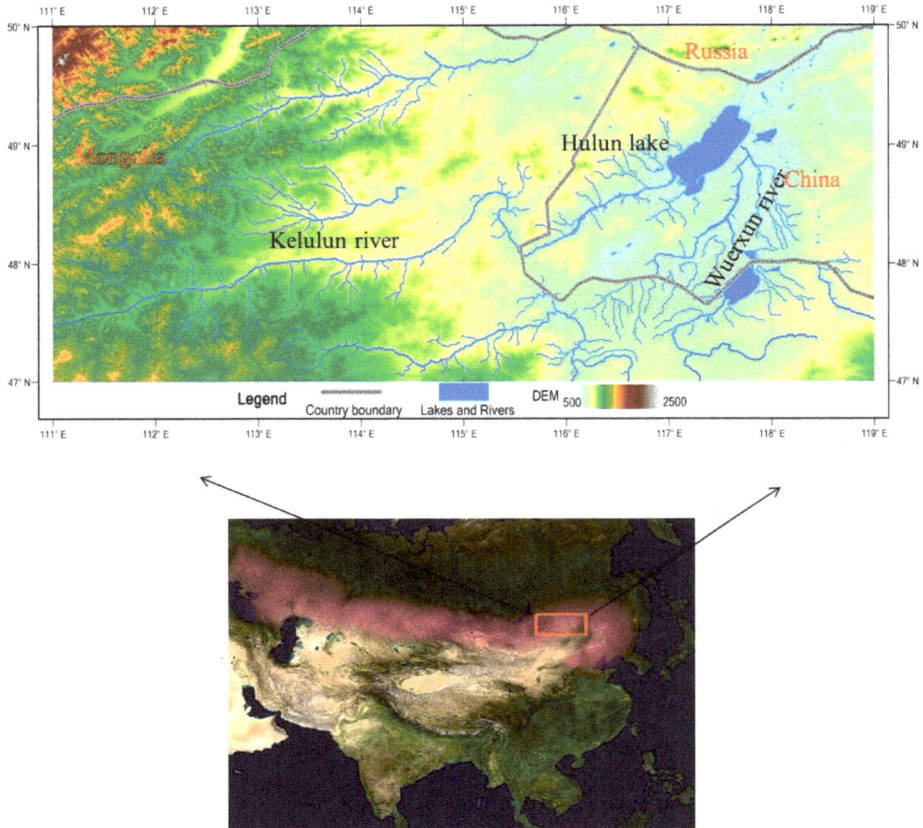

Figure 1. The geographic location and Shuttle Radar Topography Mission (SRTM)-based elevation map of the study area (red color region in the down figure shows the Eurasian Steppe zone).

The area has a mid-temperate semi-arid continental climate with the dominant mid-temperate zone characterized by drastic changes in winter and summer seasons. The average annual rainfall is about 293 mm, mainly concentrated in unfrozen season (from May to October). The average annual temperature ranges from $-13\,^{\circ}$C in winter to 12.3 $^{\circ}$C in summer, the average annual evaporation is around 249 mm, and the average annual relative humidity is 49%. The semi-arid climate with the strong winds is increasing the vulnerability of this area to desertification. Figure 2 shows the vegetation cover types of the study area. About 89.9% and 7.9% of the study area is occupied by annual grass-vegetation and forest (deciduous needle leaf, deciduous broadleaf, evergreen needle leaf and annual broadleaf vegetation), respectively. The surface water bodies cover about 2.2% of study area, with a major water bodies—Hulun Lake having surface area 2307 km^2 and Beier Lake with surface area 609 km^2. As shown in Figure 1, there are many inland rivers in the study area, but only two rivers, Kelulun and Wuerxun river, with annual discharge about 7×10^8 and 5.5×10^8 m^3, flow into Hulun lake which is the main drainage outlet in this area. Recently, however, the annual average discharge of these two rivers is less than 2×10^8 m^3. The fluctuations in water levels of Hulun lake can be used as an indicator of wet and dry conditions in the study area.

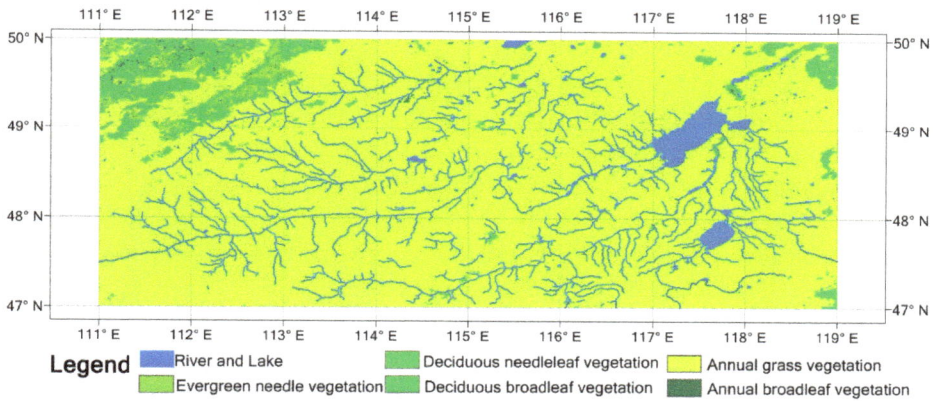

Figure 2. Vegetation map of study area.

3. Materials

3.1. Precipitation Data

The monthly precipitation for the period of 2002–2012 is obtained from Tropical Rainfall Measuring Mission (TRMM) [16]. TRMM products have been used in a number of studies of Inner Mongolia and surrounding precipitation, where they have been found to be adequate when compared with ground observations [17,18]. The product used in this study is referred as the TRMM and other precipitation dataset (denoted as 3B43). It is derived not only from TRMM sensors but also a number of other satellites and ground based rain gauged data. Monthly observed precipitation data (2002–2012) for five stations near the Hulun Lake are employed in this analysis to evaluate the applicability of satellite derived precipitation in study area.

3.2. Surface Air Temperature Data

This study uses a surface air temperature dataset namely Global Historical Climatology Network version 2 and the Climate Anomaly Monitoring System (GHCN + CAMS) which is a station observation-based global land monthly mean surface air temperature dataset at 0.5×0.5 degree resolution for the period of 1948 to the present. When compared with several existing observation-based land surface air temperature data sets, the preliminary results show that the quality of this new GHCN + CAMS land surface air temperature analysis is reasonably good and the new dataset can capture most common temporal-spatial features in the observed climatology and anomaly fields over both regional and global domains [19].

3.3. Lake Water Level Data

Monthly water level data for the Hulun Lake for the period of 2002–2012 is obtained from Hydroweb dataset [20]. The dataset is developed by Laboratoire d'Etudes en Oceanographie et Geode'sie Spatiale, Equipe Geodesie, Oceanographie, et Hydrologie Spatiales (LEGOS/GOHS) in Toulouse, France. It provides time series of water levels of large rivers, about 150 lakes and reservoirs, and wetlands around the world using the merged Topex/Poseidon, Jason-1and 2, ENVISAT and Geosat Fellow-On (GFO) data. Recent study has showed that the accuracy of water level data from Hydroweb was very high with R^2 range from 0.96 to 0.99 compared with *in situ* data in US, Netherlands and Ethiopia [11].

3.4. Satellite Imagery Data

MODIS Terra surface reflectance product (Mod09A1) [21] is employed to map and monitor spatial and temporal variations in water surface of the Hulun Lake from 2002 to 2012. Images on which snow covers the lake surface and surrounding region in winter time are not selected, because it is difficult to retrieve lake surface area in those scenes. Besides that, we also use several scenes of LANDSAT TM/ETM+ data [22] with spatial resolution of 30m to validate lake surface area derived from Mod09 A1.

3.5. GRACE TWS Data

TWS was derived from the latest version monthly GRACE gravity solutions (RL05) generated by the Center for Space Research at the University of Texas at Austin [23], from August 2002 through December 2012. Each solution consists of sets of spherical harmonic (Stokes) coefficients, C_{lm} and S_{lm}, to degree l and order m, both size less than or equal to 60. We calculated these coefficients by combining GRACE data with ocean model output as Swenson *et al.* [24] did. TWS calculation and the post processing method used here were similar with Duan *et al.* [25] with two Fan filter [26] radiuses 500 and 800 km, respectively. Finally, these coefficients were transformed into 1×1 degree gridded data that reflect vertically integrated water mass change represented by equivalent water thickness.

3.6. NDVI Data

NDVI dataset acquired from the Advanced Very High Resolution Radiometer (AVHRR) sensor aboard NOAA satellites processed by the Global Inventory Monitoring and Modeling Studies (GIMMS) at the National Aeronautics and Space Administration (NASA) [27]. The database ranges from July 1981 to December 2013 at a spatial resolution 8 km^2. The data are composited over approximately 15 day periods with the maximum value compositing technique, which minimizes the influences of atmospheric aerosols and clouds. This study analyzes the *NDVI* trend for the period from 2002 to 2012.

4. Methods

4.1. Water Resource Spatial-Temporal Series Analysis

Empirical Orthogonal Functions (EOF) analysis is a widely and easily used statistical method for analyzing large multidimensional datasets. When applied to a space-time dataset, EOF analysis can be used to decompose the observed variability into a set of spatial change patterns (EOFs), which are statistically independent and spatially orthogonal to the others, and a set of times series called time coefficients (PCs) that describes the time evolution of the particular EOFs. Together, the EOFs and PCs can be combined to reconstruct the variability in the original dataset. Basically, the goal of EOF analysis is to transform an original set of variables into a substantially smaller set of uncorrelated variables, which can reflect most of the information of the original dataset. It also has the ability to isolate various processes mixed in observation data [28]. The EOF has recently become a popular tool in various science areas such as meteorology, geology, and geography [29]. In this study, EOF analysis is applied to study both the spatial and temporal changes of precipitation and TWS.

4.2. Lake Water Surface Area Estimation

Several land cover classification methods can be used for delineating water bodies from multi-temporal satellite imagery to date from conventional unsupervised methods to more advanced artificial neural networks and support vector machine classifier [12,30]. The Modified Normalized Difference Water Index (*MNDWI*) method proposed by Xu [31] has been widely applied and proved efficient to retrieve water surface. The *MNDWI* is a band ratio index between Green (correspond to band 4 of MOD09A1 imagery) and Shortwave Infrared (*SWIR*, correspond to band 6 of MOD09A1 imagery) spectral bands that enhances water features. MNDWI is defined as:

$$MNDWI = \frac{Green - SWIR}{Green + SWIR} \tag{1}$$

Following other studies [32,33], we set the threshold for *MNDWI* to zero. *MNDWI* values > 0 represent water bodies and < 0 non-water cover types. Water features have positive *MNDWI* values because of their higher reflectance in the Green band than in the SWIR band while non-water features (soil and vegetation) have negative *MNDWI* values due to their low reflectance in the Green band than the SWIR band. However, some parts of Hulun Lake with the average depth 5.7 m are very shallow, usually less than 1 m and many aquatic plants grow out of water surface. This makes the *MNDWI* values negative in some grid located inside the lake. Here, we combine the *NDVI* value to eliminate that effect. We decide if *NDVI* < 0 or *MNDWI* > 0 and only one rule satisfies, then it is classified as water body. To validate the results, we estimate the water body from several scenes of LANDSAT imagery by traditional manual digitization, which is time consuming but has high accuracy.

4.3. NDVI Variation Trend Analysis Method

The Theil-Sen Median trend analysis, Mann-Kendall [34] are used to study the vegetation covered regions of our study area, namely, the temporal variation characteristics of the *NDVI* of the pixel covered region with *NDVI* values greater or equal to 0.1. The Theil-Sen trend analysis method can be effectively combined with the Mann-Kendall test. These are important methods for detecting the trend of long time series data, and this combination has been gradually used to analyse the long time series of vegetation reflecting the variation in trends of each pixel in a time series.

The Theil-Sen Median trend analysis is a robust trend statistical method, and it calculates the median slopes between all $n \cdot (n - 1)/2$ pair-wise combinations of the time series data. It is based on non-parametric statistics and is particularly effective for the estimation of trends in small series. The slope of Theil-Sen Median can represent the increase or decrease in the *NDVI* over the 11 years between 2002 and 2012 on a pixel scale. It is calculated by:

$$TS_{NDVI} = median \left(\frac{NDVI_m - NDVI_n}{m - n} \right), \ 2002 \le n < m \le 2012 \tag{2}$$

where, TS_{NDVI} refers to the Theil-Sen median, and $NDVI_m$, $NDVI_n$ represent the *NDVI* values for years of m and n, in case of $TS_{NDVI} > 0$, the *NDVI* shows a rising trend, otherwise, the *NDVI* presents a decreasing trend.

The Mann-Kendall test measures the significance of a trend. It is a non-parametric statistical test, and it has the advantage that samples do not need to follow certain distributions and is free from the interference of outliers. It has been broadly used to analyse the trends and variations at sites with hydrological and meteorological time series. Recently, this method has been applied to detection of vegetation variations over long time periods. The calculation algorithm is as follows:

It is assumed that $NDVI_m$, m stands for time series from 2002 to 2012. The statistics of Z is defined as:

$$Z = \begin{cases} \frac{S-1}{\sqrt{s(S)}} & , s > 0 \\ 0 & , s = 0 \\ \frac{S+1}{\sqrt{s(S)}} & , s < 0 \end{cases} \tag{3}$$

where, $S = \sum\limits_{n=1}^{t-1} \sum\limits_{m=n+1}^{t} sgn(NDVI_n - NDVI_m)$, $sgn(NDVI_n - NDVI_m) = $

$$\begin{cases} 1 & , NDVI_n - NDVI_m > 0 \\ 0 & , NDVI_n - NDVI_m = 0 \\ -1 & , NDVI_n - NDVI_m < 0 \end{cases},$$

$$s(s) = \frac{t(t-1)(2t+5)}{18} \tag{4}$$

where, $NDVI_m$ and $NDVI_n$ stands for the $NDVI$ values of the pixels m and n; t is the length of the time series; sgn is a sign function; and the Z statistic is located in the range of $(-\infty, +\infty)$. A given significance level, $|Z| > \mu_{1-\alpha/2}$, signifies that the times series shows significant variations on the level of α. Generally, the value of α is 0.05. In this study, we choose $\alpha = 0.05$, means that we measure the significance of the $NDVI$ trend over period from 2002 to 2012 on pixel scale at a confidence level of 0.05.

5. Results and Discussion

5.1. Precipitation and Temperature Variation Analysis

The numbers of rain-gauge stations in the study area are limited. Hence, we used TRMM monthly data to analyze precipitation trends in this study. To confirm the feasibility of TRMM data, observed precipitation from five rain gauge stations (Table 1) located in the vicinity of the study area are used. The correlation between two data sets for respective grids is observed to be in the range of 0.74–0.94 over the period of 2002–2012 as shown in Table 1. This confirms the applicability of TRMM data for precipitation trend analysis in this study. Studies by Yatagai *et al.* [17] and Chen *et al.* [18] also validated the applicability of TRMM data in this region.

After applying EOF to TRMM data, we found three dominant EOFs and PCs in study area (as shown in Figure 3). EOF1 and PC1 represent about 65% of total variance of precipitation, which shows superposition of annual and seasonal variability. The EOF1 is found positive throughout the study area with high values in central part highlighting uniform changing pattern over study area. A significant decreasing trend is observed over a period of 2002–2007, however an increasing trend from 2008 to 2012 can be seen from PC1. Low negative PC1 values corresponding to summer season from 2003 to 2007 indicates five consecutive years of below average rainfall, which induced a very serious drought. Using a linear regression, we found an average precipitation decline of 23.1 mm/year and increase of 18.2 mm/year for the periods of 2002–2007 and 2008–2012, respectively. We do not interpret the second and third mode of EOF on precipitation variation (*i.e.*, EOF2, PC2 and EOF3, PC3 here, respectively), since the temporal pattern change is not obvious, and it accounts for only 13% and 6% of variance in precipitation, respectively.

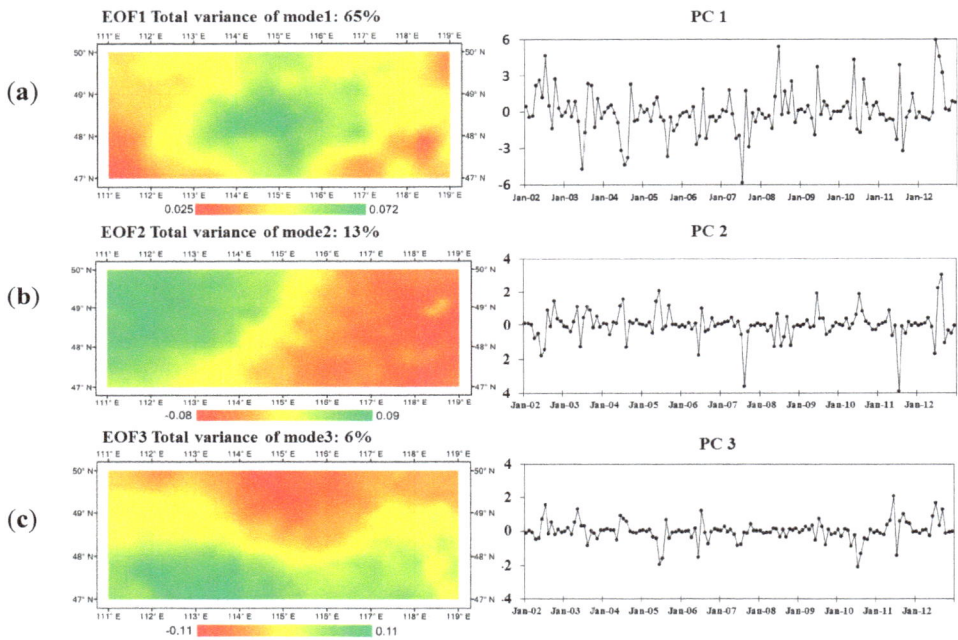

Figure 3. EOF decomposition of precipitation changes derived from TRMM satellite data over study area. EOF patterns are shown in left side and corresponding unit-less temporal patterns (PCs) in right side. (**a**): The first change mode of precipitation changes; (**b**): The second change mode of precipitation changes; (**c**): The third change mode of precipitation changes.

We also analyzed the warm (May–October) and cold (November–April) season average temperature over study area as shown in Figure 4. Average temperature of warm and cold season shows opposite change pattern against precipitation. Rising temperature may have caused more evapotranspiration, then further exacerbated water storage depletion and drought.

Table 1. Correlation between gauged precipitation with tropical rainfall measuring mission (TRMM) precipitation in respective grids.

No.	Station Name	Latitude	Longitude	R^2
1	Xinyouqi	48.67° N	116.82° E	0.75
2	Xinzuoqi	48.21° N	118.27° E	0.74
3	Manzhouli	49.57° N	117.43° E	0.82
4	Hailaer	49.22° N	119.75° E	0.92
5	Aershan	47.17° N	119.93° E	0.94

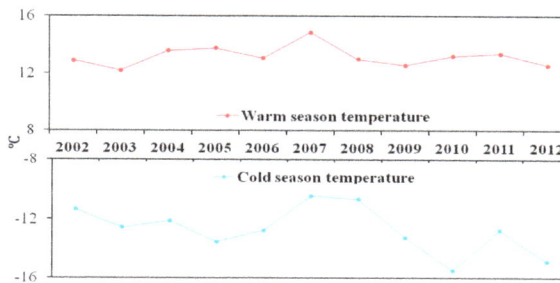

Figure 4. Temperature trend over study area from 2002 to 2012.

5.2. Water Storage Change

Figure 5 shows the overall EOF results of water storage variability in spatial and temporal scale over the study area. EOF analysis result provides us the general understanding of water storage change condition. EOF1 and PC1 represent 88% of total variance in water storage changes and EOF2 and PC2 represent only 10%. For EOF1, all values are positive, which means all places have same change pattern with high significance in the central and north part of study area, while the corresponding PC1 shows the dominant trend. By a linear regression, we found an average water storage decline of 14.2 mm/year and an increase of 3 mm/year in the study area for the periods of 2002 to 2007 and 2008 to 2012, respectively. EOF2 delineates a spatial east-west dipole structure and PC2 shows an increasing trend in the southwest corner and decreasing trend in the east of study area over period of 2008 to 2012. Overall, the water storage first reduced sharply (2002 to 2007) and then restored slightly (2008–2012), especially in the central and north part of study area (EOF1).

EOF2 and PC2 (Figure 3) representing 13% of total rainfall change pattern shows the similar east west dipole structure of the TWS pattern 2 (as shown in Figure 5). It is found that EOF1/2 and PC1/2 of precipitation (Figure 3) is very consistent with EOF1/2 and PC1/2 of TWS (Figure 5). This explains that the precipitation is one of the major driving factors behind water storage changes. Similar trends in precipitation and water storage changes observed over whole study area (as shown in Figure 6), which indicates a very sharp decreasing trend over the period of 2002–2007. However, for the period 2007–2012, in spite of the increasing precipitation trend, TWS did not show a significant increasing trend as that of precipitation but increased slightly. There may be two reasons for that: first, as we mentioned in the study area section, water income from the two rivers flow to this region has been lower than usual recently; second, actual evapotranspiration has increased because of above-normal vegetation development (as we will explain in Section 5.4).

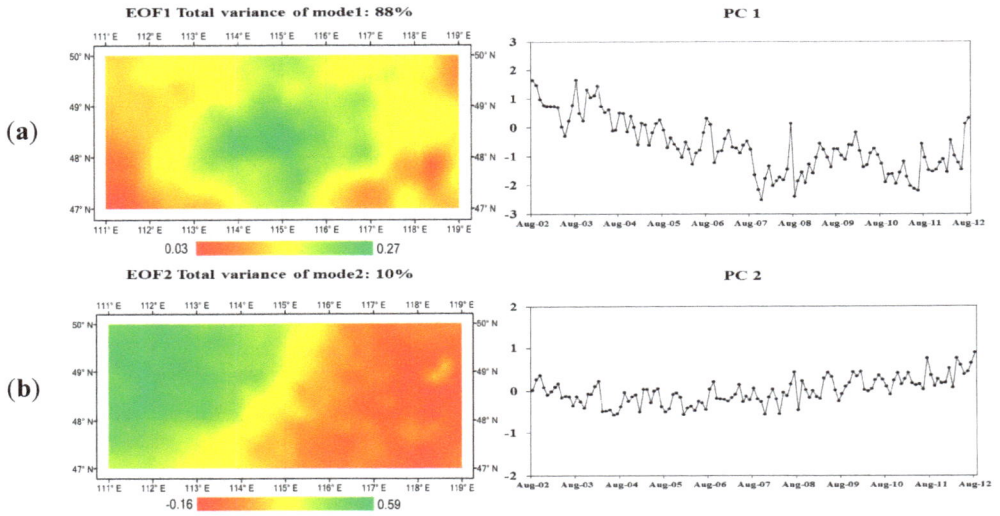

Figure 5. EOF decomposition of TWS changes over the study area. EOF patterns are shown on the left side and corresponding unit-less temporal patterns (PCs) are shown on the right side. (**a**): The first change mode of TWS changes; (**b**): The second change mode of TWS changes.

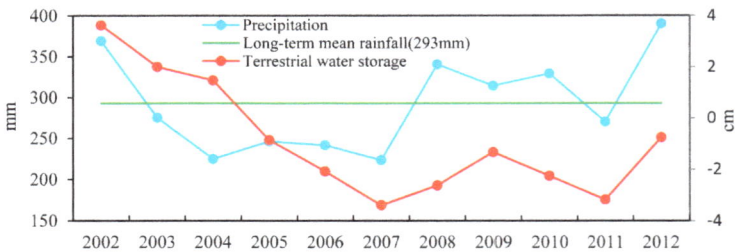

Figure 6. Annual precipitation and water storage changes series from 2002 to 2012.

44

5.3. Lake Response to Water Storage Change

As mentioned in the study area introduction, Lake Hulun drains in most parts of the study area, so its water volume change is a good indicator of water resource variation for this region. Table 2 shows the comparison between MODIS derived and LANDSAT digitalized results of lake surface area. It shows a very close relation with R^2 value as high as 0.95 and very low absolute and relative errors. It demonstrates the validity of MODIS derived lake surface area with high accuracy in unfrozen season (May–October). Figure 7 gives the lake level and lake water area time series with their correlation. Both shows rapid declination from 2002 to 2009 with about three meters of water level drop and 400 km^2 of shrinkage in lake surface area, respectively, and remained stable after 2009. This temporal change pattern is not consistent with water storage and precipitation change. Although precipitation increased after 2007, lake volume still decreased till 2009 and did not show obvious rise after that. For this phenomenon, there may be several possible reasons. Firstly, because of drought, Lake Hulun got less inflow from the upper reaches of two main rivers (Kelulun and Wuerxun) during 2002 to 2007. According to local news, these two rivers dried up from September 2007 and did not discharge water into the lake for almost for one year [35]. Secondly, people and livestock suffering from drought, which has driven water scarcity around the lake may have withdrawn more water from it than a normal year. Finally, the most important point, a large proportion of precipitation may have been contributed to recover soil moisture deficit and depleted groundwater levels due to consecutive years of droughts. This might have delayed the river inflow and groundwater discharge to the lake. In summary, when drought attacks this region, it needs more time and water to recover to the normal state even after enough rainfall.

Table 2. Comparison of Moderate Resolution Imaging Spectro-radiometer (MODIS) and LANDSAT derived Hulun lake area.

LANDSAT		MODIS		Absolute Error (km^2)	Relative Error (%)
Date	Lake Surface Area (km^2)	Date	Lake Surface Area (km^2)		
1 July 2000	2306.6	4 July 2000	2290.1	−16.4	−0.71%
6 September 2001	2236.4	7 September 2001	2186.3	−50.2	−2.24%
8 August 2002	2154.4	6 August 2002	2221.2	66.7	3.10%
27 August 2003	2114.2	30 August 2003	2106.9	−7.3	−0.35%
13 August 2004	2002.5	13 August 2004	2058.2	55.7	2.78%
16 August 2005	1977.6	14 August 2005	1948.8	−28.7	−1.45%
26 July 2006	1938.6	29 July 2006	1942.9	4.3	0.22%
29 July 2007	1907.2	29 July 2007	1902.3	−4.9	−0.26%
25 August 2008	1837	13 August 2008	1862.1	25.1	1.36%
5 October 2009	1791.5	1 October 2009	1772.8	−18.7	−1.04%

45

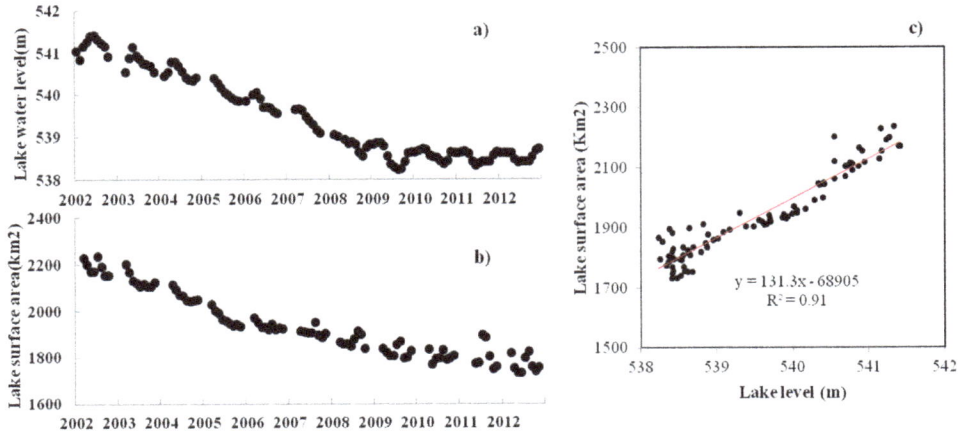

Figure 7. (a) Lake water level; (b) lake water surface area; and (c) correlation between lake water level and surface area.

5.4. Vegetation Response to Water Resource Change

Average *NDVI* distribution over the period of 2002–2012 is shown in Figure 8a. It can be seen that the areas with high *NDVI* (0.4–0.5) are located in northwestern part. *NDVI* values in central and south part of study area, and in the vicinity of lake grassland region are relatively low (0.2–0.3). The variations in trends of *NDVI* can be effectively captured by the Theil-Sen median trend analysis and the Mann-Kendall test to reflect the spatial distribution of vegetation responses to water resource changes. Because regions with a TS_{NDVI} of 0 strictly do not exist, we made the following classifications according to the real conditions of the TS_{NDVI}. Regions with a TS_{NDVI} from −0.0005 to 0.0005 are categorized as stable regions, regions with TS_{NDVI} larger than or equal to 0.0005 are categorized improved regions and regions with TS_{NDVI} less than −0.0005 are categorized as degraded areas. Moreover, significance test results of the Mann-Kendall test, at the confidence level of 0.05, are determined as significance variations ($Z > 1.96$ or $Z < −1.96$) or insignificant variations ($−1.96 \leq Z \leq 1.96$). Through combining the classification results of the Theil-Sen median trend analysis and the MK test, it is comparable with the data of trend variations of the *NDVI*. The results are summarized into five classes as shown in Table 3. It shows the regions with vegetation condition improvement, regions with stable vegetation condition, and regions with vegetation degradation, which account for 70%, 13.5% and 16.5%, respectively.

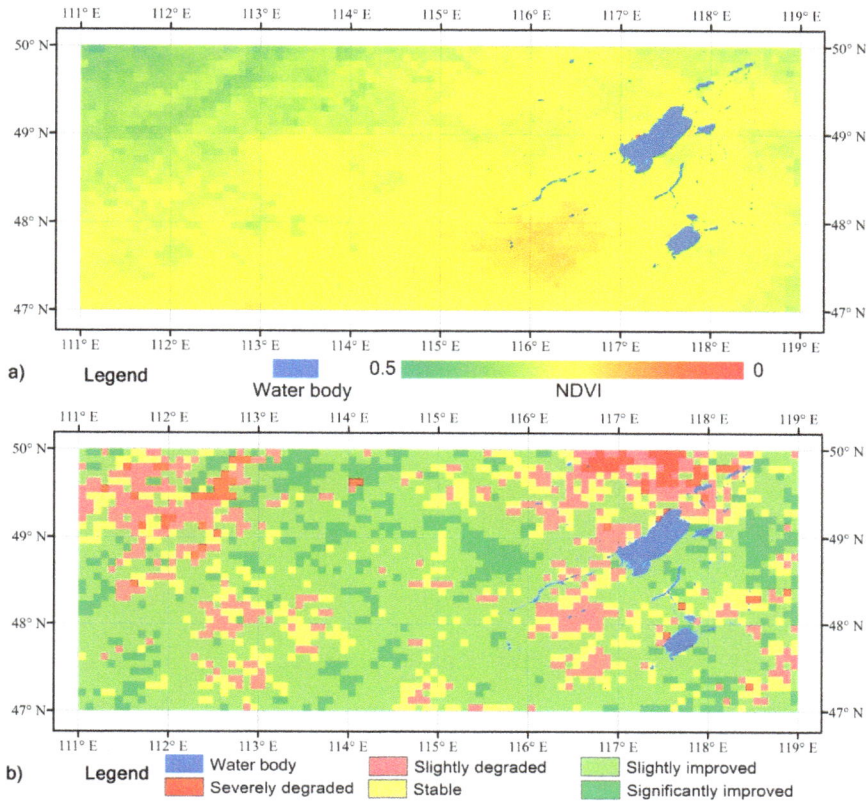

Figure 8. Spatial distribution of (**a**) average *NDVI* and (**b**) trends of inter-annual *NDVI* change over 2002 to 2012.

Table 3. Changing trend of normalized difference vegetation index (NDVI) in the study area.

TS_{NDVI}	Z	*NDVI* Trend	Area (%)
<-0.0005	<-1.96	Severely degraded	2.69%
<-0.0005	$-1.96–1.96$	Slightly degraded	13.80%
$-0.0005–0.0005$	$-1.96–1.96$	Stable	13.51%
≥ 0.005	$-1.96–1.96$	Slightly improved	57.35%
≥ 0.005	≥ 1.96	Improved	12.64%

As shown in Figure 8b, the region with improved vegetation condition is far larger than the regions with degrading trend, and mainly scattered in the central part of the study area. This indicates resilience of grassland vegetation to droughts. The decreasing precipitation from 2002 to 2007 had not much impact on vegetation in the central part. It can be seen that the vegetation condition recovered quickly after the withdrawal of drought and the increase in precipitation in the following years. Hence, average *NDVI* over 2002–2012 showed an increasing trend. This shows the close relationship between vegetation conditions and precipitation as found by other researchers [36]. It is obvious to detect such vegetation conditions with a decreasing trend distributed in the northwest mountain areas and Lake Hulun surrounding areas (especially, in the northern and western part, Figure 2 shows that these are the evergreen and deciduous forest areas). Decrease in water resources had serious negative and long-term impacts on vegetation conditions. Even the precipitation increased after 2007, vegetation conditions did not recover to its normal state. We can categorize those areas as ecologically vulnerable regions, where more protective measures and effective management are needed. Possible causes about vegetation degradation in the north of Lake Hulun may have also affected by anthropogenic activities. Since it became the more important trading port between China and Russia after 2000, human activities have been more frequent than before. In addition, rapid urbanization affected vegetation conditions in this area.

6. Conclusions

In the present study, water resource changes over Hulun Lake region derived from monthly GRACE and TRMM products were analyzed. The EOF results from both GRACE and TRMM showed decreasing trends in water storage changes and precipitation over 2002 to 2007 and increasing trends after 2007 to 2012. Water storage and precipitation changes in spatial and temporal scale showed a very consistent pattern. Further analysis proved that water storage changes were mainly caused by precipitation and temperature changes in this region. Based on the general understanding about water resource variations, we checked the response of Hulun Lake. Results indicated that lake level and lake surface area both declined during 2002 to 2009, with about three meters of water level drop and 400 km^2 shrinkage in lake surface area, respectively, and then remained stable after 2009 even though precipitation had recovered back to pre-2002 level. We can infer that water resource conditions needed more time and precipitation to recover from a long term drought in this typical semi-arid region. Furthermore, the vegetation response to water resource variations reflected that vegetation resilience to drought in most regions was high, forests were less resilient to drought than grasslands. Drought did not bring serious negative implications on vegetation growing conditions. Only 16.5% of the study area which is located in the northern and western sections of Hulun Lake and northwest mountain areas showed vegetation degradation. These areas that are categorized as ecological vulnerable regions need more protection and effective management in the future. Finally, this study demonstrated the feasibility of estimating water resource variation on the spatial–temporal scale and its impact on eco-environment using freely available remote sensing data in a sparsely gauged semi-arid area, which can also be adapted to other regions. Such spatiotemporally distributed analysis at the regional and basin level is particularly important considering that most of the water management and eco-environmental protection also take place at these scales.

Acknowledgments: The authors would like to express their sincere gratitude to the Ministry of Education, Culture, Sports, Science and Technology (Monbukagakusho: MEXT) and University of Yamanashi, Japan for providing financial assistance to undertake this study.

Author Contributions: Then manuscript was primarily written by Shaowei Ning, with Parmeshwar Udmale contributing to its preparation and English check. Hiroshi Ishidaira and Yutaka Ichikawa supervised the research and critically reviewed the draft.

Conflicts of Interest: The authors declare no conflict of interests.

References

1. Long, D.; Scanlon, B.R.; Longuevergne, L.; Sun, A.-Y.; Fernando, D.N.; Save, H. GRACE satellites monitor large depletion in water storage in response to the 2011 drought in Texas. *Geophys. Res. Lett.* **2013**, *40*, 3395–3401.

2. Huffman, G.J.; Bolvin, D.T.; Nelkin, E.J.; Wolff, D.B.; Adler, R.F.; Gu, G.; Yang, H.; Kenneth, P.B.; Erich, F.; Stocker, E.F. The TRMM multi-satellite precipitation analysis (TMPA): Quasi-global, multiyear, combined-sensor precipitation estimates at fine scales. *J. Hydrometeorol.* **2007**, *8*, 38–55.

3. Wang, X.; Gong, P.; Zhao, Y.; Xu, Y.; Cheng, X.; Niu, Z.; Luo, Z.; Huang, H.; Sun, F.; Li, X. Water-level changes in China's large lakes determined from ICESat/GLAS data. *Remote Sens. Environ.* **2013**, *32*, 131–144.

4. Gokmen, M.; Vekerdy, Z.; Verhoef, W.; Batelaan, O. Satellite based analysis of recent trends in the ecohydrology of a semi-arid region. *Hydrol. Earth Syst. Sci. Discuss.* **2013**, *10*, 6193–6235.

5. Zhang, X.; Zwiers, F.W.; Hegerl, G.C.; Lambert, F.H.; Gillett, N.P.; Solomon, S.; Nozawa, T. Detection of human influence on twentieth-century precipitation trends. *Nature* **2007**, *448*, 461–465. PubMed]

6. Syed, T.H.; Famiglietti, J.S.; Rodell, M.; Chen, J.; Wilson, C.R. Analysis of terrestrial water storage changes from GRACE and GLDAS. *Water Resour. Res.* **2008**, *44*.

7. Fensholt, R.; Rasmussen, K. Analysis of trends in the Sahelian "rain-use efficiency" using GIMMS NDVI, RFE and GPCP rainfall data. *Remote Sens. Environ.* **2011**, *115*, 438–451.

8. Du, J.; He, F.; Zhang, Z.; Shi, P. Precipitation change and human impacts on hydrologic variables in Zhengshui River Basin, China. *Stoch. Environ. Res. Risk Assess.* **2011**, *25*, 1013–1025.

9. Moiwo, J.P.; Tao, F.; Lu, W. Analysis of satellite-based and in situ hydro-climatic data depicts water storage depletion in North China Region. *Hydrol. Process.* **2012**, *27*, 1110–1020.

10. Dorothea, D.; Richard, G. Remote sensing analysis of lake dynamics in semi-arid regions: Implication for water resource management. *Water* **2013**, *5*, 698–727.

11. Duan, Z.; Bastiaanssen, W.G.M. Estimating water volume variations in lakes and reservoirs from four operational satellite altimetry databases and satellite imagery data. *Remote Sens. Environ.* **2013**, *134*, 403–416.

12. Eckert, S.; Hüsler, F.; Liniger, H.; Hodel, E. Trend analysis of MODIS NDVI time series for detecting land degradation and regeneration in Mongolia. *J. Arid Environ.* **2015**, *113*, 16–28.

13. Jiang, W.; Yuan, L.; Wang, W.; Cao, R.; Zhang, Y.; Shen, W. Spatio-temporal analysis of vegetation variation in the Yellow River Basin. *Ecol. Indic.* **2015**, *51*, 117–126.

14. Shinoda, M.; Nachinshonhor, G.U.; Nemoto, M. Impact of drought on vegetation dynamics of the Mongolian steppe: A field experiment. *J. Arid Environ.* **2010**, *74*, 63–69.

15. Da Silva, E.C.; de Albuquerque, M.B.; de Azevedo Neto, A.D.; da Silva Junior, C.D. Drought and its consequences to plants—From individual to ecosystem. In *Responses of Organisms to Water Stress*; InTech: Rijeka, Croatia, 2013.

16. Kummerow, C.; Barnes, W.; Kozu, T.; Shiue, J.; Simpson, J. The tropical rainfall measuring mission (TRMM) sensor package. *J. Atmos. Ocean. Technol.* **1998**, *15*, 809–817.

17. Yatagai, A.; Xie, P.; Kitoh, A. Utilization of a new gauge-based daily precipitation dataset over monsoon Asia for validation of the daily precipitation climatology simulated by the MRI/JMA 20-km-mesh AGCM. *SOLA* **2005**, *1*, 193–196.

18. Chen, Y.; Velicogna, I.; Famiglietti, J.S.; Randerson, J.T. Satellite observations of terrestrial water storage provide early warning information about drought and fire season severity in the Amazon. *J. Geophys. Res. Biogeosci.* **2013**, *118*, 495–504.

19. Fan, Y.; van den Dool, H. A global monthly land surface air temperature analysis for 1948–present. *J. Geophys. Res.* **2008**, *113*.

20. LEGOS/GOHS. Available: http://www.legos.obs-mip.fr (accessed on 3 January 2013).

21. U.S. Government Computer. Available online: http://e4ftl01.cr.usgs.gov/MOLT/MOD09A1 (accessed on 14 March 2015).

22. Landsat Data Access. Available online: http://landsat.usgs.gov/Landsat_Search_and_Download.php (accessed on 20 March 2015).

23. Center for Space Research. Available online: ftp://podaac.jpl.nasa.gov/allData/grace/L2/CSR/RL05/ (accessed on 12 February 2013).

24. Swenson, S.; Chambers, D.; Wahr, J. Estimating geocenter variations from a combination of GRACE and ocean model output. *J. Geophys. Res.* **2008**, *113*.

25. Duan, X.J.; Guo, J.Y.; Shum, C.K.; Wal, W. On the postprocessing removal of correlated errors in GRACE temporal gravity field solutions. *J. Geod.* **2009**, *83*, 1095–1106.

26. Zhang, Z.-Z.; Chao, B.F.; Lu, Y.; Hsu, H.-T. An effective filtering for GRACE time-variable gravity: Fan filter. *Geophys. Res. Lett.* **2009**, *36*.

27. GIMMS, "ECOCAST", NASA. Available online: http://ecocast.arc.nasa.gov/data/pub/gimms/ (accessed on 19 April 2015).

28. Longuevergne, L.; Florsch, N.; Elsass, P. Extracting coherent regional information from local measurements with Karhunen-Loève transform: Case study of an alluvial aquifer (Rhine valley, France and Germany). *Water Resour. Res.* **2007**, *43*.

29. Perry, M.A.; Niemann, J.D. Analysis and estimation of soil moisture at the catchment scale using EOFs. *J. Hydrol.* **2007**, *334*, 388–404.

30. Song, X.; Duan, Z.; Jiang, X. Comparison of artificial neural networks and support vector machine classifiers for land cover classification in Northern China using a SPOT-5 HRG image. *Int. J. Remote Sens.* **2012**, *33*, 3301–3320.

31. Xu, H. Modification of normalised difference water index (NDWI) to enhance open water features in remotely sensed imagery. *Int. J. Remote Sens.* **2006**, *27*, 3025–3033.

32. Deus, D.; Gloaguen, R. Remote Sensing Analysis of Lake Dynamics in Semi-Arid Regions: Implication for Water Resource Management, Lake Manyara, East African Rift, Northern Tanzania. *Water* **2013**, *5*, 698–727.

33. Cai, B.F.; Yu, R. Advance and evaluation in the long time series vegetation trends research based on remote sensing. *J. Remote Sens.* **2009**, *13*, 1170–1186.

34. Hoaglin, D.C.; Mosteller, F.; Tukey, J.W. *Understanding Robust and Exploratory Data Analysis*; Wiley: New York, NY, USA, 1983.

35. National Center for Agricultural Scientific Data Sharing. Available online: http://grassland.agridata.cn/client.c?method=fgbz&menuid=106&id=59 (accessed on 26 May 2015).

36. Ding, M.; Zhang, Y.; Liu, L.; Zhang, W.; Wang, Z.; Bai, W. The relationship between *NDVI* and precipitation on the Tibetan Plateau. *J. Geogr. Sci.* **2007**, *17*, 259–268.

Hydrological Evaluation of Lake Chad Basin Using Space Borne and Hydrological Model Observations

Willibroad Gabila Buma, Sang-Il Lee and Jae Young Seo

Abstract: Sustainable water resource management requires the assessment of hydrological changes in response to climate fluctuations and anthropogenic activities in any given area. A quantitative estimation of water balance entities is important to understand the variations within a basin. Water resources in remote areas with little infrastructure and technological knowhow suffer from poor documentation, rendering water management difficult and unreliable. This study analyzes the changes in the hydrological behavior of the Lake Chad basin with extreme climatic and environmental conditions that hinder the collection of field observations. Total water storage (TWS) from the Gravity Recovery and Climate Experiment (GRACE), lake level variations from satellite altimetry, and water fluxes and soil moisture from Global Land Data Assimilation System (GLDAS) were used to study the spatiotemporal variability of the hydrological parameters of the Lake Chad basin. The estimated TWS varies in a similar pattern as the lake water level. TWS in the basin area is governed by the lake's surface water. The subsurface water volume changes were derived by combining the altimetric lake volume with the TWS over the drainage basin. The results were compared with groundwater outputs from WaterGAP Global Hydrology Model (WGHM), with both showing a somewhat similar pattern. These results could provide an insight to the availability of water resources in the Lake Chad basin for current and future management purposes.

Reprinted from *Water*. Cite as: Buma, W.G.; Lee, S.-I.; Seo, J.Y. Hydrological Evaluation of Lake Chad Basin Using Space Borne and Hydrological Model Observations. *Water* **2016**, *8*, 205.

1. Introduction

In some developing parts of the world, very limited and low quality ground water data often hinder proper water management studies [1]. Moreover, the estimation of large-scale water balance using these limited ground-based measurements is prone to inaccuracies [2]. Sometimes, obtaining these datasets from the appropriate authorities involves lengthy administrative procedures, rendering studies extremely difficult.

Some, if not all of these, are associated with the Lake Chad Basin (LCB). Its scale and lack of modern infrastructure are major challenges for data collection, analysis and management [3]. Under these circumstances, satellite gravimetric, altimetry and hydrological models have proven useful in the study of these water bodies.

The Gravity Recovery and Climate Experiment (GRACE) is a joint mission between Deutsche Forschungsanstalt fur Luft und Raumfahrt (DLR) and National Aeronautics and Space Administration (NASA) that was launched in 2002. It records the Earth's time variable gravity field with a temporal and spatial resolution usually within a few hundreds of kilometers. These products together with the products [data] from satellite altimetry, Global Land Data Assimilation System (GLDAS), and WaterGAP Global Hydrology Model (WGHM) were used for this study. Satellite and hydrologic model products have widely been used for ground-based hydrological measurements and studies, and they also serve as inputs to land surface and atmospheric models [4,5]. They are also used to verify these models.

Lake Chad Basin (LCB) extends between latitude 6° N and 24° N, and between longitude 8° E and 24° E (Table 1). It covers an area of about 2,400,000 km^2, which is equivalent to 8% of the total area of the African Continent. About 20% of this total area is the conventional basin, which is under the mandate of the Lake Chad Basin Commission (LCBC).

Table 1. Morphometric data for Lake Chad.

Parameter	Lake Chad Basin
Location	6° N and 20° N, 7° E and 25° E
Catchment area	2.4×10^6 km^2
Conventional Basin	427,500 km^2
Lake area	1350 km^2

Lake Chad itself occupies the central region of the LCB. It is a closed lake, predominantly fed by two perennial rivers (the Chari and the Logone) and an ephemeral one (the Komadugu Yobe) (Figure 1). It serves as a source of freshwater and fish, and also aids pastoral and agricultural land for a population of 30 million across the basin by offering a relatively easy and permanent access to water [6].

Increase in population, dam constructions, and irrigation development facilities during the last four decades have caused the surface area of Lake Chad to shrink from 24,000 km^2 to 1300 km^2 [7,8] (Figure 2). Studies have shown that the decrease was due to persistent drought and irrigation activities in the area [9–11].

In an attempt to manage and reduce the persistent droughts in this area, the water transfer project, whose main objective is halting the shrinkage of Lake Chad through an inflow of water coming from the Ubangi River, was introduced by the LCBC.

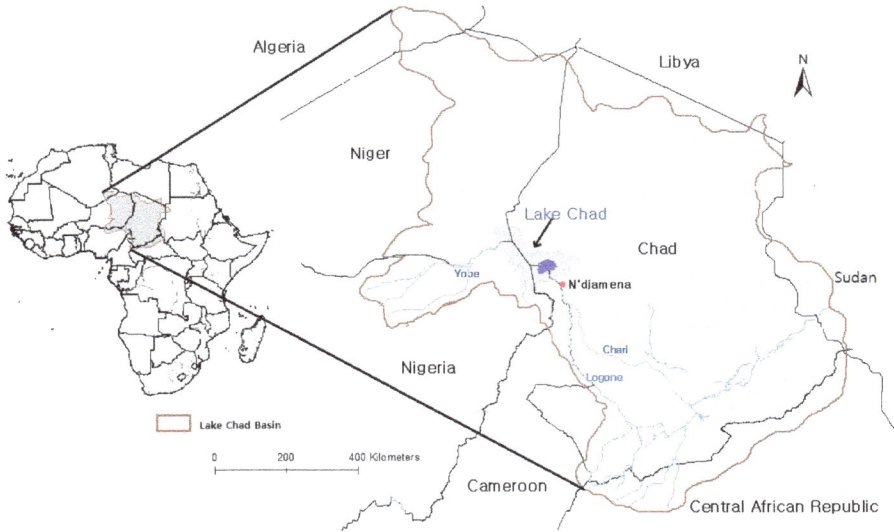

Figure 1. Locations of the Lake Chad Basin.

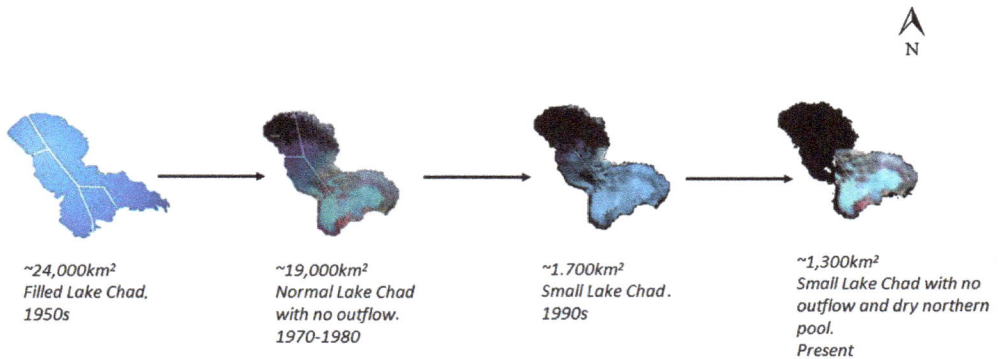

~24,000km²
Filled Lake Chad.
1950s

~19,000km²
Normal Lake Chad
with no outflow.
1970-1980

~1.700km²
Small Lake Chad.
1990s

~1,300km²
Small Lake Chad with no
outflow and dry northern
pool.
Present

Figure 2. Schematics of the state of Lake Chad from Landsat 5 images; courtesy of NASA. Modified after [12].

Water scarcity triggers food insecurity, poverty, migration and conflicts. As such, it is very important for a population or nations as a whole to secure stable and reliable water resource management techniques. A step towards this would be to understand the changes experienced by the water body in their vicinity. With limited and unreliable *in situ* data collection, understanding and documenting these changes can be very challenging and costly prompting researchers to rely on the use of satellite gravimetric, altimetry and hydrological models in monitoring water resources in such remote areas (Table 2).

Previous studies have applied remote sensing and satellite data to the Lake Chad region-some of which focused on the changes in stream flow patterns connected to the lake [9,13–15], Leblanc *et al.* [16] reported on the existence of a mega-lake Chad. They used satellite images from Landsat and Moderate Resolution Imaging Spectrometer (MODIS) for their studies. Thermal remote sensing techniques, such as the Meteosat thermal maximum composite data, was used to account for the variability of inundated areas within the lake under flooded vegetation [17]. Satellite imagery and GRACE data were used to study the regional hydrogeology and made an attempt at estimating the actual evapotranspiration over the LCB [5]. Altimetry data and ground-based information were used to predict the downstream lake and marsh heights using imperial regression techniques [18].

Few studies have used remote sensing data for investigating groundwater recharge around this area like [3,6] used satellite images (Meteosat thermal data) combined with hydrogeological data to identify the thermal change of groundwater in the depression zones, and then estimated values of recharge and discharge of the area. We try to define time series data of groundwater depending on the nature and fluctuation of this property in time and space.

In this study, we combined updated remotely sensed and hydrological model datasets. These datasets have been used to study the diverse aspects of basin hydrology within the continent [6,7,11,19–24]. Some of these point out the lack of readily available *in situ* data for these studies [5,22], some cases validated the *in situ* measurements in Lake Chad and other parts of Africa (Table 2).

Table 2. Literature review of remotely sensed data sets used in the studies of some watershed in Africa.

Study Area	Data Products			Reference
	Terrestrial Water Storage	Rainfall	Lake Height	
Lake Chad	GRACE [1]	GPCP [2]	–	[5]
Lake Chad	–	–	Sat. Alt. [3]	[25]
Lake Chad	–	NOAA [4], TRMM [5]	–	[12]
East African Great Lake	GRACE, WGHM [6]	GPCP	Sat. Alt.	[25]
Lake Victoria, Malawi and Tamganyika	GRACE	GLDAS, TRMM	Sat. Alt.	[26]
Okavango catchment	GRACE	TRMM	Sat. Alt.	[27]
Congo river basin	GRACE	GLDAS, TRMM	Sat. Alt.	[26]
Lake Victoria, Tamganyika and Malawi	GRACE, WGHM	GLDAS, TRMM	Sat. Alt.	[28]

[1] GRACE: Gravity Recovery and Climate Experiment; [2] GPCP: Global Precipitation Climatology Project; [3] Sat. Alt.: Satellite Altimetry; [4] NOAA: National Oceanic and Atmospheric Administration; [5] TRMM: Tropical Rainfall Measuring Mission; [6] WGHM: WaterGap Hydrological Model.

We utilized satellite gravimetric, altimetric and hydrological models products over the Lake Chad basin to characterize the spatiotemporal and multiscale variability in its hydrological cycle, to infer the effect of rainfall on water storage in this region, and, finally, to investigate subsurface water variations within this region and perform comparison with groundwater outputs from a global hydrological model.

2. Materials and Methods

2.1. Terrestrial Water Storage (TWS) from GRACE

GRACE is known for estimating high-precision time-varying gravity field and the changes of Earth's surface mass at a high degree of accuracy on a time scale ranging from months to a decade [29,30]. These variations are mainly due to redistribution of water mass in the surface fluid envelopes of the earth. It provides estimates of TWS, which encompasses surface water, soil moisture, groundwater and snow. However, experimental errors while using GRACE increase rapidly and concurrently as the degree of the spherical harmonic coefficients, causing inaccurate results at higher degree terms of the spherical harmonic coefficients [31,32]. Spatial averaging functions are normally used to reduce the high degree of noise in the GRACE gravity field. This provides researchers with accurate surface mass changes. An additional de-stripping averaging filter is used for suppressing the "N–S" stripping noise in the GRACE data. There is also a leakage effect, which is caused by the spatial averaging functions. This causes some signals of the GRACE mass anomalies to leak outside the region of interest. The accuracy of GRACE is high enough to detect surface mass variations corresponding to hydrological loads of 1 cm at monthly and longer time scales, with horizontal dimensions of hundreds of kilometers and larger [30].

Numerous studies on the reliability of its data sets has been carried out by comparing its TWS products to that of Land Surface Models or *in situ* land observations-India [31], the Korean Peninsula [33,34], the East African lakes [25], Mali in Africa [35]. It has also been widely used in the studies of lakes around the world [36–39].

The GRACE Level 3 (Release 05) is the latest and more accurate of GRACE products. It provides processed time variability gravity field products. These products are provided as sets of spherical harmonic coefficients averaged on a monthly scale.

For this study, we used the monthly land mass grid observations (Level 3) provided by the Center of Space Research (CSR), University of Texas, at Austin from January 2003 to December 2013. The data are available as monthly $1° \times 1°$ grids of TWS over our study area [40]. The data set was truncated at 60 degrees and smoothed with the Gaussian filter of 300 km. GRACE data enhancement techniques provided [30] were also included to improve the accuracy of these TWS estimates.

For the time frame of our study, we encountered six months of missing data during these dates: June 2011, May and October 2012, and March, August, and November 2013. For these missing datasets, the temporal linear interpolation was done between them since they were not contiguous.

2.2. Lake Height from Altimetry

Generally, satellite altimeters are nadir-pointing instruments that record the average surface "spot" height directly below the satellite as it transverses over the Earth's surface. It basically determines the distance from the satellite to a target surface by measuring the satellite-to-surface round-trip time of a radar pulse. The altimeter emits a radar wave and analyzes the return signal that bounces off the surface. The surface height is calculated as the difference between the satellite's positions on orbit with respect to an arbitrary reference surface, *i.e.*, the Earth's center is represented by a reference ellipsoid and the range between the satellite and the surface is obtained by calculating the time taken for the signal to return. From this, the measurements of the sea surface height and other characteristics of oceans, lakes, floodplains, and rivers can be obtained. A lot of information can be extracted from satellite altimetry.

Institutions like the Foreign Agricultural Services (FAS) of the United States Department of Agriculture, Hydroweb, and European Space Agency (ESA) have been making available the up-to-date and reliable user-friendly data sets.

Lakes, rivers and oceans have all been monitored over the years using these data sets [9,11,41–44]. Surface water level data sets are sometimes given in the form of graphs and tables for major water bodies based on combination of various radar altimetry sensors. These data sets are made available free of charge via web applications, such as USDA's Global Reservoir and Lake Monitor (http://www.pecad.fas.usda.gov), Hydroweb of Geodesy, Oceanography and Hydrology from Space (GOHS; http://www.legos.obs-mip.fr/), and River and Lake system provided by ESA (http://tethys.eaprs.cse.dmu.ac.uk). Repeat track methods used in the derivation of time series of the lake surface height variation uses the reference lake height profile. This is derived from averaging all height profiles across the lake within a given time span. This effectively smoothens out any varying effects of tide and wind set-up. These resulting time series of height variations are expected to have an accuracy of about 20 cm root mean square (RMS) for lakes with minimal tides and limited dynamic variability.

For our study area, the satellite altimetry data has widely been used in the studies of the lake in which *in situ* datasets were compared with these altimetry products. The results showed accurate water level variations for Lake Chad in the two data sets [11,40,41]. Altimetry missions with a 10-day repeat track, such as TOPEX/Poseidon (1992–2006), Jason-1 (2001–2013), and Jason-2 (2008–present), or those with a 35-day repeat track, such as ERS-2 (1995–2000) and ENVISAT (2002–2010), can be used to extract lake height variations. ENVISAT altimetry estimates were used for this study.

2.3. Soil Moisture from GLDAS

Operated by NASA and the National Oceanic and Atmospheric Administration (NOAA), GLDAS is a land surface simulation system that aims to ingest satellite and ground-based observational data products, using advanced land surface modeling and data assimilation techniques, in order to generate optimal fields of land surface state (e.g., soil moisture and surface temperature) and flux (e.g., evaporation and sensible heat flux) products [4]. It adopts four advanced land surface models (LSMs): The Community Land Model (CLM), Mosaic, Noah, and Variable Infiltration Capacity (VIC). GLDAS executes spatial resolutions globally at both 0.25° and 1.0°, with temporal resolutions of three hours and monthly products since 1979. GLDAS data are widely used for land-surface flux simulations. As such, the simulation accuracy using GLDAS dataset is largely contingent upon the accuracy of the GLDAS dataset. The data are available from the Goddard Earth Sciences Data and Information Services Center (GES DISC).

In this study, we used Noah 1.0° grid data which has four layers of vertical soil moisture. The monthly average soil moisture is computed as the sum of all the layers [45].

2.4. Groundwater Estimates from the WaterGap Hydrological Model

The WaterGAP Global Hydrological Model (WGHM) is a submodel of the global water use and availability model WaterGAP 2.2 It computes groundwater recharge, surface runoff and river discharge as well as storage variations of water in canopy, rivers, soil, lakes, wetlands, groundwater and snow at a spatial resolution of 0.5° [46].

WGHM is based on the best global data sets currently available, and it is able to simulate variations in water bodies. It computes the water storage in the snow pack, rooted soil zone, groundwater, on vegetation surfaces, and in surface water reservoirs. Here, the simulated estimates provided by [47] were used. In order to obtain a reliable estimate of water availability, they tuned the model against the observed discharge at 1235 gauging stations, which represent 50% of the global land area and 70% of the actively discharging area. In Africa, most basins north of the equator do not perform well [47]. Detailed information about the modelling concept and its corresponding assumptions can be found in [48].

Model outputs assessment performance was not carried out in the LCB due to limited data availability. However, the Chari and Komadougou Yobe river basins, which predominantly feed the lake, were included in the calibration scheme of this region. Too much water was modeled for both basins. The Chari-Logone river system, which supplies most of the water into the southern part of LC, has a Nash-Sutcliffe efficiency of around 0.6, which is quite good. On other hand, inland water bodies in Africa showed a good match with the WGHM model output-for instance, the East African great lakes as reported by [46].

For this study, outputs from the WaterGap 2.2a model forced with precipitation from the Global Precipitation Climatology Centre (GPCC) and data from the European Center for Medium-Range Weather Forecast (ECMWF) integrated forecast system were used. These outputs include; global-scaled gsssroundwater storage, total water storage, baseflow, and groundwater recharge (diffused and below surface water bodies). There is no data available after 2009 (Table 3). This can be found on the website (https://www.uni-frankfurt.de/49903932/7_GWdepletion) [49].

Table 3. Summary of data sets used for this study.

Variable	Dataset	Resolution		Period
		Spatial	Temporal	
Terrestrial Water Storage	GRACE	$1° \times 1°$	1 month	2003–2013
Lake Height	Sat. Alt.	$1° \times 1°$	30 days	2003–2013
Rainfall	GLDAS	$1° \times 1°$	1 month	2003–2013
Soil moisture	GLDAS	$1° \times 1°$	1 month	2003–2013
Groundwater	WGHM	$0.5° \times 0.5°$	1 month	2003–2009

2.5. Data Processing

2.5.1. Variability in TWS and Lake Height

Seasonal-Trend Decomposition Procedure based on Loess (STL) method is a filtering procedure that decomposes a time series into its additive components of variation (trend, seasonal and the remainder components) by the application of Loess smoothing models [50]. This was used to model GRACE monthly storage variations as well as the time series of altimetric lake height.

In brief, the steps performed during STL decomposition are as follows:

1. Cycle-subseries smoothing: series are built for each seasonal component, and smoothed separately.
2. Low-pass filtering of smoothed cycle-subseries: the subseries are put together again, and smoothed.
3. Detrending of the seasonal series.
4. Deseasonalizing the original series, using the seasonal component calculated in the previous steps; and Smoothing the deseasonalized series to get the trend component.

In R statistical software, the STL algorithm is available through the *stl* function. We use it with its default parameters. The degrees for the Loess fitting are $d = 1$ in steps (iii) and (iv), and $d = 0$ in step (ii).

The parameter values must be chosen by the data analyst. We assume each observation X_i in time series is the sum of these components:

$$X_i = T_i + S_i + I_i \tag{1}$$

where T_i = Trend, S_i = Seasonality and I_i = Interannual components.

Often, six parameters determine the degree of smoothing in trend and seasonal components. For detailed information on method and parameters, consult [50] paper on STL methods. For our study, these parameters are:

$n_{(p)}$: The number of observations in each seasonal cycle, = 12 months (yearly periodicity with monthly data);

$n_{(i)}$: The number of passes through the inner loop (usually set to equal one or two) = 1 month;

$n_{(o)}$: The number of robustness iterations of the outer loop (Values qual one or two) = po robustness while a zero value has no robustness iteration) = 5 months;

$n_{(l)}$: The span of the loess window for the low-pass filter (computed as the next odd number to $n_{(p)}$) = 13 months;

$n_{(s)}$: The smoothing parameter for the seasonal component, = 12 months (seasonal length is same as the periodic length);

$n_{(t)}$: The smoothing parameter for the trend component, = 22 months.

$$n_{(t)} \geqslant \left\lceil \frac{1.5 n_{(p)}}{1 - 1.5 n_{(s)}^{-1}} \right\rceil \times 2 \tag{2}$$

For this analysis, R statistical software was used [51]. It is a free software environment and a programming language for statistical computing and graphics. It is widely used among statistics and data miners for developing statistical software and data analysis. MS excel was also used for subsequent data representation and analysis.

2.5.2. Subsurface Water Volume Change

Subsurface water volume (Groundwater + Soil moisture) was investigated. GRACE data provides changes in total water storage, which includes Lake water storage (LS), Snow water equivalent storage (SWES), soil moisture storage (SMS), and groundwater storage (GWS) within the basin. With satellite and model-based estimates of LS and SMS, subsurface water volume can be estimated. SWES was ignored for our study area since this area is humid.

Estimates of the subsurface water changes was evaluated using the following disaggregation equation,

$$\Delta SSW = \Delta SM + \Delta GW = \Delta TWS - \Delta LS \qquad (3)$$

Here, ΔSSW = Subsurface Water, ΔGW = Groundwater, ΔLS = Lake water, ΔTWS = Terrestrial Water Storage, ΔSM = Soil Moisture.

In an attempt to express ΔSSW and ΔLS in terms of volume, both were multiplied by the LCB area and Lake Area, respectively.

3. Results and Discussions

3.1. TWS and Altimetry Lake Height

Based on our study period, the STL trend of the time series of monthly GRACE TWS shows a decrease in average TWS of the Lake Chad basin (Figure 3) for the periods 2003–2005 and 2009–2010 with the latter being the lowest water estimates at −0.54 cm/year. There is an increase in TWS concentrations from 2006 to 2008, and 2010 through 2013, with the latter being the highest storage estimates of 0.69 cm/year.

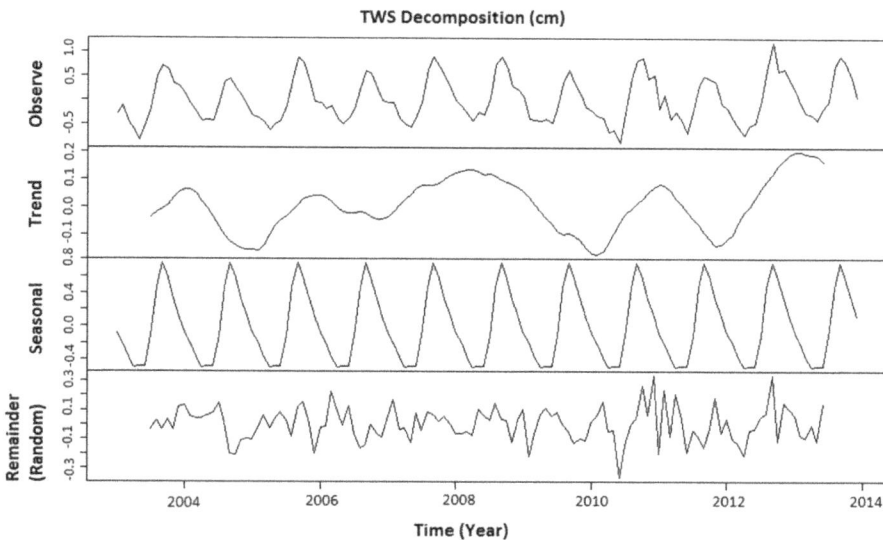

Figure 3. STL decomposition of the time series of monthly GRACE TWS.

The STL trend of the time series of altimetric lake height (Figure 4) shows a decrease in lake level from 2003 through 2005 and a steady increase until after 2008 with an average height of about 0.3 m/year. From this point, it begins to slope down towards 2010. From 2010 to 2012, the Lake experiences its lowest height averaging to about 0.23 m/year. Different rates are shown in Table 4.

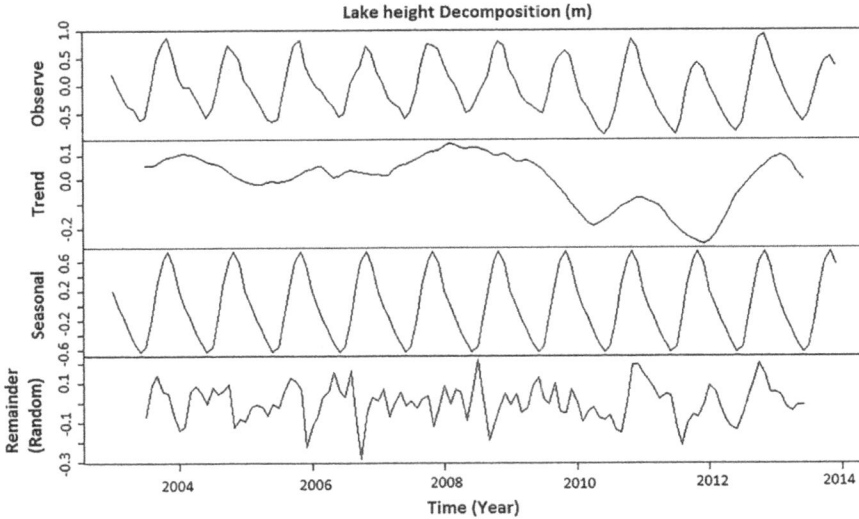

Figure 4. STL decomposition of the time series of monthly lake altimetric height.

Table 4. STL fitted trend of the time series of TWS and lake height.

Period	TWS (cm)	Lake Height (m)
2003–2005	0.25	0.32
2006–2009	0.84	0.62
2010–2013	0.38	−0.78

The STL decomposition plot of the monthly TWS estimates and lake altimetric height shows a similar pattern with their seasonal components suggesting an annual increase from the months of July–September as well as a decrease from October–June. This implies the Lake's height follows the seasonal pattern of the rainfall cycle around this area. There is a correlation (>80%) between them which points out the similarity in their pattern (Figure 5a).

Their seasonal component suggests an average annual increase in September and a main annual drop exists in November. This is due to the rainfall regime that exists over the Lake region. This relationship will be discussed in Section 3.2. The largeness and uniform size of the seasonal cycle of Lake Chad means that, over the years, Lake Chad has a fast water renewal process.

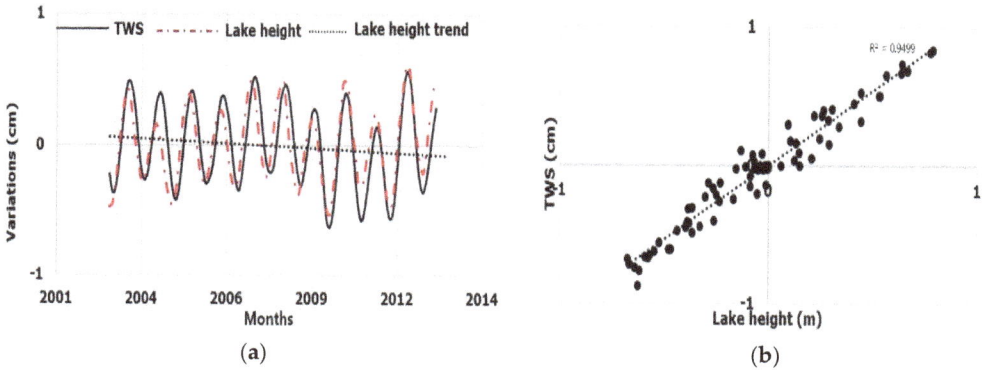

Figure 5. (**a**) Represents TWS and lake height after smoothing with a six-month window; (**b**) Represents the autocorrelation between TWS and lake height.

3.2. TWS and Rainfall

Rainfall estimates from GLDAS were compared to GRACE TWS over the LCB. During the wet season (July–September), there is an increase in seasonal pattern of rainfall over the study region with 2012 having the maximum annual averages. Figure 6 shows a comparison between the time series of the monthly estimated GLDAS rainfall and the change in GRACE TWS and, as expected, both curves show a good agreement during most of the study period in terms of pattern.

Based on Figure 6, we can clearly see the existence of a phase shift between GLDAS rainfall and GRACE TWS. This phase shift is about a month and a half. Their lagged correlation was also high (>0.9). From trend analysis, rainfall that precedes TWS increases throughout the study period. Its seasonal cycle goes ahead to confirm this phase shift that exists between rainfall and TWS in this region.

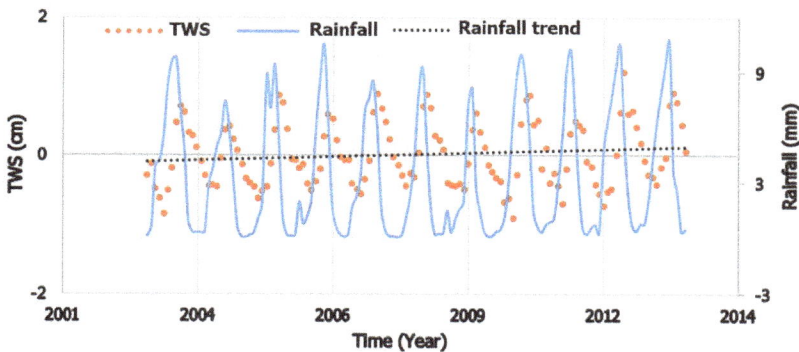

Figure 6. Comparisons between the yearly estimates of rainfall and TWS.

Figure 7 also confirms the bimodal rainfall regime that exists within this region with most of its heavy rainfall occurring between July–September and shorter rains from October–December.

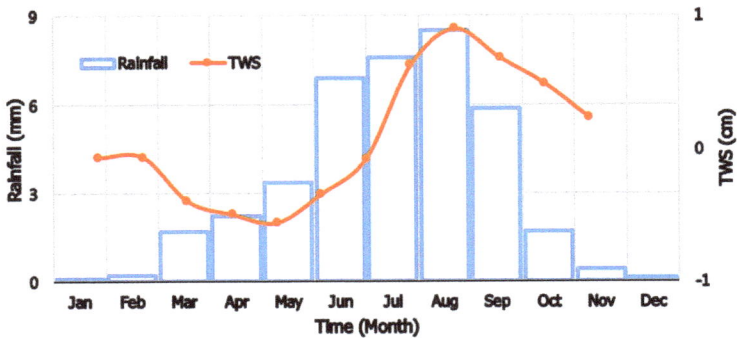

Figure 7. Comparisons between the monthly average rainfall and TWS.

3.3. TWS and Soil Moisture

Terrestrial water storage (TWS) of the Lake Chad Basin and the lake's surface water volume change from altimetry are critically analyzed. Due to its improved precision, GRACE can detect gravitational changes within an area of 200,000 km^2 or larger [44]. Hence, it is suitable for our study area which is 427,500 km^2.

Figure 8 shows a good agreement between altimetric lake water volume with GRACE basin water volume with a correlation of about 73%. There is also a good agreement in their seasonal cycles (Figure 9). We can infer that surface water of Lake Chad governs the volume of water in the basin area.

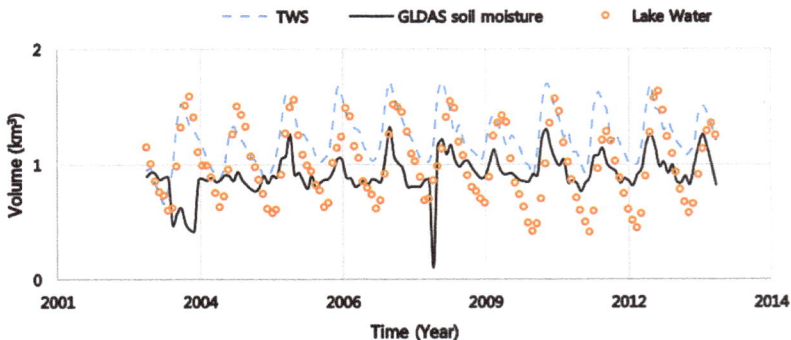

Figure 8. Effect of altimetric lake water and soil moisture on TWS.

GLDAS outputs of soil moisture estimates can also be seen in Figure 8. It has a weak agreement with basin water volume with a correlation of about 35%. Hence, soil moisture content plays little or no role in the variability in water volume stored in the Lake Chad Basin. It is solely governed by surface water of the Lake Chad.

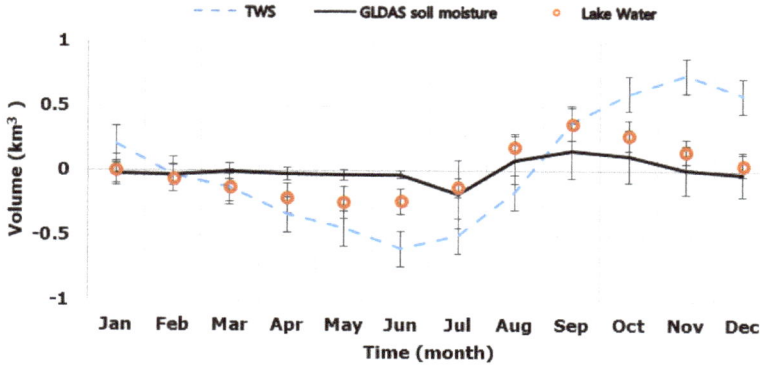

Figure 9. Monthly averaged seasonal cycle of the Lake Chad. Error bars represent the standard deviation for each month.

Additionally, investigated were the changes in subsurface water volume, which comprises of groundwater and soil moisture (Section 2.5.2). In early 2003 and 2007, there is a dissimilarity between volume estimates of WGHM groundwater and subsurface water estimates. The results obtained were compared with WGHM outputs (Figure 10). It shows two peak periods at approximately mid-2005 and 2006. Both curves have a somewhat similar pattern with a correlation coefficient of about 47% between them.

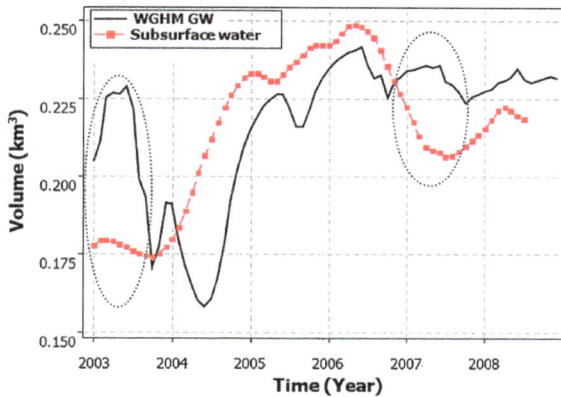

Figure 10. Changes in subsurface water (groundwater and soil moisture) and WGHM GW outputs. The annual signal is removed and data are smoothed with a six-month window.

4. Conclusions

In this study, we see how the use of GRACE TWS data along with other data sources could be of great value in the hydrological studies of remote areas. We investigated hydrological variability associated within the LCB from January 2003 to December 2013. GRACE TWS and altimetric lake volume showed a similar pattern for the majority of the study period. After comparing TWS with soil moisture content, we found that the discharge in this area is governed by surface water of the lake. This is the first attempt in which these remotely sensed datasets were used to study the varying patterns of the lake's hydrology. Deriving this characteristic spatiotemporal analysis of surface mass anomalies across the LCB, future improvements can be made in the management of water resources in this area.

For much of the study period, GRACE TWS variations within the basin show a similar pattern of variation as the averaged lake height variation from altimetry. A trend analysis showed increasing precipitation with maximum annual average increase in August but decreasing water level in the lake from altimetry with the minimum annual average occurring in 2012 with a value of 0.2 m. Our study also showed that altimetry-based volume with TWS from GRACE provides information on soil moisture and groundwater. This could help in detecting subsurface water storage changes in relation to climate variability or anthropogenic activities especially in situations where *in situ* measurements are not available. This could also lead to new research prospects, where researchers could try to find out the main cause behind the lake's decreasing trend.

This characterization can help in the proposed Water Transfer Project from the Ubangi River to Lake Chad in a number of ways. For instance, the primary objective of this project is to halt the shrinkage of Lake Chad through an inflow of water coming from the Congo basin. The association between LC level and precipitation will enable managers to plan for the total water volume that can be released and retained based on future forecasts.

Acknowledgments: This research was supported by a grant (15AWMP-B079625-02) from the Water Management Research Program funded by the Ministry of Land, Infrastructure and Transport of Korean government and the Basic Science Research Program through the National Research Foundation of Korea (NRF) funded by the Ministry of Education (NRF-2015R1D1A1A09060690).

Author Contributions: Willibroad G. Buma developed the methodology and conducted the work under the supervision and review of Sang-Il Lee. Jae Young provided technical assistance and helped in developing the manuscript.

Conflicts of Interest: The authors declare no conflict of interest.

References

1. Swenson, S.; Wahr, J. Monitoring the water balance of Lake Victoria, East Africa, from space. *J. Hydrol.* **2009**, *370*, 163–176.

2. Allen, R.; Joseph, J.; Kevin, L.; Ahjond, S. Adaptive management for a turbulent future. *Environ. Manag.* **2011**, *92*, 1339–1345.

3. Leblanc, M.; Leduc, C.; Razack, M.; Lemoalle, J.; Dagorne, D.; Mofor, L. Applications of remote sensing and GIS for groundwater modelling of large semiarid areas: Example of the Lake Chad Basin, Africa. *IAHS Publ.* **2003**, *278*, 186–192.

4. Rodell, M.; Houser, P.R.; Jambor, U.; Gottschalck, J.; Mitchell, K.; Meng, C.-J.; Arsenault, K.; Cosgrove, B.; Radakovich, J.; Bosilovich, M.; *et al.* The global land data assimilation system. *Am. Meteorol. Soc.* **2004**, *85*, 381–394.

5. Boronina, A.; Ramillien, G. Application of AVGRR imagery and GRACE measurements for calculation of actual evapotranspiration over the Quaternary aquifer (Lake Chad basin) and validation of groundwater models. *Hydrol. J.* **2008**, *348*, 98–109.

6. Leblanc, M.; Favreau, G.; Tweed, S.; Leduc, C.; Razack, M.; Mofor, L. Remote sensing for ground water modelling in large semiarid areas: Lake Chad basin, Africa. *Hydrol. J.* **2007**, *15*, 97–100.

7. Gao, H.; Bohn, T.J.; Podest, E.; McDonald, K.C.; Lettenmaier, D.P. On the causes of the shrinking of Lake Chad. *Environ. Res. Lett.* **2011**, *6*, 034021.

8. Singh, A.; Diop, S.; M'mayi, P.L. Africa's Lakes: Atlas of Our Changing Environment (Nairobi: UNEP). Available online: www.unep.org/dewa/AfricaAfricaAtlas/ (accessed on 10 May 2016).

9. Campbell, R.W. Earthshots: Satellite Images of Environmental. Available online: http://earthshots.usgs.gov/ (accessed on 10 May 2016).

10. Coe, M.T.; Foley, J.A. Human and natural impacts on the water resources of the Lake. *Geophys. Res. Atmos.* **2001**, *106*, 3349–3356.

11. Birkett, C.M. Synergistic remote sensing of Lake Chad: Variability of basin inundation. *Remote Sens. Environ.* **2000**, *72*, 218–236.

12. Churchill, O.; Belay, D.; Sium, G. Characteristics of Lake Chad Level Variability and Links to ENSO, precipitation and river discharge. *Sci. World J.* **2014**, *13*, 145893.

13. Li, K.Y.; Coe, M.T.; Ramankutty, N.; De Jong, R. Modeling the Hydrological Impact of Land-Use Change in West Africa. *J. Hydrol.* **2007**, *337*, 258–268.

14. Li, B.; Rodell, M.; Zaitchik, B.F.; Reichle, R.H.; Koster, R.D.; Tonie, M.A. Assimilation of Grace Terrestrial Water Storage into a Land Surface Model: Evaluation and Potential Value for Drought Monitoring in Western and Central Europe. *J. Hydrol.* **2012**, *446–447*, 103–115.

15. Le Coz, M.; Delclaux, F.; Genthon, P.; Favreau, G. Assessment of Digital Elevation Model (Dem) Aggregation Methods for Hydrological Modeling: Lake Chad Basin, Africa. *Comp. Geosci.* **2009**, *35*, 1661–1670.

16. Leblanc, M.; Favreau, G.; Maley, J.; Nazoumou, Y.; Leduc, C.; Stagnitti, F.; van Oevelen, P.; Delclaux, F.; Lemoalle, J. Reconstruction of Megalake Chad using Shuttle Radar Topographic Mission data. *Palaeogeogr. Palaeoclimatol. Palaeoecol.* **2006**, *239*, 16–27.

17. Leblanc, M.; Razack, M.; Dagorne, D.; Mofor, L.; Jones, C. Application of Meteosat thermal data to map soil infiltrability in the central part of the Lake Chad basin, Africa. *Geophys. Res. Lett.* **2003**, *30*.

18. Coe, M.T.; Birkett, C.M. Calculation of river discharge and prediction of lake height from satellite radar altimetry: Example for the Lake Chad basin. *Water Resour. Res.* **2004**, *40*.

19. Isiorho, S.A.; Matisoff, G. Groundwater seepage and its implication on the water resources planning and management in the Chad Basin. *Water Resour. Res.* **1989**, *1*, 210–215.

20. Isiorho, S.A.; Matisoff, G. Groundwater recharge from Lake Chad. *Limnol. Oceanogr.* **1990**, *35*, 931–938.

21. Leduc, C.; Sabljak, S.; Taupin, J.D.; Marlin, C.; Favreau, G. Estimation de la recharge de la nappe quaternaire dans le Nord-Ouest du bassin du lac Tchad (Niger oriental) à partir de mesures isotopiques. *Earth Planet. Sci. Lett.* **2000**, *330*, 355–361.

22. Goni, I.B. Tracing stable isotope values from meteoric water to groundwater in the southwestern part of the Chad basin. *Hydrogeol. J.* **2006**, *14*, 742–752.

23. Ngounou, N.B.; Jacques, M.; Jean, S.R. Groundwater Recharge from Rainfall in the Southern Border of Lake Chad in Cameroon. *World App. Sci. J.* **2007**, *2*, 125–131.

24. Lemoalle, J.; Leblanc, M.; Bader, J.C.; Tweed, S.; Mofor, L. Thermal remote sensing of water under flooded vegetation: New observations of inundation patterns for the Small Lake Chad. *Hydrol. Res.* **2011**, *404*, 87–98.

25. Becker, M.; LLovel, W.; Cazenave, A.; Guntner, A.; Crétaux, J.F. Recent hydrological behaviour of the East African great lakes region inferred from GRACE, satellite altimetry and rainfall observations. *Comptes Rendus Geosci.* **2010**, *342*, 223–233.

26. Jean-Paul, B.; Jacques, H.; Caroline, D.L. Retrieval of Large-Scale Hydrological Signals in Africa from GRACE Time-Variable Gravity Fields. *Pure Appl. Geophys.* **2012**, *169*, 1373–1390.

27. Milzow, C.; Krogh, P.E.; Bauer-Gottwein, P. Combining satellite radar altimetry, SAR surface soil moisture and GRACE total storage changes for hydrological model calibration in a large poorly gauged catchment. *Hydrol. Earth Syst. Sci.* **2011**, *15*, 1729–1743.

28. Hassan, A.A.; Jin, S. Lake level change and total water discharge in East Africa Rift Valley from satellite-based observations. *Global Planet Change* **2014**, *117*, 79–90.

29. Tapley, B.D.; Bettadpur, S. The gravity recovery and climate experiment: Mission overview and early results. *Geophys. Res. Lett.* **2004**, *31*.

30. Wahr, J.; Swenson, S.; Zlotnicki, V.; Velicogna, I. Time-variable gravity from GRACE: First results. *Geophys. Res. Lett.* **2004**, *31*.

31. Swenson, S.; Chambers, D.; Wahr, J. Estimating geocenter variations from a combination of GRACE and ocean model output. *J. Geophys. Res. Solid Earth* **2008**, *113*.

32. Rodell, M.; Velicogna, I.; Famiglietti, J.S. Satellite-based estimates of groundwater depletion in India. *Nature* **2009**, *460*, 999–1002.

33. Lee, S.I.; Kim, J.S.; Lee, S.K. Estimation of average terrestrial water storage changes in the Korean Peninsula using GRACE satellite gravity data. *J. Korea Water Resour. Assoc.* **2012**, *8*, 805–814.

34. Lee, S.I.; Seo, J.Y.; Lee, S.K. Validation of terrestrial water storage change estimates using hydrologic simulation. *J. Water Resour. Ocean Sci.* **2014**, *4*, 5–9.

35. Henry, C.M.; Allen, D.M.; Huang, J. Groundwater storage variability and annual recharge using well-hydrograph and GRACE satellite data. *Hydrogeol. J.* **2011**, *19*, 741–755.

36. Joseph, L.A.; Sharifi, M.A.; Ogonda, G.; Wickert, J.; Grafarend, E.W.; Omulo, M.A. The Falling Lake Victoria Water Level: GRACE, TRIMM and CHAMP Satellite Analysisof the Lake Basin. *Water Resour. Manag.* **2008**, *22*, 775–796.

37. Song, C.; Huang, B.; Ke, L. Modeling and analysis of lake water storage changes on the Tibetan Plateau using multi-mission satellite data. *Remote Sens. Environ.* **2013**, *135*, 25–35.

38. Longuevergne, L.; Wilson, C.R.; Scanlon, B.R.; Crétaux, J.-F. GRACE water storage estimates for the Middle East and other regions with significant reservoir and lake storage. *Hydrol. Earth Syst. Sci.* **2013**, *17*, 4817–4830.

39. Tourian, M.J.; Elmi, O.; Chen, Q.; Devaraju, B.; Roohi, Sh.; Sneeuw, N. A spaceborne multisensor approach to monitor the desiccation of Lake Urmia in Iran. *Remote Sens. Environ.* **2015**, *156*, 349–360.

40. NASA MEaSUREs Program. Available online: http://grace.jpl.nasa.gov (accessed on 10 May 2016).

41. Crétaux, J.-F.; Birkett, C.M. Lake studies from satellite radar altimetry. *Comptes Rendus Geosci.* **2006**, *338*, 1098–1112.

42. Kouraev, A.V.; Semovski, S.V.; Shimaraev, M.N.; Mognard, N.M.; Légresy, B.; Remy, F. Observations of Lake Baikal ice from satellite altimetry and radiometry. *Remote Sens. Environ.* **2007**, *108*, 240–253.

43. Kropacek, J.; Braun, A.; Kang, S.C.; Feng, C.; Ye, Q.H.; Hochschild, V. Analysis of lake level changes in Nam Co in central Tibet utilizing synergistic satellite altimetry and optical imagery. *Int. J. Appl. Earth Obs. Geoinf.* **2012**, *17*, 3–11.

44. Duan, Z.; Bastiaanssen, W.G.M. Estimating water volume variations in lakes and reservoirs from four operational satellite altimetry products and satellite imagery data. *Remote Sens. Environ.* **2013**, *134*, 403–416.

45. Goddard Earth Sciences Data and Information Services Center. Available online: http://grace.jpl.nasa.gov (accessed on 10 May 2016).

46. Döll, P.; Kaspar, F.; Bernhard, L. A global hydrological model for deriving water availability indicators: Model tuning and validation. *J. Hydrol.* **2003**, *270*, 105–134.

47. Döll, P.; Muller, H.S.; Schuh, C.; Portmann, F.T. Global-scale assessment of groundwater depletion and related groundwater abstractions: Combining hydrological modeling with information from well observations and GRACE satellites. *Water Resour. Res.* **2014**, *50*, 5698–5720.

48. Güntner, A.; Stuck, J.; Döll, P.; Schulze, K.; Merz, B. A global analysis of temporal and spatial variations in continental water storage. *Water Resour. Res.* **2007**, *43*.

49. Data source for the groundwater depletion. Available online: https://www.uni-frankfurt.de/49903932/7_GWdepletion (accessed on 10 May 2016).

50. Cleveland, R.B.; Clevelan, W.S.; McRae, J.E.; Thacker, E.L. STL: A seasonal-trend decomposition procedure based on loess. *J. Off. Stat.* **1990**, *6*, 3–73.

51. A language and environment for statistical computing. Available online: http://www.Rproject.org/ (accessed on 10 May 2016).

Water Discharge and Sediment Load Changes in China: Change Patterns, Causes, and Implications

Chong Jiang, Linbo Zhang, Daiqing Li and Fen Li

Abstract: In this research, monthly hydrological and daily meteorological data were collected across China for the period 1956–2012. Modified Mann–Kendall tests, double mass curve analysis, and correlation statistics were performed to identify the long-term trends and interrelation of the hydrometeorological variables and to examine the influencing factors of streamflow and sediment. The results are as follows: (1) In the last 60 years, the streamflow in northern China has shown different decreasing trends. For the southern rivers, the streamflow presented severe fluctuations, but the declining trend was insignificant. For the streamflow in western China, an increasing trend was shown. (2) In the northern rivers, the streamflow was jointly controlled by the East Asian monsoon and westerlies. In the southern rivers, the runoff was mainly influenced by the Tibet–Qinghai monsoon, the South Asian monsoon, and westerlies. (3) Sediment loads in the LCRB (Lancang River Basin) and YZRB (Yarlung Zangbo River Basin) did not present significant change trends, although other rivers showed different degrees of gradual reduction, particularly in the 2000s. (4) Underlying surface and precipitation changes jointly influenced the streamflow in eastern rivers. The water consumption for industrial and residential purposes, soil and water conservation engineering, hydraulic engineering, and underlying surface changes induced by other factors were the main causes of streamflow and sediment reduction.

Reprinted from *Water*. Cite as: Jiang, C.; Zhang, L.; Li, D.; Li, F. Water Discharge and Sediment Load Changes in China: Change Patterns, Causes, and Implications. *Water* **2015**, *7*, 5849–5875.

1. Introduction

Global warming caused by human-induced emissions of greenhouse gases is accelerating the global hydrological cycle [1]. The accelerated hydrological cycle is in turn altering the spatial-temporal patterns of precipitation, resulting in increased occurrences of precipitation extremes that cause increased occurrences of floods and droughts in many regions of the world [2], including China [3–5]. As a vital natural resource, water is fundamental for the sustainable development of the economy, ecosystem, and biodiversity. Therefore, water security and related implications for ecosystem and river diversity, particularly the variability and availability of regional water resources under the influences of climatic change and human activities, have been discussed in recent years [6–8]. Much attention has been given to water resource changes and their effects on the economic society by the international community. For example, the Intergovernmental Panel on Climate Change (IPCC) has reviewed changes in the global hydrological cycle and has assessed the impacts of climatic change on water resources [9]. Many countries, such as the United Kingdom, have addressed the impact of climatic change on water resource variation [10]. The climatic changes of China are controlled mainly by winter and summer monsoons [11]. Generally, precipitation in southwest China is greater than that in northwest China; these patterns are controlled mainly by the monsoon system and the effects of topography [12]. Rainy seasons in eastern China hinge on progress and retreat of the East Asian monsoon. Detailed information on the evolution of summer Asian monsoons and the associated propagation of rain belts has been reported by Ding [13].

River sediment is an important aspect of land surface processes and global change research. River sediment generation, transportation, and river delta response have become important aspects of Earth system science. In 1968, Holeman [14] investigated global sediment discharge by using global hydrological data; further research was conducted by Holland [15]. Walling and Fang [16] investigated the temporal variation of 145 rivers by using long-term data (longer than 25 years) in Asia, Europe, and North America. They reported that the sediment discharge in more than 50% of analyzed rivers presented upward or downward trends, the latter of which was dominant. However, in the remaining 50%, the sediment flux essentially remained stable [17]. A study of the sediment load in Russia showed that of the 20 rivers flowing into the Arctic Ocean, 35% showed increasing trends, 60% presented declining trends; only 5% remained stable [18]. Similar research was conducted by Liu [19], Subramainian [20], and Siakeu [21] for major rivers of Asia, India, and Japan, respectively. This research revealed that human activities, particularly reservoir and dam construction, were the main causes of sediment flux reduction.

Many researchers investigated the sediment and streamflow change in major rivers of China. The Yangtze River Basin (YARB) [22], Yellow River Basin (YRB) [23], Huai River Basin (HURB) [24], Liao River Basin (LRB) [25], and Songhua River Basin (SRB) [26] showed different degrees of decreasing trends. However, the sediment flux in western rivers such as the Yarlung Zangbo River Basin (YZRB) and Lancang River Basin (LCRB) remained stable or increased slightly. The Yellow and Yangtze rivers are two of the largest rivers in China and therefore receive more attention. Yang [27] considered that reservoir and dam construction was the main reason of sediment reduction in the Yangtze River. Miao [23] reported that reservoir construction, reduced precipitation, soil and water conservation projects jointly induced sediment reduction in the Yellow River. So far, the focus in China has been mostly on regional water and sediment resources. On the basis of instrument records of streamflow and sediment, many scholars conducted research in different river basins in China to reveal different changing patterns of runoff and sediment. Most of these studies are based on the monitoring data of individual rivers without considering the impacts of water and sand resource consumption of the economic society within the river basin. Regarding spatial scale, most studies completed thus far are based on one river basin; few are based on a national scale. An acceptable evaluation of water and sediment resources requires sufficient hydrometeorological data and extensive data-driven analysis, which is the motivation for the current study. Further, possible causes of precipitation and streamflow and sediment resource variation need to be investigated, and the related implications should be discussed.

Therefore, the objectives of this study are to (1) investigate streamflow and sediment changes in major rivers in China; (2) determine the streamflow changes and their relationship with precipitation, monsoons, and water consumption for industrial and residential purposes; (3) determine regional sediment load changes and their relationship with hydraulic engineering, soil and water conservation engineering, and underlying surface changes induced by human activities; and (4) discuss the relationship between sediment and runoff changes and their relationship with specific events, the implications of which will also be discussed. The primary goal of this study is to evaluate the impact of climate change and human activities on streamflow and sediment load and to provide basic information for water and soil resources management in this region.

2. Data and Methodology

2.1. Data Collection and Processing

In this study, annual precipitation data from 725 rain gauge stations for the period 1951–2012 were obtained from the National Climate Center (NCC) of the China Meteorological Administration (CMA). The quality of meteorological data was firmly controlled [28]. To guarantee the accuracy of the results, the data was preprocessed as follow before the analysis. The observational data of missing data years of more than 5 years (including 5 years) were excluded. The time series data of partial relocation stations were unified, and the remaining missing observation data were completed with a linear regression method and adjacent station interpolation to ensure the integrity of the time series. The missing data in 725 stations only accounted for less than 5% of total data amount. The regional averages refer to the arithmetic mean value of the stations within a region. Annual precipitation average was, thus, calculated from these records using the Thiessen Polygon method for each river basin.

The consecutive monthly data of streamflow and sediment yields from the 30 gauge stations were collected for the same period from Ministry of Water Resources (MWR). The hydrologic data from the 30 gauge stations listed in Table 1 were used to analyze changes in streamflow and sediment load. Figure 1 and Table 1 provide information on the station location and associated drainage area. It is worth to address is that, the Xinjiang Inland River Basin and the Hexi Inland River Basin (HIRB) in northwestern China are composed of many tributaries. Therefore, to reflect the overall change in streamflow, we summed the streamflow in tributary data to represent the runoff of the entire basin. In southwestern China, although many large rivers are present, we selected as study objects only two river systems, the LCRB and YZRB, considering data availability.

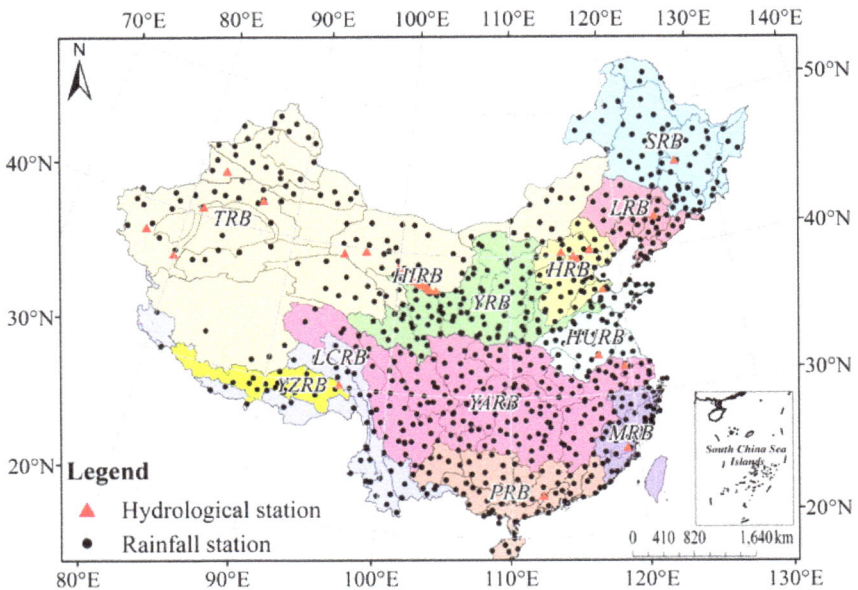

Figure 1. Distribution of 12 major basins in China showing gauge stations (red triangles) and meteorological stations (black dots) used in this study.

The daily discharge was computed from the water level by using previously calibrated discharge-water level curves. Water was sampled at fixed intervals, and suspended sediment concentration was obtained by measuring water samples in the laboratory. All the measurements of water level, discharge, SSC followed national standards issued by the Ministry of Water Conservancy, and were printed in the China Gazette of River Sedimentation [29]. Sediment loads refers to the suspended fraction only, whereas bedload was excluded due to its difficulty in field sampling. Measurement of the sediment loads was on the basis of standard procedures [30,31].

Errors in calculating sediment load were introduced through the low frequency of sampling, rather than continuous monitoring, which is likely to underestimate sediment load during peak hours. The monthly and annual streamflow and sediment load at the gauging stations were derived from the daily measured data. The accuracy and consistency of all the data used in this study have been checked out by the corresponding agencies before their release. In instances when discharge data were missing, we used discharge data from similar rainfall conditions at other times as a replacement, and the missing data in 30 stations only accounted for less than 3% of total data amount.

Table 1. Summary of gauging stations and hydrological characteristics in the 12 major basins in China. DA is the drainage area.

ID	Gauge Station	Basin (Abbreviation)	Location	Longitude (N°)	Latitude (E°)	DA (km²)	Time Span
1	Lijin	Yellow River Basin (YRB)	Mainstream	118°15′	37°29′	752,032	1952–2012
2	Datong	Yangtze River Basin (YARB)	Mainstream	117°03′	32°37′	1,705,383	1950–2012
3	Wujiadu	Huai River Basin (HURB)	Mainstream	117°23′	32°54′	121,330	1950–2011
4	Zhangjiafen	Hai River Basin (HRB)	Bai River	116°10′	39°48′	8506	1954–2011
5	Xiahui		Chao River	117°18′	40°22′	5340	1961–2011
6	Shixiali		Sanggan River	114°43′	40°16′	23,944	1952–2011
7	Xiangshuipu		Yang River	109°40′	38°01′	14,507	1952–2011
8	Yanling		Yongding River	115°49′	40°01′	43,674	1952–2011
9	Haerbin	Songhua River Basin (SRB)	Mainstream	126°32′	45°48′	389,769	1955–2012
10	Tieling	Liao River Basin (LRB)	Mainstream	123°43′	42°13′	120,764	1954–2012
11	Gaoyao	Pearl River Basin (PRB)	Xi River	112°27′	23°01′	351,535	1957–2011
12	Zhuqi	Southeast Rivers Basin (Min River Basin, MRB)	Mainstream	119°06′	26°08′	54,500	1950–2011
13	Nuxia	Southwestern Rivers Basin (Yarlung Zangbo River, YZRB)	Mainstream	95°05′	31°17′	191,235	1956–2009
14	Xiangda	Southwestern Rivers Basin (Lancang River Basin, LCRB)	Mainstream	96°28′	32°12′	17,909	1956–2012
15	Dajingxia Reservoir	Hexi Inland River Basin (HIRB)	Dajing River	103°24′	37°28′	68,300	1961–2010
16	Gulang		Gulang River	102°52′	37°27′		1961–2010
17	Huangyanghe Reservoir		Huangyang River	102°44′	37°35′		1961–2010
18	Zamusi		Zamu River	102°34′	37°42′		1961–2010
19	Nanying Reservoir		Jinta River	102°31′	37°48′		1961–2010
20	Jiutiaoling		Xiying River	102°03′	37°52′		1961–2010
21	Shagousi		Dongda River	101°55′	37°58′		1961–2010
22	Xidahe Reservoir		Xida River	101°23′	38°03′		1961–2010
23	Changmapu		Shule River	96°51′	39°49′		1961–2010

Table 1. *Cont.*

ID	Gauge Station	Basin (Abbreviation)	Location	Longitude (N°)	Latitude (E°)	DA (km²)	Time Span
24	Dangchengwan		Dang River	94°53′	39°30′		1961–2010
25	Yingluoxia		Hei River	99°55′	38°57′		1961–2010
26	Binggou		Beida River	101°56′	37°54′		1961–2010
27	Kaqun	Xinjiang Inland	Yeerqiang River	76°54′	37°59′	50,248	1957–2011
28	Tongguziluoke	River Basin	Yulongkashi River	79°55′	36°49′	14,575	1957–2011
29	Yanqi	(Tarim River	Kaidu River	86°34′	42°02′	22,516	1957–2011
30	Alaer	Basin, TRB)	Mainstream	81°19′	40°32′	127,900	1957–2011

1:100,000-scale land use maps in 1985 and 2010 were respectively obtained from the Earth System Science Data Sharing Platform and The Remote Sensing Monitoring and Assessment of Decadal Changes of National Eco-environment (2000–2010) project group. The digital elevation model (DEM), the Monitoring Report of Soil and Water Loss in China, and other maps were obtained from the Earth System Science Data Sharing Platform.

2.2. Methodology

2.2.1. Mann–Kendall Test for Monotonic Trend

To analyze the long-term trends of hydrometeorological variables, the non-parametric Mann–Kendall test [32,33] was applied. This method has been widely used to detect trends in climate and streamflow time series [34]. In the Mann–Kendall test, the null hypothesis H_0 states that $x_1, ..., x_n$ are samples of n independent and identically distributed random variables with no seasonal change. The alternative hypothesis H_1 for a two-sided test defines the distributions of x_k and x_j as non-identical for all $k, j \leq n$; with $k \neq j$. The test statistic S is given as

$$S = \sum_{i=1}^{n-1} \sum_{k=i+1}^{n} \text{sgn} \left(x_k - x_i \right) \tag{1}$$

$$\text{sgn} = \begin{cases} +1 & \theta > 0 \\ 0 & \text{if} \quad \theta = 0 \\ -1 & \theta < 0 \end{cases} \tag{2}$$

If the dataset is independent and identically distributed, the mean of S will be zero, and the variance of S will be:

$$\text{var}\,(S) = \frac{\left[n\,(n-1)\,(2n+5) - \sum\limits_{j=1}^{m} t_j\,(t_j-1)\,(2t_j+5) \right]}{18} \tag{3}$$

where n is the number of data points, t is the extent of a given time, m is the number of tied groups, and t_j is the number of data points in the j-th group. A tied group is a set of data points having the same value. A normalized test statistic Z can be computed on the basis of S as:

$$
Z = \begin{cases}
\frac{S-1}{\sqrt{\text{var}(S)}} & S > 0 \\
0 & S = 0 \\
\frac{S+1}{\sqrt{\text{var}(S)}} & S < 0
\end{cases}
\tag{4}
$$

When the significance levels are set at 0.01, 0.05, and 0.1, $|Z_\alpha|$ is 2.58, 1.96, and 1.65, respectively. At a certain significance level, if $|Z| > |Z_\alpha|$, the null hypothesis H_0 is rejected. That is, the trend is significant at the set level of significance. Otherwise, no significant trend exists.

In the Mann–Kendall test, the slope estimated by using the Theil–Sen estimator [35,36] is usually considered to detect the monotonic trend and to indicate the variable quantity in the unit time. It is a robust estimate of the magnitude of a trend and has been widely used to identify the slope of a trend line in a hydrological or climatic time series [37]. The estimator is given as:

$$
\beta = Median \left(\frac{x_j - x_l}{j - l} \right) \forall 1 < l < j
\tag{5}
$$

where $1 < l < j < n$, β is the median overall combination of record pairs for the entire dataset and is resistant to extreme observations. A positive β denotes an increasing trend, and a negative β indicates a decreasing trend.

2.2.2. Modified Mann–Kendall Test (Mann–Kendall Test with Trend-Free Pre-whitening)

The Mann–Kendall test assumes that the series is independent and the series is not robust against autocorrelation. However, certain hydrological time series may frequently display statistically significant serial correlation. This may lead to a disproportionate rejection of the null hypothesis of no trend, whereas the null hypothesis is actually true. Therefore, the effect of serial correlation is a major source of uncertainty in testing and interpretation trends. To eliminate the influence of serial correlation, "pre-whitening" was proposed by Von Storch [38] to remove the lag one serial correlation (r_1) from the time series. This method has been applied in an increasing number of studies [23,26,39].

To determine whether the observed dataset is serially correlated, the significance of the lag-1 serial correlation (r_1) should be tested at the 0.10 significance level. r_1 is calculated by using the following Equation [23,26]:

$$r_k = \frac{\frac{1}{n-k}\sum_{i=1}^{n-k}(x_i - \overline{x})(x_{i+k} - \overline{x})}{\frac{1}{n}\sum_{i=1}^{n}(x_i - \overline{x})^2} \tag{6}$$

If $\frac{-1-1.645\sqrt{n-2}}{n-1} \leq r_1 \leq \frac{-1+1.645\sqrt{n-2}}{n-1}$, the time series is assumed to be independent at the 0.10 significance level and can be subjected to the original Mann–Kendall test. Otherwise, the effect of serial correlation should be removed from the time series by pre-whitening prior to application of the Mann–Kendall test. The Mann–Kendall test is then used to detect trends in the residual series. The new time series is obtained as [40].

$$x_i' = x_i - (\beta \times i) \tag{7}$$

The r_1 value of this new time dataset is calculated and used to determine the residual series as:

$$y_i' = x_i' - r_1 \times -x_{i-1}' \tag{8}$$

The value of $\beta \times i$ is added again to the residual dataset as:

$$y_i = y_i' + (\beta \times i) \tag{9}$$

The y_i series is then subjected to trend analysis.

2.2.3. Double Mass Curve

Double mass curve analysis is a simple and practical visual method widely used in the study of the consistency and long-term trend test of hydrometeorological data [41]. This method was first used to analyze the consistency of precipitation data in Susquehanna watershed, Pennsylvania, USA [42]; a theoretical explanation was later reported [43]. The theory of the double mass curve is based on the fact that a plot of two cumulative quantities during the same period exhibits a straight line if the proportionality between the two remains unchanged; the slope of the line represents the proportionality. This method can smooth a time series and suppress random elements in the series; thus, it can show the main trends of the time series. In the last 30 years, Chinese scholars analyzed the effects of soil and water conservation measures and land use/cover changes on streamflow and sediment by using this method and have achieved good results [44]. In the present study, double mass curves of sediment *versus* streamflow were plotted for the different periods to detect the relationship change before and after transition years. The appearance of the inflection point denotes that the relationship between the sediment and streamflow begins to change significantly [44].

3. Variation of Streamflow and Sediment Load

3.1. Overall Change of Streamflow

In this study, we defined the northern rivers and southern rivers as those north and south of the eastern monsoon zone in China, respectively. Therefore, the northern rivers include the YRB, HRB, LRB, SRB, and HURB; the southern rivers include the Pearl River Basin (PRB), YARB, and Min River Basin (MRB); and the western rivers include the Tarim River Basin (TRB), HIRB, LCRB, and YZRB. Based upon the years of average runoff (Table 2), the order of runoff in the southern rivers was YARB (8944.8×10^8 m^3) > PRB (2166.4×10^8 m^3) > MRB (532.7×10^8 m^3); that in the northern rivers was SRB (404.7×10^8 m^3) > YRB (299.2×10^8 m^3) > HURB (266.9×10^8 m^3/a) > LRB (29.3×10^8 m^3) > HRB (18.1×10^8 m^3); and that in the western rivers was YZRB (312.5×10^8 m^3) > LCRB (247.3×10^8 m^3) > TRB (157.3×10^8 m^3) > HIRB (44.0×10^8 m^3). It should be noted that Xiangda Station, which represents the LCRB, is located at the source area of LCRB; therefore, the streamflow was smaller than that in the entire basin. Actually, the average streamflow in the downstream region of the LCRB was 740.5×10^8 m^3 [45].

Table 2. Average annual streamflow and sediment load.

Basins	Water Discharge (10^8 m^3)	Sediment Load (10^4 t)
HRB	18.1	795.1
HURB	266.9	881.4
YRB	299.2	76,655
LRB	29.3	1112.9
SRB	404.7	598.8
MRB	532.7	573.3
PRB	2166.4	6274.6
YARB	8944.8	40791
HIRB	44.0	–
TRB	157.3	–
YZRB	312.5	1710.4
LCRB	247.3	341.0

Figure 2a,b show the cumulative curve of the streamflow in the major rivers. The cumulative curves of YARB, PRB, MRB, LCRB, YZRB, TRB, and HIRB, presented linear increasing trends with essentially no fluctuation or inflection point. Among them, the runoff in the HIRB showed a significant increasing trend at 0.16×10^8 m^3/a, $P < 0.001$. That of other basins fluctuated near the mean level, as shown in Figure 3;

the interdecadal anomalies are shown in Table 3. For the northern rivers, the cumulative curve of streamflow in HURB presented complicated changes. The overall direction was a straight line, indicating severe fluctuation on the interannual scale, and no obvious trend was found. The other northern rivers generally presented convex curves or lines, indicating that runoff in these rivers showed decreasing trends (Figure 3). As shown in Table 4, the order of decrease rate was YRB (-7.25×10^8 m^3/a, $P < 0.001$) > SRB (-3.32×10^8 m^3/a, $P < 0.001$) > HRB (-0.69×10^8 m^3/a, $P < 0.001$) > LRB (-0.48×10^8 m^3/a, $P < 0.001$).

3.2. Overall Change in Sediment Load

Figure 2c,d show the cumulative curve of the sediment load in the major rivers. No obvious convex state was presented in those of LCRB and YZRB, which means the sediment discharge variation had no significant trend. The cumulative curve of sediment load in other rivers showed obvious convex shapes, which denote that the sediment had different degrees of gradual reduction. In particular, after 2000, the decrease was between 59.1% and 98.7% (Table 3). In the southern rivers, the decrease in the 2000s was between 59.1% and 63.4%, with YARB showing the largest value. In the northern rivers, the decrease was between 59.4% and 98.7%; LRB and HRB were reduced by 97.6% and 98.7%, respectively (Table 3).

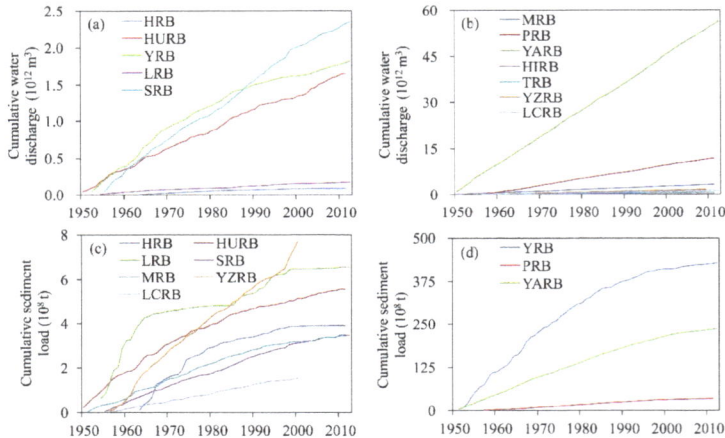

Figure 2. Cumulative curves of (**a**), (**b**) water discharge and (**c**), (**d**) sediment discharge in major basins in China, during 1950–2012.

3.3. Pattern of Changes in Streamflow and Sediment Load

Figures 3 and 4 show respectively the temporal variation and double mass curves of sediment load and streamflow in the 12 major rivers in China. On the whole, the variation of streamflow *versus* sediment can be divided into three categories. In the first, the streamflow was stably maintained, but the sediment load was reduced (HURB, YARB, MRB, PRB). In the second, the streamflow and sediment load were both reduced (YRB, LRB, HRB, SRB). In the third, both water and sediment discharge remained stable (LCRB, YZRB).

Table 3. Interdecadal anomalies of water discharge and sediment load. The reference value is the average value during the 1950s to 1960s.

Basins	Change in Water Discharge (%)				Change in Sediment Load (%)			
	1970s	1980s	1990s	2000s	1970s	1980s	1990s	2000s
HRB	−18.6	−55.9	−56.4	−84.6	−35.0	−77.5	−77.2	−98.7
HURB	−27.8	0.1	−45.5	−6.5	−43.4	−46.8	−73.7	−69.6
YRB	−36.4	−41.5	−71.2	−67.5	−26.0	−47.3	−67.8	−88.7
LRB	−56.9	−36.8	−27.3	−62.8	−90.2	−80.3	−59.9	−97.6
SRB	−31.6	−2.9	−4.6	−43.4	−21.3	7.7	−19.6	−59.4
MRB	−1.7	−2.6	4.2	−5.4	−1.6	−13.4	−49.0	−63.4
PRB	11.2	−4.0	13.3	−6.4	10.9	14.7	4.4	−59.1
YARB	−6.1	−0.9	5.8	−5.7	−13.0	−10.9	−29.7	−63.1
YZRB	−6.6	−14.2	−4.5	5.2	−14.9	−24.0	−5.2	−
LCRB	−1.4	19.3	1.7	14.1	5.7	39.4	−14.4	−

Table 4. Results of Sen's slope estimator and the Z value by using linear regression and the Mann–Kendall test, respectively.

Basins	Water Discharge (10^8 m^3)			Sediment Load (10^4 t)		
	Slope	Z	Significance	Slope	Z	Significance
HRB	−0.69	−6.96 **	$P < 0.001$	−46.96	−6.73 **	$P < 0.001$
HURB	−1.50	−1.49	$P > 0.1$	−25.35	−5.42 **	$P < 0.001$
YRB	−7.25	−6.01 **	$P < 0.001$	−2300	−6.58 **	$P < 0.001$
LRB	−0.48	−3.2 *	$P < 0.01$	−55.33	−4.90 **	$P < 0.001$
SRB	−3.32	−3.37 **	$P < 0.001$	−8.44	−4.11 **	$P < 0.001$
MRB	−0.53	−0.58	$P > 0.1$	−9.74	−4.10 **	$P < 0.001$
PRB	−2.04	−0.98	$P > 0.1$	−77.49	−3.31 **	$P < 0.001$
YARB	−5.72	−0.36	$P > 0.1$	−6000	−6.73 **	$P < 0.001$
HIRB	0.16	2.81 *	$P < 0.01$	−	−	−
TRB	0.10	0.80	$P > 0.1$	−	−	−
YZRB	0.20	0.40	$P > 0.1$	6.12	0.47	$P > 0.1$
LCRB	0.61	1.58	$P > 0.1$	−0.96	−0.67	$P > 0.1$

Note: "*" and "**" mean the correlation coefficient reach the significance level of 0.01 and 0.001, respectively.

3.3.1. Streamflow Remained Stable and Sediment Load Reduced

In the HURB, the streamflow and sediment discharge experienced continuous declines before 1978 (Figures 3a and 4a). Since the early 1980s, and particularly in the 1990s, an obvious decrease in sediment discharge occurred. This result was due mainly to the effects of soil and water conservation projects in the HURB with areas of 1.53×10^4 km^2 and 1.04×10^4 km^2 in the 1980s and 1990s, respectively. Shi [24] reported that the soil erosion amount of the small watershed in the upper reaches of the Huai River was reduced 77%–85% after implementation of engineering projects.

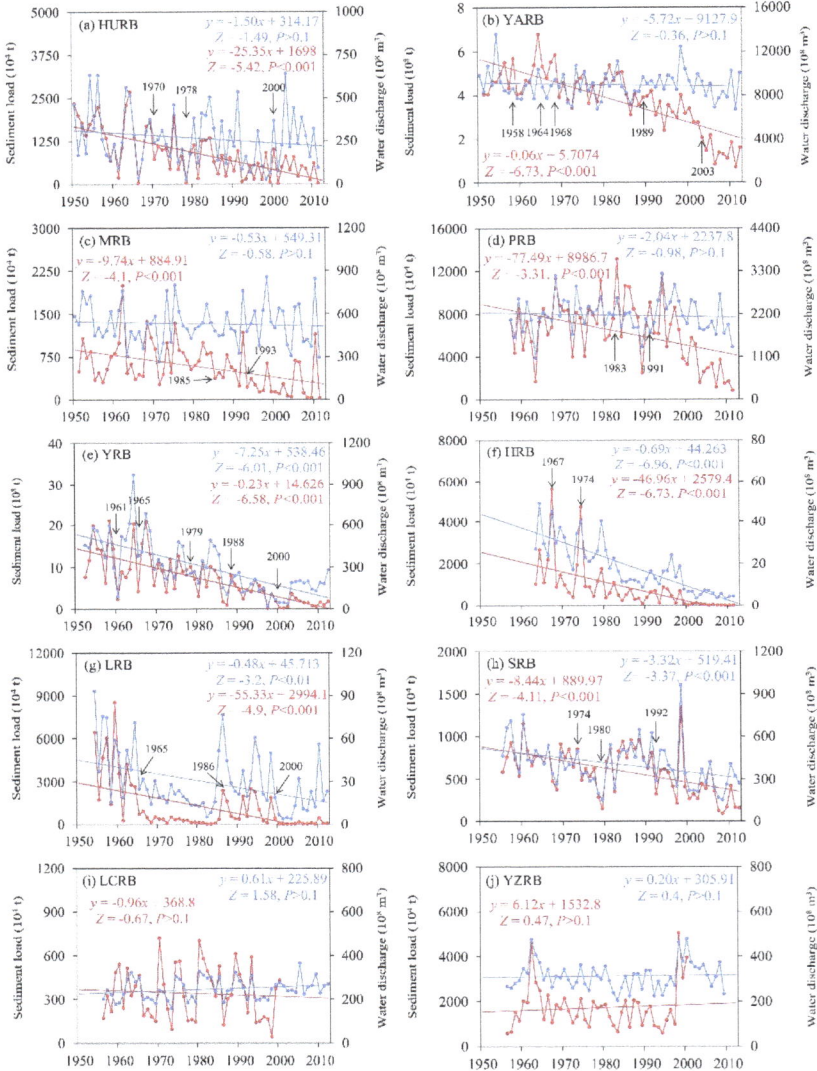

Figure 3. Interannual variability of streamflow and sediment load in major basins in China during 1950–2012. Blue and red lines represent changes in water discharge and sediment discharge, respectively.

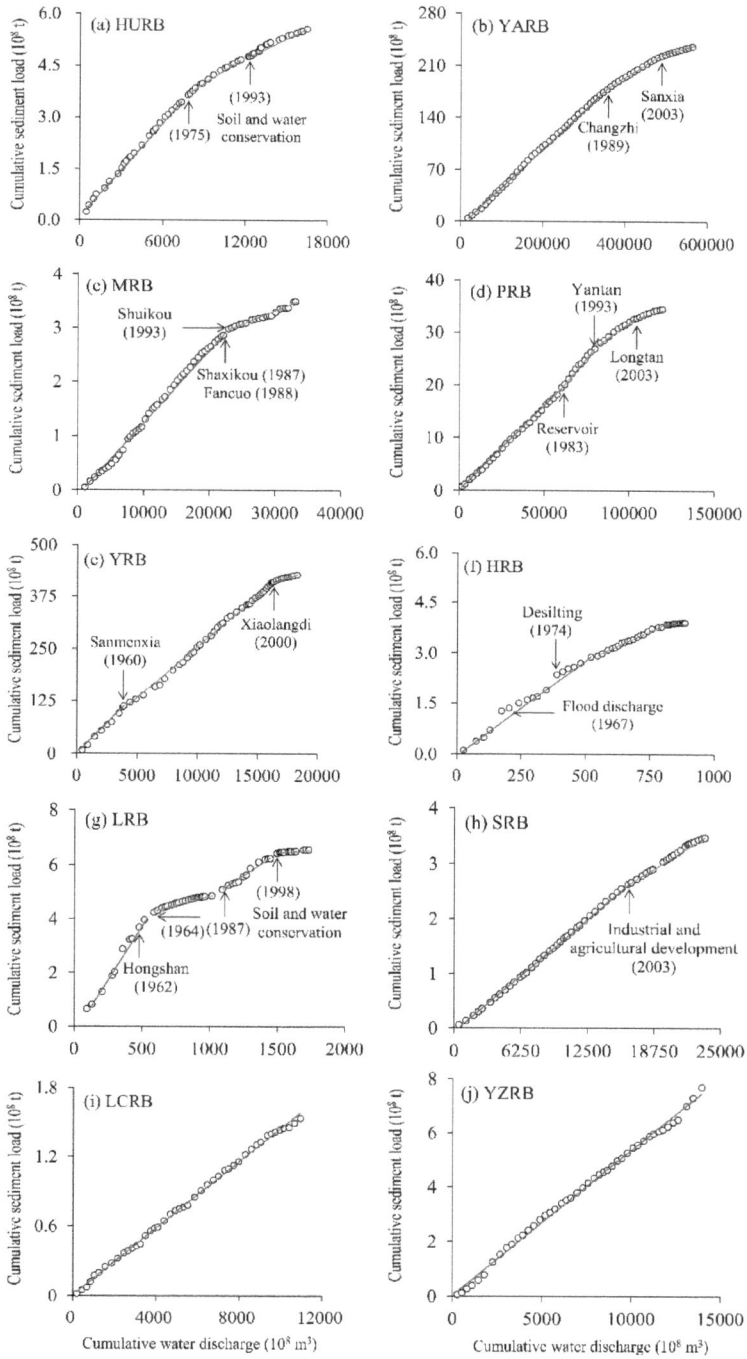

Figure 4. Double mass curve of streamflow *versus* sediment load.

In the YARB, the streamflow variation was relatively stable; however, the sediment load decreased significantly (Figures 3b and 4b). The sediment yields in 1956, 1958, and 1963–1968 were obviously high, which is likely attributed to large-scal edeforestation activities such as "Devastating Forests for Arable Land". After 1969, the sediment load value suddenly dropped. This result is likely attributed to the Danjiangkou Reservoir operation that began in 1968, which intercepted a large amount of sediment. After 1989, the sediment yield decreased again, which could have been caused by three aspects. Firstly, a series of ecological engineering projects was conducted since 1982, particularly the Changzhi Project launched in 1989. Secondly, the reservoirs were constructed in the mainstream and tributaries of the Yangtze River. Thirdly, the amount of sand dredging increased annually. The total amount in 1990–2002 was 5×10^7 t. After 2003, the sediment load decreased again, which was mainly caused by operation of the Sanxia Reservoir in 2003. In 2012, sedimentation in Sanxia Reservoir reached 14.37×10^8 t.

For MRB, as observed by Zhuqi Station, sediment load changed with streamflow variation before 1985, although the amplitude of the sediment was larger than that of the streamflow (Figures 3c and 4c). After 1985 and 1993, we detected two distinct decreasing processes obviously related to reservoir construction such as the Shaxikou hydropower station in 1987, Fancuo Dam in 1988, and the Shuikou hydropower station in 1993. In particular, the sedimentation of Shuikou Dam accounted for 86% of the total amount of sediment in the MRB. An additional factor is that the annual sand dredging amount in Mawei near Shuikou Dam was 1×10^7 t after the late 1980s. The streamflow in Gaoyao Station of PRB was stable, although the sediment load decreased significantly (Figures 3d and 4d). Before 1983, the sediment load fluctuated with streamflow change and maintained a consistent pace. In 1983–1991, the sediment load increased obviously, which may related to the construction of multiple reservoirs. Since 1994, the sediment load presented a sharp decline, which is mainly attributed to the sediment-retaining functions of reservoirs and dams such as Yantan Reservoir in the Hongshui River (operated in 1993) and Longtan Reservoir (operated in 2003).

3.3.2. Streamflow and Sediment Load Reduced Together

In the YRB, HRB, LRB, and SRB, the water and sediment discharge showed clear downward trends. In YRB, as observed by Linjin Station, the sediment loads in 1961–1965, 1980–1988, and 2000–2012 were lower than the mean level (Figures 3e and 4e). The low point in 1961 reflected the operation of Sanmenxia Reservoir, which began to retain water and sediment. In 1965, the operation mode of this reservoir changed to begin storing clear water and releasing muddy sediment; therefore, the sediment load recovered slightly. After 1980, the sediment load reduced again, with the following possible causes: Firstly, the water diversion project began operation, which reduced the streamflow and sediment discharge. Secondly, the soil and water conservation projects in the upstream function are functioning efficiently; thus, the water discharge is reduced in the lower reaches. Thirdly, the rainfall is concentrated mainly in the midstream region with less soil loss area; thus, the sediment is reduced downstream. In the early 2000s, the sediment and streamflow reduced again, mainly due to the water diversion projects and water reservoirs operating in the upstream region, such as Xiaolangdi Reservoir in 1999.

In the HRB, the sediment and streamflow after the 1980s was maintained at relatively low levels, the sediment load approached zero since 2000 (Figures 3f and 4f). The abnormally high sediment load in 1967 and 1974 was due to the desilting effect of Guanting Reservoir during the flood season through high rainfall and runoff. In the LRB, the sediment and water discharge presented periodic changes (Figures 3g and 4g). Before the 1960s, strong rainfall caused severe soil erosion, thereby inducing sediment load increases. After 1964, the sediment load decreased significantly due to the operation of Hongshan Reservoir in 1962. Until 1999, the sedimentation in Hongshan Reservoir reached 9.41×10^8 m^3, accounting for 58% of the total storage capacity. Severe fluctuation of sediment and streamflow occurred in 1985–2000, which was caused mainly by rainfall.

In the SRB, the streamflow and sediment discharge decreased at the same pace (Figures 3h and 4h). On the interdecadal scale, the streamflow and sediment discharge experienced "low-high-low" alternating variation processes in the 1970s to 1990s, which were mainly affected by precipitation. In addition, agricultural and industrial development since the beginning of the 1960s also accelerated the water consumption for industrial and residential purposes. Although some large- and medium-sized reservoirs are located in the upstream regions of Songhua River, the reciprocal relationship between runoff and sediment has been relatively good, and no obvious anomalies have been detected. This result occurred essentially because the vegetation coverage in the source area is relatively high, and density of population is low; thus, the river is seldom disturbed by human activities.

3.3.3. Both Streamflow and Sediment Discharge Remained Stable

Both streamflow and sediment discharge in the LCRB and YZRB maintained stability. No significant upward or downward trends were detected, and fluctuations of streamflow *versus* sediment have been essentially consistent (Figure 3i,j and Figure 4i,j). Actually, the vegetation coverage is good, the population density in the YZRB and LCRB is relatively low, and the level of economic development and social construction is also relatively low. Thus; the streamflow and sediment are seldom affected by human activities, and presented a good correlation among rainfall, streamflow, and sediment.

4. Influencing Factors of Streamflow Variation

4.1. Precipitation

In general, climate change is mainly characterized by temperature and precipitation variability. Precipitation drives runoff and hence directly influences the discharge of a river. Figure 5 shows precipitation changes in the major rivers and in the entire country that occurred during the past 60 years. China's precipitation experienced a "decrease, increase, decrease, increase, decrease" pattern. The 1950s was a rainy decade; the average rainfall reached 820.3 mm, which was the highest value recorded in several decades. After 1960, precipitation began to decrease. Levels were low in the 1980s and increased slightly in the 1990s. After 2000, the overall trend obviously reduced. The SRB and LRB experienced a rainy period in the 1950s that decreased in the 1960s and 1970s. The 1980s and 1990s presented slight increases that decreased again in the 2000s. Precipitation in the YRB essentially showed a decreasing trend by decade with lowest values occurring in the 1990s. After 2000, however, the level increased slightly. Rainfall in the TRB and HIRB increased by decade; however, the YZRB and LCRB presented decreasing trends since the 1960s. For the southern rivers including YARB, MRB, and PRB, changes in precipitation were consistent. Levels were lowest in the 1950s, increased in the 1960s and 1970s, and fell slightly in the 1980s before returning to the less-rainfall stage in 2000.

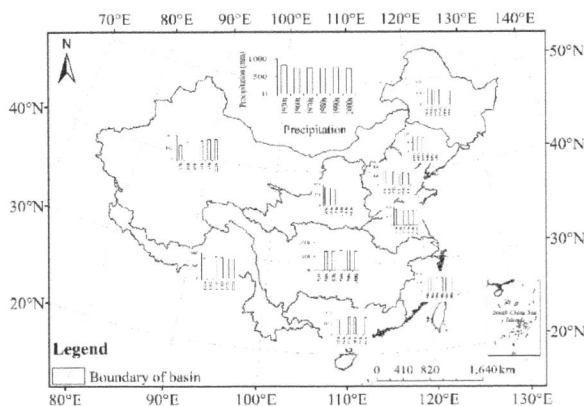

Figure 5. Interdecadal rainfall changes in the main basins in China during the 1950–2012.

Since the Reform and Opening-up policy was implemented in 1979, the demand for water resources increased substantially, and the streamflow in some rivers showed different degrees of decreasing trend (Figure 3). Therefore, we divided the streamflow sequence into two periods of before and after 1980 to explore the relationship between streamflow and precipitation (Figure 6). The subsection fitting lines of precipitation *versus* runoff represent the runoff depth produced by rainfall in 1950–1979 and 1980–2010, respectively. Assuming that the precipitation had no significant change on the basin scale, changes in the relationship of precipitation and streamflow can reflect the influence of the underlying surface on the original hydrological process. In the northern rivers, the fitting lines of rainfall-runoff in the HRB (Figure 6a), YRB (Figure 6c), and SRB (Figure 6e) moved downward during 1980–2010 comparing with that occurring in 1950–1979.

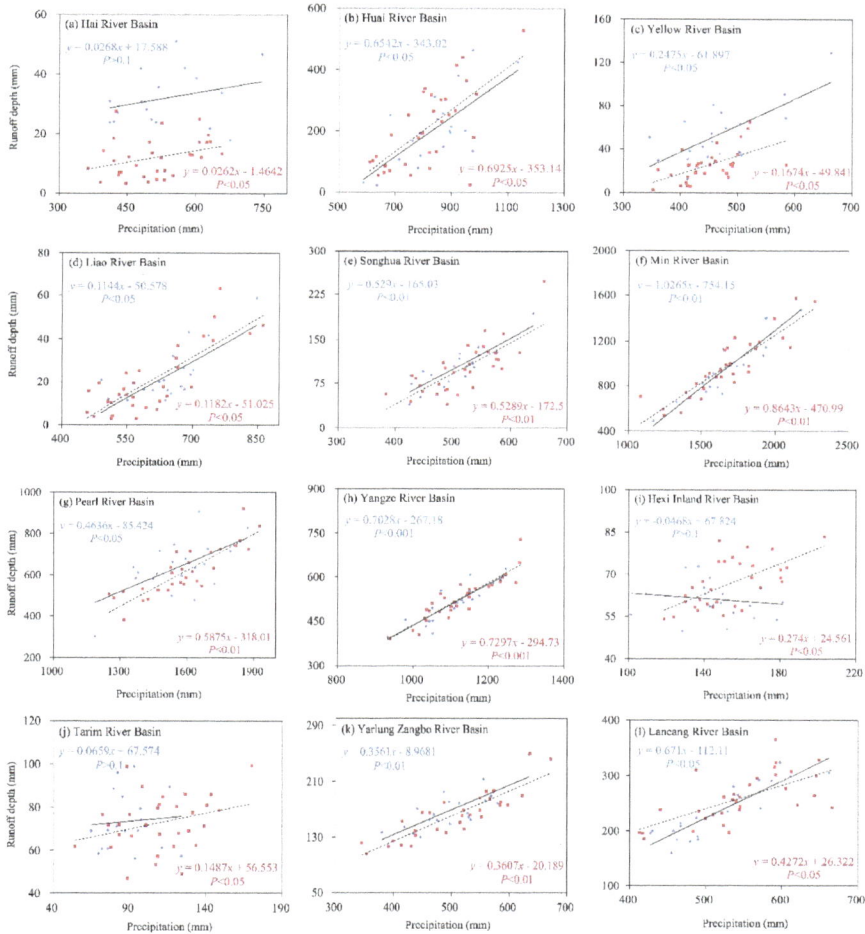

Figure 6. Subsection fitting lines of precipitation *versus* streamflow in 1950–1979 *versus* those in 1980–2012. Solid and dotted lines represent lines in 1950–1979 and 1980–2012, respectively. Blue and red points represent data of 1950–1979 and 1980–2012, respectively.

This downward movement denote the capacity of runoff yield decline. For the HURB (Figure 6b) and LRB (Figure 6d), the fitting lines moved upward, which illustrates that the capacity of runoff yield was enhanced. In the southern rivers, the runoff capacity of the PRB declined after 1980 (Figure 6g), whereas the change of MRB was not obvious (Figure 6f). The YARB also remained stable (Figure 6h). The streamflows of the TRB and HIRB in the northwest and the LCRB and YZRB in the southwest, originating from Tibet–Qinghai Plateau, were recharged by rainfall in addition to runoff from glacier and snow melt water. In this condition, the rainfall-runoff fitting lines could not fully reflect the changes in runoff yield ability. Nevertheless, the fitting lines can still be used to determine the role of rainfall change on runoff. The slope of the regression line in the HIRB during 1950–1979 was a negative value (Figure 6i), which shows that the water supply in addition to rainfall decreased. In 1980–2010, however, this part of the water supply increased. Similarly, the runoff yield in the TRB (Figure 6j) and the YZRB (Figure 6k) decreased, whereas the change in the LCRB was not obvious (Figure 6l). Changes in the rainfall–runoff relationship showed that the underlying surface affected the hydrological process, which will be further discussed in subsequent sections.

4.2. Monsoons

To determine the possible causes of streamflow variation, we examined Asian monsoon indices, including the East Asian monsoon index (EAMI), South Asian monsoon index (SAMI), Tibet–Qinghai Plateau monsoon index (TPMI), West Pacific subtropical high (WPSH), and westerly index (WI). The Asian monsoon indices were created by Li [46,47], and WI was derived from National Centers for Environmental Prediction (NCEP) reanalysis data. The TPMI was calculated from NCEP 600 mb height reanalysis data based on the method given by Wang [48].

Figure 7 shows the cumulative anomaly curves of runoff *versus* climatic indexes and their correlation coefficients. The streamflow in the northern rivers correlated positively with the EAMI and negatively with the WI, including the HRB (Figure 7a,b), HURB (Figure 7c,d), YRB (Figure 7e,f), LRB (Figure 7g,h), and SRB (Figure 7i,j). Among the northern rivers, the correlation coefficient between the YRB and EAMI streamflow was the highest ($R = 0.82$, $P < 0.001$). However, the cumulative anomaly curves of streamflow against the WI presented an opposite phase ($R = -0.83$, $P < 0.001$). The results showed that streamflow in the YRB was jointly affected by the East Asian monsoon and westerlies, which also reflects the interaction of westerlies and the East Asian monsoon. The relationship of streamflow in the HRB, HURB, LRB, and SRB and climatic indices was essentially the same as that in the YRB. However, the correlation was not as strong as that with the YRB, which may be related to the locations of the East Asia monsoon and westerly belt, as well as the influence of human activities on natural runoff. In the southeastern rivers, the runoff of the

MRB was affected mainly by the West Pacific subtropical high ($R = 0.46$, $P < 0.01$) and the Tibet–Qinghai Plateau Monsoon ($R = 0.38$, $P < 0.01$) (Figure 7k,l). The PRB (Figure 7m,n) is adjacent to the South China Sea and is also located in the westerly belt and the East Asian monsoon zone. Therefore, its runoff was influenced by several monsoon systems and therefore lacked direct correlation with a specific monsoon index; the closest relationships were with the SAMI ($R = 0.24$, $P < 0.1$) and TPMI ($R = -0.55$, $P < 0.01$). The case of the YARB (Figure 7o,p) was similar to that of the PRB. The runoff was influenced by several monsoon systems dominated by the WI ($R = 0.64$, $P < 0.001$) and TPMI ($R = 0.39$, $P < 0.01$). Ma [49] also found a similar phenomenon, although deeper indications require further investigation. In northwestern China, runoff in the HIRB (Figure 7q,r) and TRB (Figure 7s,t) correlated positively with that in the TPMI and WI, indicating that streamflow changes in these areas were controlled mainly by the Tibet–Qinghai Plateau monsoon and westerlies.

A deeper indication is that the TRB and HIRB were located in the central area of Eurasia, which could have hardly been reached by the Northwest Pacific summer monsoon. Therefore, the westerlies and Tibet–Qinghai Plateau monsoon became the dominant factors. In southwestern China, the YZRB originated from the Tibet–Qinghai Plateau hinterland. Its runoff was therefore subjected to the Tibet–Qinghai plateau monsoon, and the correlation coefficient between them was as high as 0.76 ($p < 0.001$) (Figure 7u,v). In contrast, because the YZRB was near the India Ocean, the runoff and particularly the midstream was influenced by the South Asia monsoon. In the LCRB (Figure 7w,x), although its headwater was in the Tibet–Qinghai Plateau monsoon zone, the runoff was mainly affected by the westerlies and East Asian monsoon.

4.3. Water Consumption for Industrial and Residential Purposes

Although climate change is important, particularly the impact of precipitation change on runoff, water consumption for industrial and residential purposes also plays important roles in runoff change [50–52]. Long-term statistics of water consumption are not currently available, which makes detailed analysis difficult. The "China Water Resources Bulletin (2004–2012)", reported that the YARB had the largest annual water consumption (Figure 8a). Aside from a slight reduction in water consumption in the HRB, the other basins all presented increasing trends in 2004–2012 (Figure 8a). Water for agriculture and industry accounted for the largest proportion at more than 80% of the total water consumption (Figure 8b,c). For long term trends, the "China Statistical Yearbook (1950–2009)" reported that the population, gross domestic product (GDP), and gross agricultural products showed significantly increasing trends (Figure 9). In particular, definite increases in water consumption are expected by the former two sectors. Thus, it can be inferred that the water consumption during the past 60 years also showed sharp increasing trends.

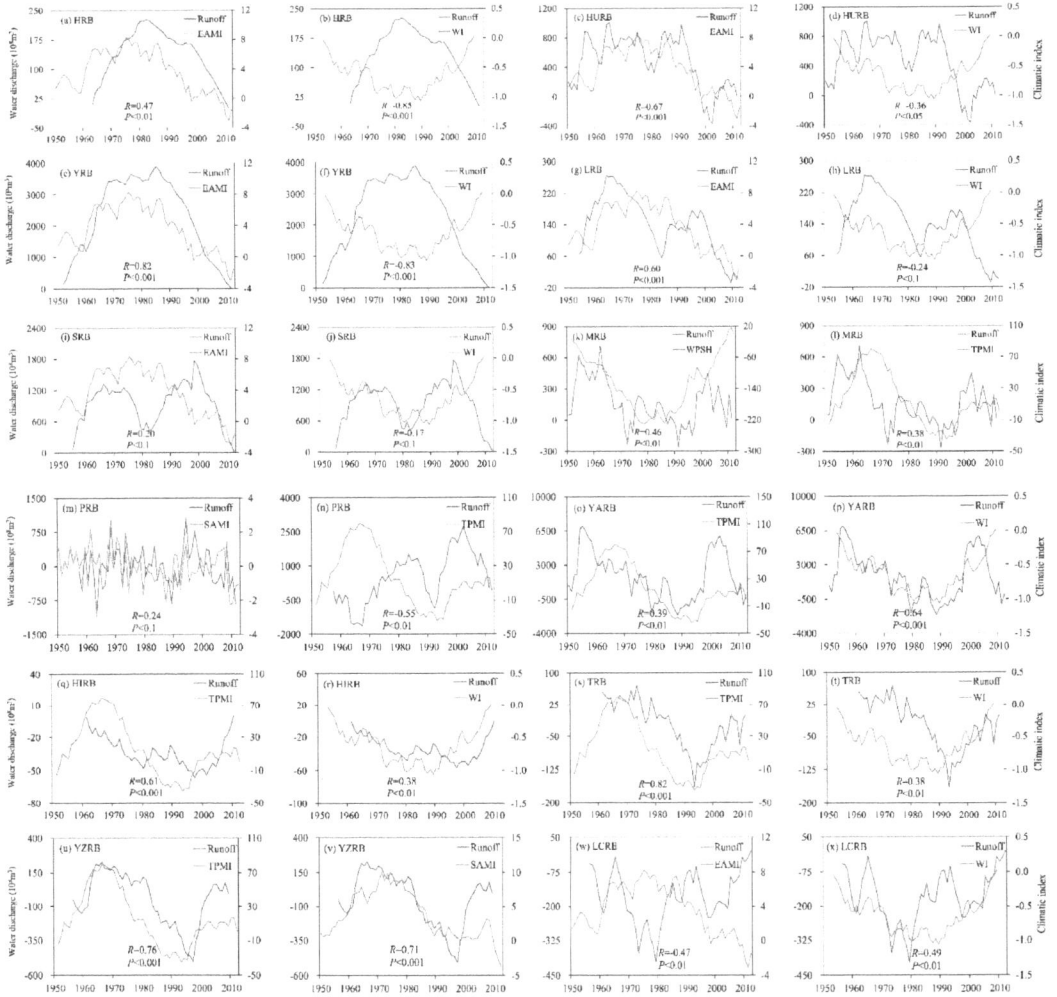

Figure 7. Cumulative anomaly curves of streamflow *versus* monsoon intensity indices in the main basins of China. EAMI, SAMI, WI, WPSH and TPMI represent the East Asian monsoon index, westerly index, West Pacific subtropical high, and Tibetan Plateau monsoon index, respectively. *R* and *P* represent the correlation coefficient and significance level, respectively.

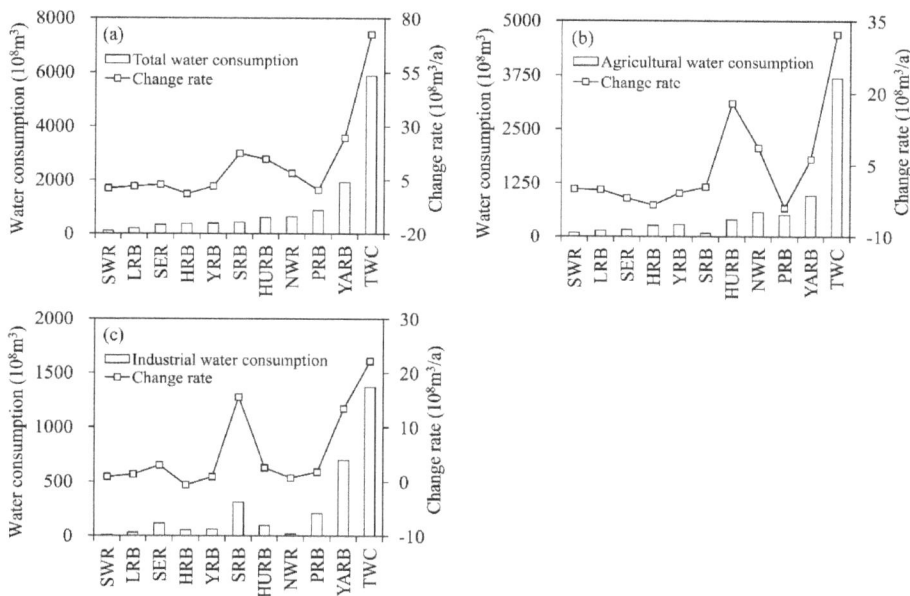

Figure 8. Interannual variability of water consumption and its change rate in the 10 major basins and in the whole country (TWC). (**a**), (**b**), and (**c**) are respectively gross water consumption, agricultural water consumption, and industrial water consumption.

4.4. Land Use/Land Coverage Change

The statistics of land use change from 1985 to 2010 are listed in Table 5. National land use change during the past 30 years shows obvious spatial differences and can be divided into two periods: from the end of the 1980s to 2000 and from 2000 to 2010. The characteristics of land use change during the first period indicate that farmland, residential land, industrial land, paddy fields, and water areas increased rapidly, and ecological land was greatly reduced. In 2000–2010, growth in farmland, residential land, paddy fields, and water areas slowed, and ecological land decreased slightly. Three possible causes can be considered. Firstly, construction land expansion was the main cause of the farmland decrease in traditional agricultural areas. In addition, large-scale ecological engineering promoted decreases in farmland. Another aspect, annual accumulated temperature increases in the northern arid zone, created suitable climatic conditions for reclamation in some forest zones, and a large area of dry land was converted to paddy fields. Secondly, implementation of the Reform and Opening-up policy resulted in rapid urbanization processes that occupied farmland and ecological land, particularly in eastern China. Thirdly, before 2000, the reclamation of grassland and forests caused soil degradation, which induced severe soil erosion. After 2000, the 'Grain for Green' program was implemented, which significantly increased the areas of grassland and forests in the ecologically fragile areas of western China.

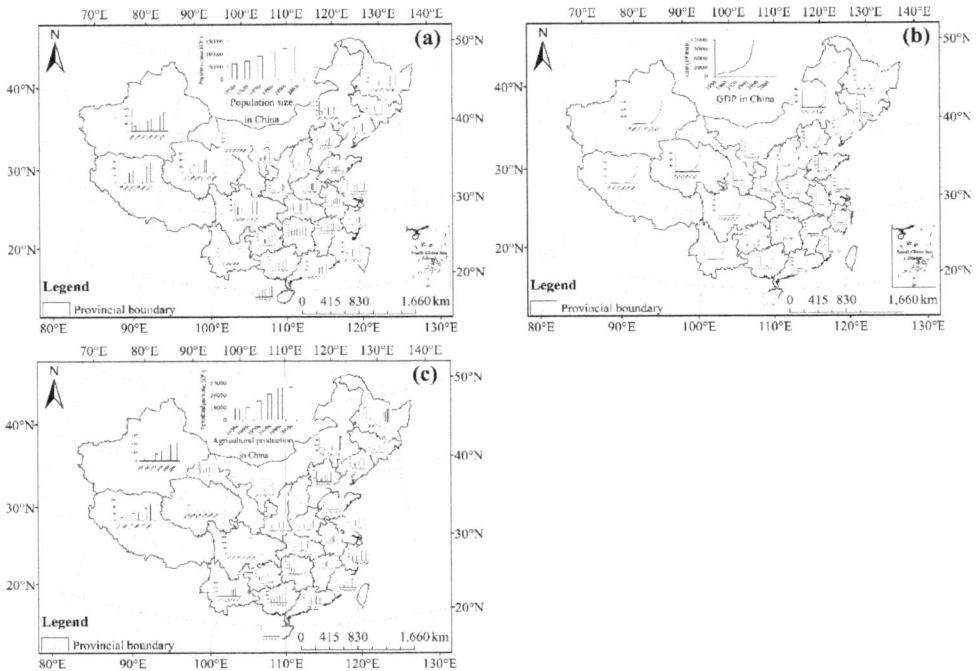

Figure 9. (**a**) Interannual variability of population; (**b**) gross domestic product (GDP); and (**c**) agricultural products in 1950–2012 in China.

Table 5. Area of land use conversion in China during the two periods of 1980s–2000 and 2000–2010.

Transfer Types	Dryland– Paddy Field	Grassland– Forest	Others– Waters	Others– ConstructionLand	Forest– Farmland	Forest– Grassland	Grassland– Farmland	Grassland– Forest	Waters– Others
1980–2000	173.15	115.80	80.81	177.63	174.73	81.14	345.76	104.73	62.40
2000–2010	138.12	142.36	114.55	378.24	37.54	41.85	197.38	88.79	83.27

The potential hydrological effect of land use/land coverage change has been a highly contested issue. Firstly, the high intensity of human activities in the city have resulted in accelerated soil compaction and crusting processes, thereby decreasing the infiltration rate and capacity of soil water storage and resulting in surface runoff increases [22,53]. Secondly, the runoff coefficients of forest areas and grassland are relatively small. When the forests and grassland converted to arable land, the flow yield will therefore increase. Thirdly, in southern China, a certain degree of lake reclamation occurred, which inhibited the process of evaporation and infiltration and resulted in runoff increases.

5. Influencing Factors of Sediment Load Variation

5.1. Hydraulic Engineering and Soil and Water Conservation Engineering

5.1.1. Soil and Water Conservation Engineering

China began to conduct soil and water conservation projects in the early 1950s, which included the construction of silt dams and terraced fields and the planting of trees and grass [54,55]. In the middle reaches of the Yellow River, for example, the most severe soil erosion in China occurred. The statistics of soil and water conservation engineering projects conducted in this area in 1979, 1989, 1996, and 2011 are plotted in Figure 10a.

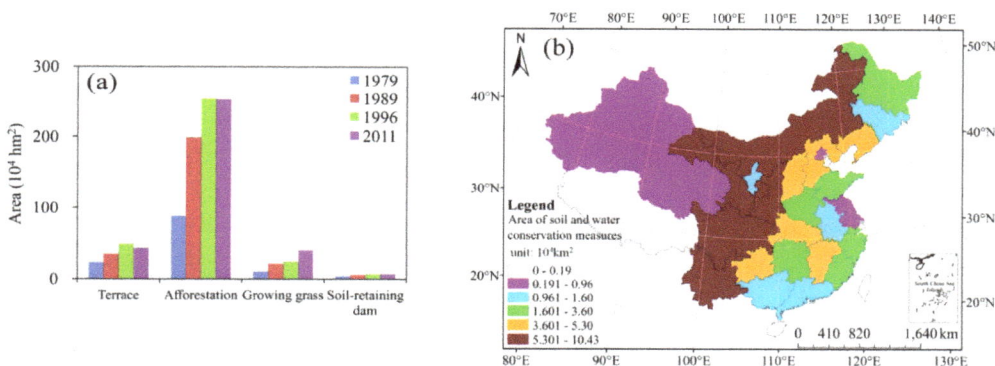

Figure 10. Statistics of soil and water conservation engineering areas in (**a**) the middle stream of the Yellow River basin (YRB) and (**b**) the national distribution.

All types of engineering showed an increasing trend, with the largest area in afforestation. On the national scale, the total area of present soil and water conservation engineering projects is 99.2×10^4 km^2, and the area of silt dams is 925.6 km^2 [56]. The largest distribution is in western China, which incurred severe soil erosion (Figure 10b). During the past 60 years, the runoff in some large rivers declined significantly (Figure 3), largely due to the impact of soil and water conservation engineering.

5.1.2. Reservoir and Dam Construction

By the end of 2012, the total number of large, medium, and small reservoirs in mainstreams and tributaries was 97,543, and a capacity of 8255×10^8 m^3 [56]. These reservoirs regulated the runoff and sediment change processes and reduced the sediment load in the downstream regions. The statistics of the "Water Resources Yearbook in 2006" [57] indicate that by the end of 2006, the numbers of large and medium reservoirs in the YARB were 149 and 1115, respectively, which ranked as highest. For the southern rivers including the YARB, PRB, HURB, and SERB, water resources were abundant. Thus, there were many reservoirs in these basins, and the ratios of total reservoir capacity (TRC) to average annual flow (AAF) was below 70%. However, in the northern rivers including the LRB, SRB, and HRB, the water resources are relatively limited; thus, the TRC/AAF was above 100%. In the two largest basins (YRB and YARB), the total amount of sedimentation in all of the reservoirs in the YARB during 1991–2005 was 17958×10^4 t. For the Sanmenxia and Xiaolangdi reservoirs, sedimentation in 1960–2012 and 1997–2012 was 64.108×10^8 m^3 and 27.625×10^8 m^3, respectively [56].

5.1.3. Water and Sediment Diversion Projects

To resolve the uneven distribution of water resources and to ease the high demand of local water resources, China has been conducting water and sediment diversion projects such as the "South-to-North Water Diversion Project" and the "Yellow River-to-Tianjin Water Diversion Project". After implementation of these projects, the streamflow and sediment load decreased in water supply areas and increased in water demand areas. In the "South-to-North Water Diversion Project", the amount of water diversion from the midstream region of the Yangtze River accounted for a very small proportion of the average streamflow, particularly in the wet seasons, and had little impact in abundant water. In the dry seasons, however, the water diversion project reduced the streamflow in the downstream region. Therefore, the saltwater traced back into the estuary of Yangtze River, and the sediment load was reduced downstream.

In the Yellow River, for example, the total amount of water diversion in the lower reaches was 3665.2×10^8 m^3 in 1958–2002, and the annual water diversion was 89.4×10^8 m^3, accounting for 23.1% of the runoff observed at Huayuankou Station during the same period and directly influencing streamflow and sediment changes in the lower reaches [23,55]. The amount of water and sediment diversion in the downstream region of the Yellow River in 2011 is shown in Table 6. These factors accounted for 15.6% and 13.6% of water and sediment discharge, respectively, during the same period.

Table 6. Amount of water and sediment diversion in the downstream reaches of the Yellow River in 2011.

Channel Segment	Xixiayuan–Huayuankou	Huayuankou–Jiahetan	Jiahetan–Gaocun	Gaocun–Sunkou	Sunkou–Aishan	Aishan–luokou	Luokou–Lijin	Lower Reaches of Lijin	Total
Length (km)	109.8	100.8	77.1	118.2	63.9	101.8	167.8	110.0	849.4
Amount of water diversion (10^8 m^3)	4.19	14.57	17.31	11.50	8.5	21.07	22.08	4.66	103.88
Amount of sediment diversion (10^4 t)	62.8	111.3	286.6	242.7	266.3	800.2	464.4	94.5	2328.8

5.2. Other Human Activities

The impacts of human activities on soil and water loss and sediment discharge in rivers include direct and indirect effects. Direct impacts include the erosion-transport-accumulation process induced by human activities such as farmland reclamation, mining, and road construction. The indirect impacts were mainly caused by destruction of vegetation, which accelerated the occurrence and development of soil erosion. With the development of society and the economy, both the strength and breadth of human activities were greater than those in the past. These activities include increases in population, the scale of reclamation, mining, and other infrastructure construction.

6. Impacts of Water and Sediment Discharge Reduction on Utilization of Sediment Resources

6.1. Effects of River Regulation and Flood Control

In the YRB, the sediment load in the lower reaches of Yellow River was greatly reduced in recent decades. Particularly after 2000, the streamflow and sediment load observed at Huayuankou Station were reduced by 52.5% and 61.7%, respectively, compared with those observed in the 1950s to 1960s. Reducing water and sediment discharges has relieved flood pressure and has influenced river regulation to some degree. Dike reinforcement with silt remains a very important project in the lower reaches of the Yellow River [58]. Decreases in sediment concentration by flooding and the threat of beach inundation by flooding have resulted in significant increases in the cost of dike reinforcement with silt.

6.2. Effects of Flood Irrigation and Soil Improvement

The sediment of the Yellow River, particularly flood sediment, is an effective material for soil improvement because it reduces salinity and alkalinity and improves land fertility. Until the early 1990s, the area of soil improvement was 23.2 × 10^4 hm^2 in the lower reaches of the Yellow River [58]. Soil improvement by silt is usually conducted during the flood season, when the sediment concentration is relatively high. However, this effect will be limited under the reduction of water and sediment resources.

6.3. Effects of Land and Wetland Formation

In the estuary area of the Yellow River, approximately 64% of sediment is precipitated on land and in shallow water, which elevates the riverbed and delta. The Linjin Station observation reported reductions of 88.7% and 67.5% in sediment and water discharge, respectively, compared with values reported in the 1950s and 1960s. Therefore, the land formation rate has decreased. For example, the 23.6 km^2/a land formation rate in 1855–1954 decreased to 8.6 km^2/a in 1992–2001 [58]. In addition, soil, water, and sediment are the main components of wetlands. Therefore, sediment and water discharge reduction will directly affect the quality and formation rate of wetlands.

7. Conclusions

The results of the present study are summarized in the following points:

1. During the past 60 years, the streamflow in northern China, including the HRB, HURB, YRB, LRB, and SRB, showed different decreasing trends. That in the southern rivers, including MRB, PRB and YARB, presented severe fluctuations, although the declining trend did not reach significant levels. For the streamflow in the TRB, HIRB, YZRB, and LCRB, increasing trends were presented. The runoff yield capacity was weakened in the HRB, YRB, SRB, and PRB and enhanced in the LRB and HURB. That in the MRB and YARB remained stable.

2. In the northern rivers, runoff correlated positively with the EAMI and negatively with the WI. In the southern rivers, runoff was mainly influenced by the Tibet–Qinghai monsoon, South Asian monsoon, and westerlies. That in the HIRB and TRB was controlled mainly by the Tibet–Qinghai monsoon and westerlies. Runoff in the YZRB was controlled by the South Asian monsoon and the Tibet–Qinghai monsoon, whereas that in the LCRB was influenced mainly by the East Asian monsoon and westerlies.

3. Sediment loads in the LCRB and YZRB did not present significant change trends. However, sediment loads in other rivers exhibited varying degrees of gradual reduction, the greatest of which was in the 2000s.

4. Underlying surface and precipitation changes jointly influenced the runoff in eastern rivers. Water consumption for industrial and residential purposes, soil and water conservation engineering projects, hydraulic engineering, and underlying surface changes induced by other factors were the main causes of runoff and sediment reduction.

Acknowledgments: This research was jointly funded by a key consulting project from the Chinese Academy of Engineering (2014-XZ-31), the National Natural Science Foundation of China (41301632) and the Chinese Research Academy of Environmental Sciences special funding for basic scientific research (2014-YKY-003).

Author Contributions: Chong Jiang, Fen Li and Linbo Zhang conceived the research idea and designed the methodology, Chong Jiang and Fen Li performed the analyses, and Chong Jiang, Daiqing Li, Fen Li, and Linbo Zhang wrote the paper.

Conflicts of Interest: The authors declare no conflict of interest.

References

1. Alan, D.Z.; Justin, S.; Edwin, P.M.; Bart, N.; Eric, F.W.; Dennis, P.L. Detection of intensification in global and continental scale hydrological cycles: Temporal scale of evaluation. *J. Clim.* **2003**, *16*, 535–547.
2. Easterling, D.R.; Meehl, G.A.; Parmesan, C.; Changnon, S.A.; Karl, T.R.; Mearns, L.O. Climate extremes: Observations, modeling, and impacts. *Science* **2000**, *289*, 2068–2074.
3. Qiu, J. China drought highlights future climate threats. *Nature* **2010**, *465*, 142–143.
4. Zhang, J.Y.; He, R.M.; Qi, J.; Liu, C.S.; Wang, G.Q.; Jin, J.L. A new perspective on water issues in north China. *Adv. Water Sci.* **2013**, *24*, 303–310.
5. Zhang, Y.J.; Hu, C.H.; Wang, Y.G. Analysis on variation characteristics and influential factors of runoff and sediment of Liaohe River Basin. *Yangtze River* **2014**, *45*, 32–35.
6. World Water Assessment Programme (WWAP). *The United Nations World Water Development Report 4: Managing Water under Uncertainty and Risk*; World Water Assessment Programme: Paris, France, 2012.
7. Vörösmarty, C.J.; McIntyre, P.B.; Gessner, M.O.; Dudgeon, D.; Prusevich, A.; Green, P.; Glidden, S.; Bunn, S.E.; Sullivan, C.A.; Liermann, C.R.; *et al.* Global threats to human water security and river biodiversity. *Nature* **2010**, *467*, 555–561.
8. Intergovernmental Panel on Climate Change (IPCC). *Managing the Risks of Extreme Events and Disasters to Advance Climate Change Adaptation*; Cambridge University Press: Cambridge, UK, 2012.
9. Climate Change 2013: The Physical Science Basis, Summary for Policy Makers, Technical Summary and Frequently Asked Questions. Available online: https://www.ipcc.ch/pdf/assessment-report/ar5/wg1/WG1AR5_SummaryVolume_FINAL.pdf (accessed on 13 October 2015).
10. Matthew, B.C.; Nigel, W.A. Adapting to climate change impacts on water resources in England-An assessment of draft water resources management plans. *Glob. Environ. Chang.* **2011**, *21*, 238–248.
11. Domroes, M.; Peng, G. *The Climate of China*; Springer: Berlin, Germany, 1988.
12. Zhai, P.M.; Zhang, X.B.; Wan, H.; Pan, X.H. Trends in total precipitation and frequency of daily precipitation extremes over China. *J. Clim.* **2005**, *18*, 1096–1108.
13. Ding, Y. *Monsoons Over China*; Kluwer Academic Publishers: Amsterdam, the Netherland, 1994.

14. Holeman, J.N. The sediment yield of major rivers of the world. *Water Resour. Res.* **1968**, *4*, 737–747.

15. Holland, H.D. River transport to the oceans. In *The Ocean Lithosphere*; John Wiley & Sons: New York, NY, USA, 1981.

16. Walling, D.E.; Fang, D. Recent trends in the suspended sediment loads of the world rivers. *Glob. Planet. Chang.* **2003**, *39*, 111–126.

17. Milliman, J.D. Delivery and fate of fluvial water and sediment to the sea: A marine geologist's view of European rivers. *Sci. Mar.* **2001**, *65*, 121–132.

18. Bobrovitskaya, N.N.; Kokorev, A.V.; Lemeshko, N.A. Regional patterns in recent trends in sediment yields of Eurasian and Siberian rivers. *Glob. Planet. Chang.* **2003**, *39*, 127–146.

19. Liu, C.; Wang, Y.G.; Sui, J.Y. Analysis on variation of seagoing water and sediment load in main rivers of China. *J. Hydraul. Eng.* **2007**, *38*, 1444–1452.

20. Subramainian, V. Sediment load of Indian rivers. *Curr. Sci.* **1993**, *64*, 928–930.

21. Siakeu, J.; Oguchi, T.; Aoki, T.; Esakid, Y.; Jarviee, H.P. Change in riverine suspended sediment concentration in central Japan in response to late 20th century human activities. *Catena* **2004**, *55*, 231–254.

22. Wang, Y.G.; Liu, X.; Shi, H.L. Variation and influence factor of runoff and sediment in the lower and middle Yangtze River. *J. Sed. Res.* **2014**, *29*, 38–47.

23. Miao, C.Y.; Ni, J.R.; Borthwick, A.G.L. Recent changes of water discharge and sediment load in the Yellow River basin. *Prog. Phys. Geogr.* **2010**, *34*, 541–561.

24. Shi, H.L.; Hu, C.H.; Wang, Y.G.; Tian, Q.Q. Variation trend and cause of runoff and sediment load variations in Huaihe River. *J. Hydraul. Eng.* **2012**, *43*, 571–579.

25. Zhang, J.Y.; Zhang, S.L.; Wang, J.X.; Li, Y. Study on runoff trends of the six larger basins in China over the past 50 years. *Adv. Water Sci.* **2007**, *18*, 230–234.

26. Miao, C.Y.; Yang, L.; Liu, B.Y.; Gao, Y.; Li, S.L. Streamflow changes and its influencing factors in the mainstream of the Songhua River basin, Northeast China over the past 50 years. *Environ. Earth Sci.* **2011**, *63*, 489–499.

27. Yang, S.L.; Zhao, Q.Y.; Belkin, I.M. Temporal variation in the sediment load of the Yangtze River and the influences of the human activities. *J. Hydrol.* **2002**, *263*, 56–71.

28. Gao, G.; Chen, D.L.; Xu, C.Y.; Simelton, E. Trend of estimated actual evapotranspiration over China during 1960–2002. *J. Geophys. Res.* **2007**, *112*, D11120.

29. International Research and Training Center on Erosion and Sedimentation (IRTCES) China. *Gazette of River Sedimentation 2011*; China Water Power Press: Beijing, China, 2012.

30. Lu, X.X.; Higgitt, D.L. Sediment yield variability in the upper Yangtze, China. *Earth Surf. Proc. Land* **1999**, *24*, 1077–1093.

31. Hassan, M.A.; Church, M.C.; Xu, J.X.; Yan, Y.X. Spatial and temporal variation of sediment yield in the landscape: Example of Huanghe (Yellow River). *Geophys. Res. Lett.* **2008**, *35*, L06401.

32. Mann, H.B. Non-parametric tests against trend. *Econometrica* **1945**, *13*, 245–259.

33. Kendall, M.G. *Rank Correlation Methods*; Griffin: London, UK, 1975; pp. 1–200.

34. Chattopadhyay, S.; Jhajharia, D.; Chattopadhyay, G. Trend estimation and univariate forecast of the sunspot numbers: Development and comparison of ARMA, ARIMA and Autoregressive Neural Network models. *C.R. Geosci.* **2011**, *343*, 433–442.

35. Theil, H. A rank invariant method of linear and polynomial regression analysis. *Proc. R. Neth. Acad. Sci.* **1950**, *53*, 1397–1412.

36. Sen, P.K. Estimates of the regression coefficients based on Kendall's tau. *J. Am. Stat. Assoc.* **1968**, *63*, 1379–1389.

37. Jhajharia, D.; Dinpashoh, Y.; Kahya, E.; Choudhary, R.R.; Singh, V.P. Trends in temperature over Godavari River basin in Southern Peninsular India. *Int. J. Climatol.* **2013**, *34*, 1369–1384.

38. Storch, H.; Navarra, A. (Eds.) Misuses of statistical analysis in climate researchvon. In *Analysis of Climate Variability: Applications of Statistical Techniques*; Springer: Berlin, Germany, 1995; pp. 11–26.

39. Chattopadhyay, S.; Jhajharia, D.; Chattopadhyay, G. Univariate modeling of monthly maximum temperature time series over northeast India: Neural network *versus* Yule-walker equation based approach. *Meteorol. Appl.* **2011**, *18*, 70–82.

40. Kumar, S.; Merwade, V.; Kam, J.; Thurner, K. Streamflow trends in Indiana: Effects of long term persistence, precipitation and subsurface drains. *J. Hydrol.* **2009**, *374*, 171–183.

41. Xu, J.X. Recent variations in water and sediment in relation with reservoir construction in the upper Changjiang River Basin. *J. Mount. Sci.* **2009**, *27*, 385–393.

42. Merriam, C.F. A comprehensive study of the rainfall on the susquehanna valley. *Trans. Amer. Geophys. Union* **1937**, *18*, 471–476.

43. Searcy, J.K.; Hardisoni, C.H.; Langbein, W.B. *Double Mass Curves*; Geological Survey Water Supply Paper 1541-B; US Geological Survey: Washington, DC, USA, 1960.

44. Mu, X.M.; Zhang, X.Q.; Gao, P.; Wang, F. Theory of double curves and its applications in hydrology and meteorology. *J. China Hydrol.* **2010**, *30*, 47–51.

45. Li, Y.Y.; Cao, J.T.; Shen, F.X.; Xia, J. The changes of renewable water resources in China during 1956–2010. *Sci. China Earth Sci.* **2014**, *57*, 1825–1833.

46. Li, J.P.; Wu, Z.W.; Jiang, Z.H.; He, J.H. Can global warming strengthen the East Asian summer monsoon? *J. Clim.* **2010**, *23*, 6696–6705.

47. Li, J.P.; Feng, J.; Li, Y. A possible cause of decreasing summer rainfall in northeast Australia. *Int. J. Climatol.* **2011**, *31*.

48. Wang, M.C.; Liang, J.; Shao, M.J.; Shi, G. Preliminary analysis on interannual change of Tibet–Qinghai Plateau monsoon. *Plateau Meteorol.* **1984**, *3*, 76–82.

49. Ma, H.; Yang, D.W.; Tan, S.K.; Gao, B.; Hu, Q.F. Impact of climate variability and human activity on streamflow decrease in the Miyun Reservoir catchment. *J. Hydrol.* **2010**, *389*, 317–324.

50. Chen, Y.N.; Fan, Y.T.; Wang, H.J.; Fang, G.H. Research progress on the impact of climate change on water resources in the arid region of Northwest China. *Acta Geogr. Sin.* **2014**, *69*, 1295–1304.

51. Song, X.M.; Zhang, J.Y.; Zhan, C.S.; Liu, C.Z. Review for impacts of climate change and human activities on water cycle. *J. Hydraul. Eng.* **2013**, *44*, 779–790.

52. He, D.M.; Liu, C.M.; Feng, Y.; Hu, J.M.; Ji, X.; Li, Y.G. Progress and perspective of international river researches in China. *Acta Geogr. Sin.* **2014**, *69*, 1284–1294.

53. Liu, C.; Hu, C.H.; Shi, H.L. Changes of runoff and sediment fluxes of river in mainland of China discharge into Pacific Ocean. *J. Sed. Res.* **2012**, *27*, 70–75.

54. Liu, C.; He, Y.; Zhang, H.Y. Trends analysis of the water and sediment loads of the main rivers in China using water-sediment diagram. *Adv. Water Sci.* **2008**, *19*, 317–324.

55. Liu, C.; Wang, Z.Y.; Sui, J.Y. Variation of flow and sediment of the Yellow River and their influential factors. *Adv. Sci. Technol. Water Resour.* **2008**, *28*, 1–7.

56. Ministry of Water Resources of the People's Republic of China (MWRPRC). *Chinese River Sediment Bulletin 2012*; China Water Power Press: Beijing, China, 2013; pp. 1–60.

57. Ministry of Water Resources of the People's Republic of China (MWRPRC). *Chinese River Sediment Bulletin 2006*; China Water Power Press: Beijing, China, 2007; pp. 1–55.

58. Wang, Y.G.; Hu, C.H.; Shi, H.L. Variation in water and sediment resources and its influence on utilization of sediment resource in the Yellow River Basin. *J. China Inst. Water Resour. Hydropower Res.* **2010**, *8*, 237–245.

The Impact of Climate Change on the Duration and Division of Flood Season in the Fenhe River Basin, China

Hejia Wang, Weihua Xiao, Jianhua Wang, Yicheng Wang, Ya Huang,
Baodeng Hou and Chuiyu Lu

Abstract: This study analyzes the duration and division of the flood season in the Fenhe River Basin over the period of 1957–2014 based on daily precipitation data collected from 14 meteorological stations. The Mann–Kendall detection, the multiscale moving *t*-test, and the Fisher optimal partition methods are used to evaluate the impact of climate change on flood season duration and division. The results show that the duration of the flood season has extended in 1975–2014 compared to that in 1957–1974. Specifically, the onset date of the flood season has advanced 15 days, whereas the retreat date of the flood season remains almost the same. The flood season of the Fenhe River Basin can be divided into three stages, and the variations in the onset and retreat dates of each stage are also examined. Corresponding measures are also proposed to better utilize the flood resources to adapt to the flood season variations.

Reprinted from *Water*. Cite as: Wang, H.; Xiao, W.; Wang, J.; Wang, Y.; Huang, Y.; Hou, B.; Lu, C. The Impact of Climate Change on the Duration and Division of Flood Season in the Fenhe River Basin, China. *Water* **2016**, *8*, 105.

1. Introduction

Global climate change is inevitable [1], causing frequently occurring extreme weather and climate events and unevenly distributed rainfall. Fenhe River, the second largest tributary of the Yellow River in Northern China, is experiencing such changes. According to historical records, it had a 5-year-long wet season from 1963 to 1967, and had a 15-year-long dry season from 1979 to 1993. It has been suggested that the climate, especially precipitation, in the Fenhe River Basin has been affected by human activities and natural phenomena such as El Niño-Southern Oscillation and Pacific Decadal Oscillation [2]. Therefore, it is of critical importance to study the variation in the flood season and evaluate the impact of climate change.

Several studies have investigated the duration of flood seasons. Odekunle [3] and Sâmia *et al.* [4] determined the onset and retreat time of the flood seasons in Nigeria and South American monsoon areas, respectively, and both of their methods were effective. Hachigonta *et al.* [5] found that the onset date of the flood season in Zambia has a significant spatial variation. Other studies [6–10] have analyzed temporal variations in flood seasons. All of these studies indicate that the onset and retreat time of the flood season have fluctuations that are affected by natural variations over long periods.

Although numerous researchers have investigated the onset and retreat times of the flood season, few studies explore the variation of the flood season in light of climate change. In this paper, we detect the abrupt changing point of the climate based on the annual precipitation in 1957–2014 in the Fenhe River Basin, estimate the onset and retreat time of flood seasons divided by this abrupt changing point, and, finally, divide the flood season into three stages to analyze the temporal variation of each stage. Based on the results, we propose measures to alleviate the discord between flood control and flood resources utilization.

2. Study Area

The Fenhe River is the largest river in the province of Shanxi, China and also the second largest tributary of the Yellow River. The Fenhe River Basin covers an area of 39471 km^2 (110°30′–113°32′ E; 35°20′–39°00′ N), accounting for 25.3% of the area of Shanxi. The Fenhe River Basin lies between Mt. Lvliang and Mt. Taihang and irregularly distributed as a wide stripe in West-Central Shanxi. It is located in the mid-latitude continental monsoon climate zone with arid and semiarid climates. During 1957–2014, the mean annual precipitation was 472.4 mm. The drainage map of the Fenhe River Basin is shown in Figure 1.

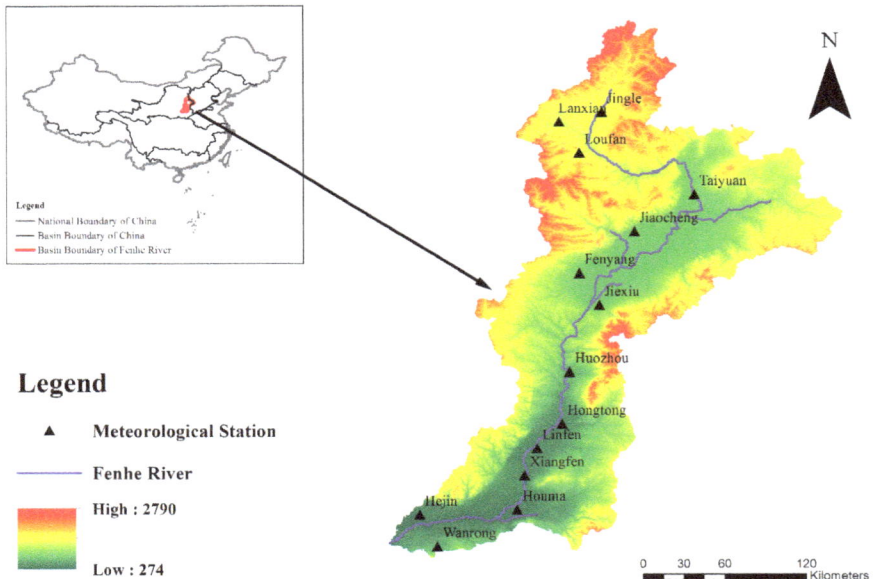

Figure 1. Plane graph of the Fenhe River Basin.

The data used for this study include daily precipitation data from 1957 to 2014 at 14 meteorological stations within the basin, and are obtained from the China Meteorological Administration [11]. Some stations have incomplete and false data records for some years. We use the inverse distance weighted method to fill and replace these missing or false data. It should be stated that the precipitation data during 1957–1979 are separated by rain and snow, but they are no longer separated and have not been since 1980. In this study, the snow data before 1980 are converted to the corresponding rainfall.

3. Materials and Methods

3.1. Climate Abrupt Change Detection

Mann–Kendall detection is a nonparametric statistical test method which is widely used in time series data trend tests [12]. Samples are not needed to obey a certain distribution in this way, which is also unacted on the interference of a few outliers. Therefore, it is suitable for hydrology, meteorology, and other non-normal distribution data. In this paper, the annual precipitation during 1957–2014 are chosen as the time series data to detect the abrupt point. The specific steps of Mann–Kendall detection are then further explored [13].

3.2. The Onset and Retreat Dates of Flood Season

Currently, the onset and retreat time of flood season are mostly defined by experience, which is inevitably subjective. The accurate detect of the two dates will provide scientific support to relevant sectors to forecast the flood season and to regulate the reservoirs for the purpose of flood control. The moving t-test is suitable for the multiscale detection of the change of flood season; the beginning of flood season is a symbol that the precipitation varies from less to more, at which time the abrupt point may exist. The situation is just the opposite when it comes to the end of flood season. Fraedrich *et al.* took the historical flood level of the River Nile, for example, and found three distinct epochs from AD 622 to AD 1470 [14]. Wang *et al.* detected the beginning and ending dates of rainy season in Dalian from 1951 to 1998 by applying the moving t-test method, which showed that it is more objective and exact than the empirical method [15]. Jiang *et al.* redistricted the coherently dry/wet episode of the Nile River and found the results were coincided with the historical disaster and famine of Egypt [16]. Other researchers also obtained ideal results by using this method [17–19]. Therefore, the multiscale moving t-test is applied in this paper to define the onset and retreat dates of flood season.

The multiscale moving t-test is used to detect the differences between two subsamples before and after the abrupt point with equivalent sample sizes, namely,

$n = n_1 = n_2$. The t-statistic of precipitation in the Fenhe River Basin can be calculated as follows:

$$t(n,i) = (\bar{x}_{i2} - \bar{x}_{i1})\, n^{1/2} \left(s_{i2}^2 + s_{i1}^2\right)^{-1/2} \tag{1}$$

where

$$\bar{x}_{i1} = \sum_{j=i-n}^{i-1} x_j / n;\ s_{i1}^2 = \sum_{j=i-n}^{i-1} \left(x_j - \bar{x}_{i1}\right)^2 / (n-1) \tag{2}$$

$$\bar{x}_{i2} = \sum_{j=i}^{j+n-1} x_j / n;\ s_{i2}^2 = \sum_{j=i}^{i+n-1} \left(x_j - \bar{x}_{i2}\right)^2 / (n-1) \tag{3}$$

The test of the multiscale abruption is realized by altering the n value. $N = 2, 3, \ldots , <N/2$, and $i = n+1, n+2, \ldots , N\text{-}n$, where N is the number of days in a year (365 or 366), and n is the timescale. The confidence level of t-test at 0.01 is approximately equivalent to the confidence level of Yamamoto and the Mann–Kendall test at 0.05. In order to make the graphics visualized and the analysis convenient, the further calculation is as follows:

$$t_r(n,i) = t(n,i) / t_{0.01}(n) \tag{4}$$

$tr(n, i)$ can be treated as the threshold to judge whether there is an abrupt point. When $tr(n, i)$ is greater than 1.0, it means that the abruption of variation trend is rising; when $tr(n, i)$ is less than -1.0, it means the abruption of variation trend is descending. The center of extreme values of the $tr(n, i)$ absolute value means that the abruption is the most prominent when the time is i and the timescale is n.

3.3. Flood Season Division

Flood season division means dividing the flood season into several stages in the light of the distinct differences and the regularity of flood characteristics during the different periods. Accurate division is the prerequisite for controlling the flood limit level, which can bring huge benefits to agricultural irrigation, power generation, etc., and alleviate the contradiction between flood control and flood resources utilization. Extensive researches on flood season division are conducted, and a mass of methods has been applied to flood season division. Liu et al. successfully divided the flood season in the Geheyan Reservoir by means of the changing point method [20]. Jin et al. used the fuzzy set analysis method with the city of Nanping and divided the flood season into three stages [21]. Many other methods are also ideal for solving this problem [22–27].

The essence of flood season division is a multidimensional orderly clustering analysis of time series. The Fisher optimal partition method is just a method of clustering of ordered samples and was used to divide seismicity period successfully by He *et al.* [28]. It has some basic characteristics: firstly, it can take multi-index into consideration; secondly, it can meet the time sequence of the flood season; finally, it can confirm the optimum number of flood season divisions. Therefore, the Fisher optimal partition method is applied in this paper to divide the flood season of the Fenhe River Basin.

We can define $\{x_1, x_2, \ldots, x_n\}$ as ordered samples, where every sample is m-dimensional vector. The symbol $B(n, k)$ indicates that n ordered samples are divided into k parts. The division can be expressed as follows:

$$P_1 = \{i_1, i_1 + 1, \ldots, i_2 - 1\}; P_2 = \{i_2, i_2 + 1, \ldots, i_3 - 1\}; \ldots P_k = \{i_k, i_k + 1, \ldots, n\}, \quad (5)$$

where division points are $1 = i_1 < i_2 < \ldots < i_k < n = i_{k+1} - 1$.

Supposing a part P contains samples $\{x_i, x_{i+1}, \ldots, x_j\}$, the mean value can be denoted as follows:

$$\overline{x_p} = \frac{1}{j - i + 1} \sum_{t=1}^{j} x_t \quad (6)$$

Therefore, the sum of the squares of deviations of the part can be denoted as follows:

$$D(i, j) = \sum_{t=i}^{j} (x_t - \overline{x_p})^T (x_t - \overline{x_p}) \quad (7)$$

The purpose of the Fisher optimal partition method is to find division points, which make the sum of the squares of deviations of every part the minimum. Consequently, the optimal division can be depicted as the following formula:

$$B(n, k) = \min \sum_{t=1}^{k} D(i_t, i_{t+1} - 1) \quad (8)$$

The theorem says that the optimal division $B(n, k)$ must be equal to the optimal division $B(n-1, k-1)$ plus the remaining part. According to the theorem above, the optimal division $B(n, 2)$ can be denoted as follows:

$$B(n, 2) = \min_{2 \leqslant i \leqslant n} \{D(1, i - 1) + D(i, n)\} \quad (9)$$

Similarly, the optimal division $B(n, k)$ can be denoted as follows:

$$B(n, k) = \min_{2 \leqslant i \leqslant n} \{B(i - 1, k - 1) + D(i, n)\} \quad (10)$$

Then, we can obtain the optimal solution by finding the division points in descending order. The division point i_k must meet the requirement of the following equation:

$$B(n,k) = \{B(i_k - 1, k - 1) + D(i_k, n)\} \tag{11}$$

Then, the part k can be obtained: $P_k = \{i_k, i_k + 1, \ldots, n\}$. Similarly, the remaining parts $P_{k-1}, P_{k-2}, \ldots, P_2, P_1$ can be obtained successively.

In this process, we need to draw the $B(n, k) - k$ curve according to the results, and then calculate the absolute value of slope of every division point $f(k)$. Next, the $f(k) - k$ curve can be drawn—if $f(k)$ is greater, the division into k parts is better than $k-1$ parts. When $f(k)$ is close to zero, there is no need for division. Generally, k is the optimal division number of parts when $f(k)$ meets the maximum, and the corner of $B(n, k) - k$ curve is also seen as the optimal division point.

4. Results

4.1. The Detection of Climate Abrupt Point

Figure 2 displays the results of Mann–Kendall detection of annual precipitation during 1957–2014. Here, UF(K) and UB(K) are statistics sequence which obey standard normal distribution. A intersection exists between UF(K) line and UB(K) line, and it just lies between the two critical lines whose confidence level are 0.05. The corresponding time of the intersection is 1974. Therefore, we can divide the period 1957–2014 into two periods, namely, 1957–1974 and 1975–2014. The annual precipitation is on the rise during the former period, while the trend declines during the latter period.

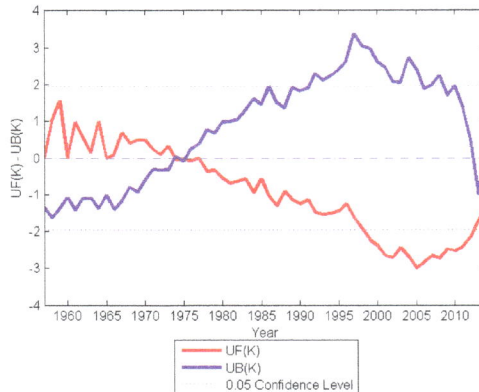

Figure 2. Mann-Kendall detection of the precipitation in the Fenhe River Basin for the period 1957–2014.

4.2. The Onset and Retreat Dates of Flood Season

Taking the year 1979 as an example, the result of the multiscale moving t-test of the Fenhe River Basin precipitation is shown in Figure 3. The color scale of the contour map shows that the larger the value, the deeper the color. The maximum value of $tr(n, i)$ is 2.2, which is greater than 1.0. The corresponding day is 165 and the corresponding timescale is 65 days. It means that the onset date of flood season in 1979 is Day 165, on which the abruption is the most conspicuous in the timescale of 65 days. Similarly, the retreat date of flood season in 1979 is Day 255, on which the abruption is the most conspicuous in the timescale of 63 days. Figure 4 shows the variation trend during 1957–2014, the red line represents the onset date of flood season in the Fenhe River Basin, and the blue line indicates the retreat date. Consequently, the annual average flood season of the two period 1957–1974 and 1975–2014 can be obtained. The onset and retreat dates of flood season during 1957–1974 are 170 and 253, respectively, and the flood season lasts 84 days long. In a similar way, the onset and retreat dates of flood season during 1975–2014 are 156 and 252, respectively, and the flood season is 13 days longer than that during 1957–1974. Out of convenience with respect to the flood season division, it was necessary to transform the timescale from day to pentad. According to the traditional Chinese calendar, a pentad is 5 days; one year contains 72 pentads. For instance, Day 170 can be converted to 4 June, which means the fourth pentad in June. Therefore, the flood season of 1957–1974 and 1975–2014 are 4 June–2 September and 1 June–2 September, respectively.

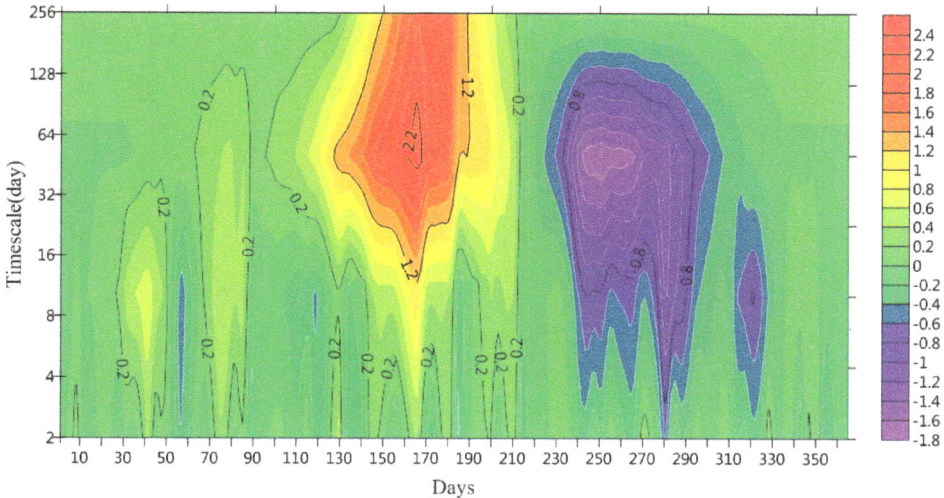

Figure 3. The multiscale moving t-test on the Fenhe River Basin daily mean precipitation in 1979.

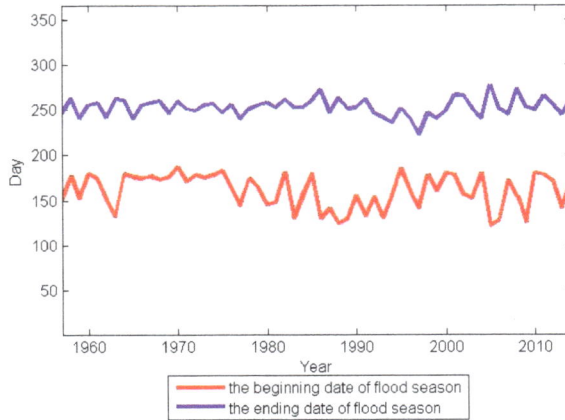

Figure 4. The onset and retreat dates of flood seasons for the period 1957–2014.

4.3. Flood Season Division

Five factors that can represent the seasonal rules of precipitation and flood are chosen as basic elements for flood season division: mean pentad precipitation, mean coefficient of variation of pentad precipitation, mean maximum 1-day-precipitation within a pentad, mean maximum 3-day-precipitation within a pentad, and number of hard rain days within a pentad. Tables 1 and 2 display the result of division of 1957–1974 and 1975–2014, respectively. Figure 5 shows the $B(n, k) - k$ curve and $f(k) - k$ curve of 1957–1974 and 1975–2014, respectively. The $f(k)$ is the maximum when k is equal to 3 in picture (*b*), and the curve $B(n, k) - k$ has a turning at the same time. Therefore, the optimal division number k is 3 during 1957–1974, which means the flood season can be divided into three stages during 1957–1974, namely, pre-flood season, main flood season, and post-flood season. The corresponding classification in Table 2 is 1–3, 4–9, 10–17, which means the pre-flood season is from 4 June to 6 June, the main flood season is from 1 July to 6 July, and the post-flood season is from 1 August to 2 September. Analogously, the flood season can be divided into three stages during 1975–2014. To be specific, the pre-flood season is from 1 July to 6 July, the main flood season is from 1 July to 6 August, and the post-flood season is from 1 September to 2 September. Figure 6 shows the length of each stage. By contrast, we can see the pre-flood season is 15 days in advance; thus, it lasts 15 days longer than before, the main flood season lasts 30 days longer than before, and the post-flood season is 30 days less, accordingly.

Table 1. Result of flood season division of the Fenhe River Basin during 1957–1974.

k	B(n,k)	f(k)	Classification
2	0.742		1–2,3–17
3	0.223	0.519	1–3,4–9,10–17
4	0.091	0.132	1–2,3–4,5–9,10–17
5	0.072	0.019	1–2,3–4,5–7,8–9,10–17
6	0.052	0.02	1–2,3–4,5–7,8,9,10–17
7	0.036	0.016	1–2,3,4,5–7,8,9,10–17
8	0.025	0.011	1–2,3,4,5,6–7,8,9,10–17
9	0.016	0.009	1–2,3,4,5,6–7,8,9,10,11–17
10	0.01	0.006	1–2,3,4,5,6,7,8,9,10,11–17
11	0.007	0.003	1–2,3,4,5,6,7,8,9,10,11–14,15–17
12	0.004	0.003	1–2,3,4,5,6,7,8,9,10,11–13,14,15–17
13	0.002	0.002	1–2,3,4,5,6,7,8,9,10,11–13,14,15–16,17
14	0	0.002	1–2,3,4,5,6,7,8,9,10,11–12,13,14,15–16,17
15	0	0	1,2,3,4,5,6,7,8,9,10,11-12,13,14,15–16,17
16	0	0	1,2,3,4,5,6,7,8,9,10,11,12,13,14,15–16,17

Table 2. Result of flood season division of the Fenhe River Basin during 1975–2014.

k	B(n,k)	f(k)	Classification
2	1.021		1–6,7–20
3	0.437	0.584	1–6,7–18,19–20
4	0.263	0.174	1–6,7–11,12,13–20
5	0.192	0.071	1–6,7–11,12,13–18,19–20
6	0.146	0.046	1–6,7–8,9–11,12,13-18,19–20
7	0.103	0.043	1–4,5–6,7–8,9–11,12,13–18,19–20
8	0.062	0.041	1–4,5–6,7–8,9–11,12,13–17,18,19–20
9	0.025	0.037	1–4,5–6,7–8,9–11,12,13–16,17,18,19–20
10	0.021	0.004	1–4,5–6,7–8,9–11,12,13,14–16,17,18,19–20
11	0.013	0.008	1–4,5–6,7–8,9–11,12,13,14,15–16,17,18,19–20
12	0.01	0.003	1–4,5–6,7–8,9–10,11,12,13,14,15-16,17,18,19–20
13	0.007	0.003	1–2,3–4,5–6,7–8,9–10,11,12,13,14,15–16,17,18,19–20
14	0.005	0.002	1–2,3,4,5–6,7–8,9–10,11,12,13,14,15–16,17,18,19–20
15	0.004	0.001	1–2,3,4,5–6,7–8,9–10,11,12,13,14,15,16,17,18,19–20
16	0.003	0.001	1–2,3,4,5–6,7,8,9–10,11,12,13,14,15,16,17,18,19–20
17	0.002	0.001	1–2,3,4,5–6,7,8,9–10,11,12,13,14,15,16,17,18,19,20
18	0.001	0.001	1–2,3,4,5–6,7,8,9,10,11,12,13,14,15,16,17,18,19,20
19	0	0.001	1,2,3,4,5–6,7,8,9,10,11,12,13,14,15,16,17,18,19,20

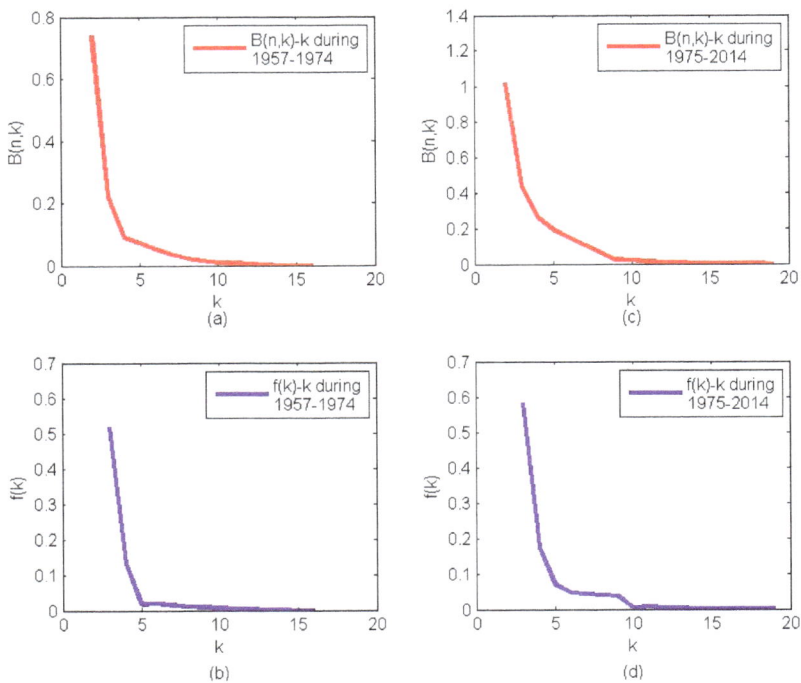

Figure 5. The $B(n, k) - k$ and $f(k) - k$ curves during 1957–1974 (**a,b**) and 1975–2014 (**c,d**).

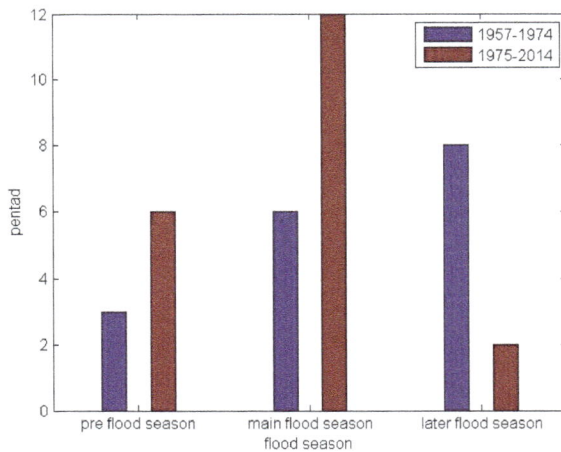

Figure 6. The bar graph of the length of each sub flood season.

5. Discussion

This research suggests that the flood season has advanced and extended from climate change since 1974, and it can be divided into three stages. Liu *et al.* also detected the abrupt point in the Fenhe River Basin, and they found that the precipitation changed abruptly in 1978 [29], which is approximately consistent with the results in this paper. The inconsistency between their article and this paper may derive from the inconsistency of the data—the period studied and the meteorological stations are different, while it is clear that the climate indeed changed significantly in the Fenhe River Basin.

With respect to the onset and retreat time of flood season, according to statistics, the precipitation during flood season account for more than 60% of the annual precipitation. Thus, the multiscale moving *t*-test method is objective and consistent with the characteristics of the precipitation. Pan *et al.* reconstructed the beginning times of flood season of the middle reaches of Yellow River according to the historical water level data from 1921 to 1950, and the results showed that they range from 6 July to 10 July. Our results indicate that the average onset dates of flood season during the period 1957–1974 and 1975–2014 in the Fenhe River Basin are 19 June and 5 June, respectively, which is inconsistent with Pan's results. Although the Fenhe River Basin lies in the middle reaches of the Yellow River, the terrain here is different from the mainstream of the Yellow River; the Fenhe River Basin is sandwiched between Mt. Lvliang and Mt. Taihang. The method and water level data used in Pan's paper are also different from the moving *t*-test method and the precipitation data used in this paper, which may have caused these differing results.

When it comes to the flood season division, Yang analyzed the diachronic distribution of storm floods and the regularity of flood season division in the Fenhe River Reservoir by recording the number of rainstorm days and the frequency of flood. The results indicated that the flood season can be divided into three stages [30], which is similar to the results of this paper. However, the Fisher optimal partition method focuses on the study from the angle of mathematics; it takes index into consideration more, and the time interval in this paper is 5 days, which is more accurate than 10 days in Yang's paper, so the result of the Fisher optimal partition method is more accurate.

Some countermeasures should be guaranteed to effectively utilize the flood water resources. We often set different flood limit levels during each stage [31]. In general, the prime task during the pre-flood season and the main flood season is flood prevention, so the flood limit levels in the pre- and main flood season should be a little lower. Further, the main purpose during the post-flood season is storing water, so the flood limit level during this period should be a little higher. As the pre-flood season has advanced 15 days and the main flood season has extended by 30 days, the flood limit level during this period should be decreased in advance and sustained longer so that the flood can flow safely through the basin. As for the shortening and delay of the post-flood season, the flood limit level should be elevated later so that flood can be transferred to water resources during this period.

Although some valuable findings have been obtained, further studies are needed to explore the flood season in greater detail. We used daily precipitation only to define the onset and retreat time of the flood season in the Fenhe River Basin. In fact, water level is also relative to the flood season, and the run-off also can impact the flood season. Future research should gather more data and consider more variables. If possible, we can take one reservoir as an example to evaluate the utilizable quantity of flood resources under climate change.

6. Conclusions

By using the Mann–Kendall detection, the multiscale moving t-test and the Fisher optimal partition methods, we present the onset and retreat dates of each stage in the Fenhe River Basin, which are characterized as the temporal variations from climate change. In the meantime, some useful conclusions of this study can be summarized as follows:

(1) The climate in the Fenhe River Basin has changed since 1974, and the annual precipitation illustrates a downward trend.
(2) The flood season in the Fenhe River Basin during 1975–2014 is 15 days longer than that during 1957–1974. Specifically, the pre-flood season is from 16 June to 30 June during 1957–1974, while it is from 1 June to 30 June during 1975–2014; the main flood season is from 1 July to 31 July during 1957–1974, while it is from 1 July to 31 August during 1975–2014; the post-flood season is from 1 August to 10 September during 1957–1974, whereas it is from 1 September to 10 September during 1975–2014.
(3) The flood limit level should be lowered at an earlier time to resist floods during the pre- and main flood season, and it should be raised at a later time so that more water for utilization can be stored after the flood season.

Author Contributions: Jianhua Wang contributed to design the methods; Yicheng Wang and Hejia Wang contributed ideas concerning the structure and content of the article; Baodeng Hou and Ya Huang contributed to data collection and processing; Chuiyu Lu drew the figures and maps; Weihua Xiao and Hejia Wang analyzed the results; Hejia Wang wrote the final manuscript.

Conflicts of Interest: The authors declare no conflicts of interest.

References

1.   International Panel of Climate Change (IPCC) Working Group I. *IPCC Fifth Assessment Report (AR5)*; IPCC: Stockholm, Sweden, 23–26 September 2013.

2. Ma, Z.G.; Shao, L.J. Relationship between dry/wet variation and the Pacific Decade Oscillation (PDO) in Northern China during the last 100 years. *Chin. J. Atmos Sci.* **2006**, *30*, 464–474.

3. Odekunle, T.O. Determining rainy season onset and retreat over Nigeria from precipitation amount and number of rainy days. *Theor. Appl. Climatol.* **2006**, *83*, 193–201.

4. Sâmia, R.G.; Alan, J.P.C.; Mary, T.K. Revised method to detect the onset and demise dates of the rainy season in the South American Monsoon areas. *Theor. Appl. Climatol.* **2015**.

5. Hachigonta, S.; Reason, C.J.C.; Tadross, M. An analysis of onset date and rainy season duration over Zambia. *Theor. Appl. Climatol.* **2008**, *91*, 229–243.

6. Ding, J.L.; Xu, Z.S.; Fei, J.F.; Qiang, X.M. Analysis of the definition of the onset and ending dates of the annually first rainy season in South China and its interannual variation characteristics. *J. Trop. Meteorol.* **2009**, *25*, 59–65.

7. Gu, R.Y.; Zhou, W.C.; Bai, M.L.; Di, R.Q.; Yang, J. Influence of climate change on ice slush period at Inner Mongolia section of Yellow River. *J. Dese. Res.* **2012**, *32*, 1751–1756.

8. Pan, W.; Fei, J.; Man, Z.M.; Zheng, J.Y.; Zhuang, H.Z. The fluctuation of the beginning time of flood season in North China during AD1766-1911. *Quatern. Int.* **2015**, *380*, 377–381.

9. Owusu, K.; Waylen, P.R. The changing rainy season climatology of mid-Ghana. *Theor. Appl. Climatol.* **2013**, *112*, 419–430.

10. Ding, L.L.; Ge, Q.S.; Zheng, J.Y.; Hao, Z.X. Variation of starting date of pre-summer rainy season in South China from 1736 to 2010. *Acta. Geogr. Sin.* **2014**, *24*, 845–857.

11. China Meteorological Administration. Available online: http://data.cma.gov.cn/ (accessed on 15 March 2016).

12. Pettitt, A.N. A non-parametric approach to the change-point problem. *Appl. Stat.* **1979**, *28*, 126–135.

13. Wei, F.Y. *Modern Climatic Statistical Diagnosis and Prediction Technology*; China Meteorological Press: Beijing, China, 1999; pp. 69–72.

14. Fraedrich, K.; Jiang, J.M.; Gerstengarbe, F.W.; Werner, P.C. Multiscale detection of abrupt climate changes: application to River Nile flood levels. *Int. J. Climatol.* **1997**, *17*, 1301–1315.

15. Wang, L.L.; Zou, Y.R.; Sui, H.Q. An objective determination of the beginning and ending date of rainy season in Dalian. *Meteorol. Mon.* **2000**, *26*, 12–16.

16. Jiang, J.M.; Fraedrich, K.; Zou, Y.R. A scanning t test of multiscale abrupt changes and its coherence analysis. *Chin. J. Geophys.* **2001**, *44*, 31–39.

17. Jiang, J.M.; Mendelssohn, R.; Schwing, F.; Fraedrich, K. Coherency detection of multiscale abrupt changes in historic Nile flood levels. *Geophys. Res. Lett.* **2002**, *29*, 112.

18. Schwing, F.; Jiang, J.M.; Mendelssohn, R. Coherency of multi-scale abrupt changes between the NAO, NPI, and PDO. *Geophys. Res. Lett.* **2003**, *30*, 325–348.

19. Gu, X.Q.; Jiang, J.M. A new application of scanning *t*-test: Partition of dry/wet episodes in the western USA. *Quat. Sci.* **2006**, *26*, 742–751.

20. Liu, P.; Guo, S.L.; Li, W.; Xiong, H.K.; Zhang, W.X.; Guo, H.J.; Xu, D.L.; Wang, Z.X. Application of changing-point method for flood season stage in Geheyan Reservoir. *J. Yangtze River. Sci. Res. Inst.* **2007**, *24*, 8–11.

21. Jin, B.M.; Fang, G.H. Application of fuzzy set analysis method on flood stage study of Nanping. *Water Power* **2010**, *36*, 20–22.

22. Gao, B.; Liu, K.L.; Wang, Y.T.; Hu, S.Y. Application of system clustering method to dividing flood season of reservoir. *Water Resour. Hydropower Eng.* **2005**, *35*, 1–5.

23. Dong, Q.J.; Wang, X.J.; Wang, J.P.; Fu, C. Application of fractal theory in the stage analysis of flood season in Three Gorges Reservoir. *Resour. Environ. Yangtze Basin* **2007**, *16*, 400–404.

24. Xie, F.; Wang, W.S. Set pair analysis and its application to the division of flood period. *S-N Water Divers. Water Sci. Technol.* **2011**, *9*, 60–63.

25. Wang, Z.Z.; Cui, T.T.; Wang, Y.T.; Yu, Z.B. Flood season division with an improved fuzzy c-mean clustering method in the Taihu Lake Basin in China. *Procedia Eng.* **2012**, *28*, 66–74.

26. Wang, Z.Z.; Wang, Y.T.; Wu, H.Y.; Cui, T.T.; Xu, H.; Zhang, Y. Novel flood season division method based on fuzzy time series-effective cluster and its application to Taihu lake basin. *J. Hydroelectric Eng.* **2012**, *31*, 29–34.

27. Chen, L.; Singh, V.P.; Guo, S.L.; Zhou, J.Z.; Zhang, J.H.; Liu, P. An objective method for partitioning the entire flood season into multiple sub-seasons. *J. Hydrol.* **2015**, *528*, 621–630.

28. He, H.W.; Zhang, A.L. The application of Fisher method to dividing seismicity period in Yunnan province. *J. Seismol. Res.* **1994**, *17*, 231–239.

29. Liu, Y.F.; Sun, H.; Yuan, Z.H.; Li, Y.R. Time series analysis of precipitation in flood season in fenhe river basin. *B. Soil Water Conserv.* **2011**, *31*, 121–125.

30. Yang, H.X. Research on the storm-flood characteristic of basin controlled by fenhe reservoir and its dividing by periods. *Shanxi Hydrotech.* **2008**, *3*, 40–41.

31. Gao, B.; Wang, Y.T.; Hu, S.Y. Adjustment and application of the limited level of reservoirs during the flood season. *Adv. Water Sci.* **2005**, *16*, 326–333.

Bivariate Drought Analysis Using Streamflow Reconstruction with Tree Ring Indices in the Sacramento Basin, California, USA

Jaewon Kwak, Soojun Kim, Gilho Kim, Vijay P. Singh, Jungsool Park and Hung Soo Kim

Abstract: Long-term streamflow data are vital for analysis of hydrological droughts. Using an artificial neural network (ANN) model and nine tree-ring indices, this study reconstructed the annual streamflow of the Sacramento River for the period from 1560 to 1871. Using the reconstructed streamflow data, the copula method was used for bivariate drought analysis, deriving a hydrological drought return period plot for the Sacramento River basin. Results showed strong correlation among drought characteristics, and the drought with a 20-year return period (17.2 million acre-feet (MAF) per year) in the Sacramento River basin could be considered a critical level of drought for water shortages.

Reprinted from *Water*. Cite as: Kwak, J.; Kim, S.; Kim, G.; Singh, V.P.; Park, J.; Kim, H.S. Bivariate Drought Analysis Using Streamflow Reconstruction with Tree Ring Indices in the Sacramento Basin, California, USA. *Water* **2016**, *8*, 122.

1. Introduction

Generally, long-term data are recommended for analyzing floods and droughts. However, although precipitation data are available from the 16th century onwards, their quality and reliability in many countries are questionable because of the methods of observation, different periods of observations, uncertainties associated with gaging sites, and temporal resolution of observations [1,2]. Many studies have, therefore, used tree-ring data as a way to acquire data for longer periods of time, up to 500 years [3].

Since Ferguson [4] correlated observed hydro-meteorological data and tree-ring data in California, many studies have utilized tree-ring data to reconstruct time series of the past. Fritts [5] did tree ring analysis and correlated the data with climate, suggesting that it can be used in water resources management [6]. Also, some studies have correlated hydro-meteorological variables with tree-ring data to reconstruct climate factors [7–10]. Some studies have reconstructed seasonal series, such as precipitation [11,12], natural hazards [13], and temperature [14], based on tree ring width data. However, most of the studies on tree rings and hydrological phenomena have focused on droughts [15–24]. This may be because annual data are normally reconstructed using tree ring data and hence it is difficult to analyze intra-annual hydrological phenomena. Long time-scale data are mainly used for droughts, because it is difficult to determine the onset or end of a drought and often the drought may last for several months or years. The long time-scale occasionally gives rise to a sample size problem [25] for drought analysis that can be overcome with long-term tree ring reconstruction [26]. Some studies have directly correlated tree rings with droughts [15,27], drought patterns or oscillations [28–30], drought index and its trend [23,31–33], return periods [34], and spatial drought characteristics [35].

Bivariate (or multivariate) analyses of drought characteristics, such as severity, duration, and arrival time, are being increasingly made [36]. These analyses have introduced multivariate drought indices, such as Multivariate Drought Index (MDI) including precipitation, runoff, evapotranspiration, and soil moisture [37]; Multivariate Standardized Drought Index (MDSI) which combines the Standardized Precipitation Index and the Standardized Soil Moisture Index [38]; and Vegetation Drought Response Index (VegDRI) which integrates climatic indicators; and satellite-derived vegetation index [39,40]. Some studies have employed conventional multivariate analysis for drought indices with PDSI [23,41] and SWSI [42,43], or bivariate frequency analysis [44,45] assuming that all variables had the same probability distribution. To overcome this restriction, the copula method has been developed [46]. For doing bivariate drought analysis by the copula method [46,47], tree ring reconstruction can be employed to our advantage.

The objective of this study, therefore, is to reconstruct the annual streamflow of the Sacramento River in California and Oregon using tree-ring width data, and use the reconstructed data for bivariate drought frequency analysis with the copula method. Selected tree-ring data were used in an artificial neural network for streamflow reconstruction and the reconstructed data were verified by comparing with actual observations. The Archimedean copula function was applied to the reconstructed streamflow data and then the return period plot of the hydrological drought in the Sacramento River basin was derived. The advantages of using the copula method have been discussed in many studies [34,36,48–55]. Since drought may last from months to years, as has happened in California, long-term reconstruction based on tree-ring data, addresses the drawback of short-term data [56].

2. Materials and Methods

2.1. Study Area and Data

Annual streamflow data for four Sacramento rivers and tree-ring data of the nearby region were employed in this study. The tree-ring data from 17 sites (Figure 1 and Table 1) in California and Oregon, which reflect standard chronologies of ring width [57], were obtained from the International Tree-Ring Databank [58]. The tree ring width data were standardized [59]. Annual streamflow is the sum of four river flows, which are the Sacramento River above Bend Bridge (SBB), the Feather River at the Lake Oroville (FTO), the Yuba River near Smartville (YRS) and the American River at Folsom (AMF), which was obtained from the California Data Exchange Center of California Department of Water Resources [60] for the period 1872 to 1977. It has a long-term mean of 18.9 MAF (million acre-feet; 1.23×10^9 m^3), median of 17.6 MAF, and maximum flow of 51.6 MAF that occurred in the year 1890. These tree ring and streamflow sites are shown in Figure 1 and Table 1.

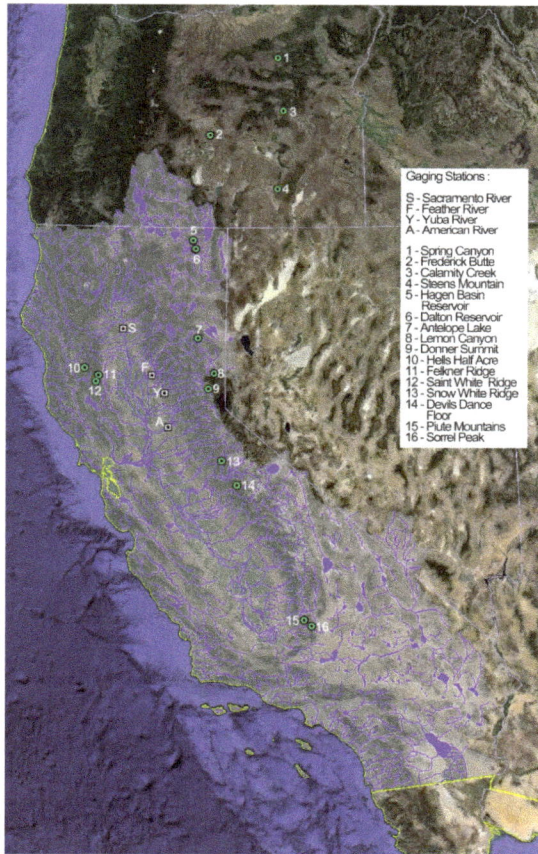

Figure 1. Study area, tree ring sites, and streamflow sites.

2.2. Drought Definition Using the Run Theory

This study qualitatively defines the hydrological drought (hereafter referred to as "drought") as a significant decrease in the availability of streamflow in the

117

river. Quantitatively, drought was defined using the run theory, which allows us to calculate drought duration, severity, and arrival time [61,62]. Thus, drought can be defined as the time when a hydro-meteorological time series x_t falls below the truncation level x_0 and that represents a hydro-meteorological event (Figure 2).

Table 1. Tree ring sites and streamflow observatory specification [58].

Category	Index in Figure 1	ID	Name	Site Location			Tree Ring Species
				Lat (degree)	Lon (degree)	Height (EL.m)	
	7	ANTEP	Antelope Lake	40.15	−120.6	1480	*Pinus jeffreyi Balf*
	7	ANTEP	Antelope Lake	40.15	−120.6	1480	*Pinus ponderosa Douglas ex C. Lawson*
	3	CALAM	Calamity Creek	43.98	−118.8	1464	*Juniperus occidentalis Hook*
	6	DALTON	Dalton Reservoir	41.62	−120.7	1531	*Pinus ponderosa Douglas ex C. Lawson*
	14	DEVILS	Devil's Dance Floor	37.75	−119.75	2084	*Pinus jeffreyi Balf*
	9	DONNER	Donner Summit	39.32	−120.35	2265	*Pinus jeffreyi Balf*
	11	FELKN	Felkner Ridge	39.5	−122.67	1494	*Pinus lambertiana Douglas*
	2	FREDER	Frederick Butte	43.58	−120.45	1494	*Juniperus occidentalis Hook*
Tree ring sites	5	HAGER	Hager Basin Reservoir	41.77	−120.75	1524	*Juniperus occidentalis Hook*
	10	HELLS	Hell's Half Acre	39.6	−122.95	1922	*Pinus jeffreyi Balf*
	8	LEMON	Lemon Canyon	39.57	−120.25	1859	*Pinus jeffreyi Balf*
	15	PIUTE	Piute Mountain	35.53	−118.43	1975	*Pinus jeffreyi Balf*
	13	SNOWHT	Snow White Ridge	38.13	−120.05	1731	*Pinus ponderosa Douglas ex C. Lawson*
	16	SORREL	Sorrel Peak	35.43	−118.28	2011	*Pinus jeffreyi Balf*
	1	SPRING	Spring Canyon	44.9	−118.93	1366	*Juniperus occidentalis Hook*
	4	STEENS	Steens Mountain	42.67	−118.92	1656	*Juniperus occidentalis Hook*
	13	STJOHN	St. White Mountain	39.43	−122.68	1555	*Pinus ponderosa Douglas ex C. Lawson*
	S	SBB	Sac. River, Abv bend bridge	40.29	−122.19	56.6	-
Flow site	F	FTO	Feather River, Oroville	39.52	−121.55	45.4	-
	Y	YRS	Yuba River, Smartville	39.24	−121.27	85.3	-
	A	AMF	American River, Folsom	38.68	−121.18	0	-

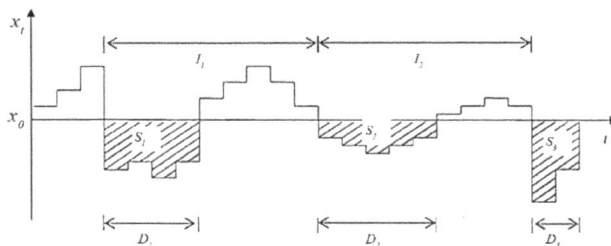

Figure 2. Drought characteristics using the run theory: $D_1, D_2, \ldots$ denote drought duration; $S_1, S_2, \ldots$ denote drought severity; $I_1, I_2, \ldots$ denote drought arrival time.

Drought events are based on the truncation level, so the selection of the level is one of the important issues for proper drought analysis. Generally, the mean value of streamflow has been widely used as the truncation level [63–68]. However, the Sacramento River streamflow shows high variability, between 5.13 and 51.65 MAF, so a median value of annual streamflow was regarded as a more reliable truncation level in this study.

118

2.3. Artificial Neural Network

The factors that influence annual hydro-meteorological behavior can be roughly classified into four groups: (i) atmospheric-climatic; (ii) geologic-geomorphic; (iii) soil-vegetation; and (iv) runoff-channel factors [69]. Tree-ring widths in a trunk of a tree are also influenced by atmospheric-climatic and soil-vegetation factors, such as precipitation, evapotranspiration, and soil moisture. This indicates that an appropriate modeling technique and tree rings that have a correlation with atmospheric-climatic factors can be used to reconstruct annual streamflow, and this study employed an Artificial Neural Network (ANN) model. ANN mimics the structure and functions of a biological neural system, in which neurons are connected through nodes [70]. After the perceptron was proposed to categorize information patterns [71], ANNs have been widely used to recognize nonlinear relationships between different variables. The ANN used in this study was comprised of three layers: the input layer that represents observed streamflow data, the output layer that produces simulated streamflow, and the hidden layer that is constituted by a network of neurons that are trained to recognize patterns from observations (Figure 3).

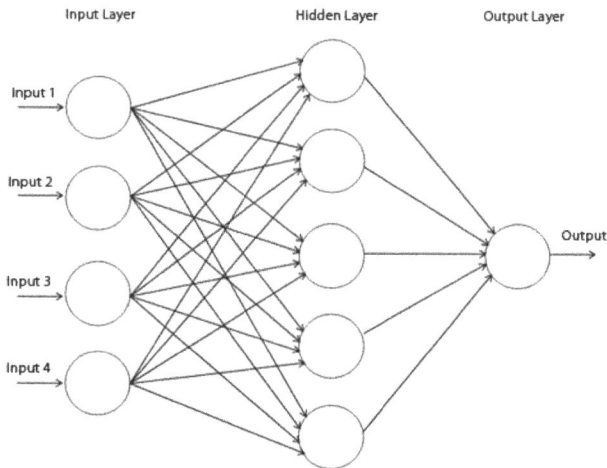

Figure 3. ANN schematization: *Input i* is the input set, *output* is the result of network delay, and each circle represents neural network [71]; each line indicates nodes between neurons that have their own connection strength.

The back-propagation algorithm was used to train the network through the adjustment of connection strength to learn about the error and optimize the neurons. It calculates the error function with respect to all the weights in the network and the gradient of error functions is fed into the optimization technique, which attempts to minimize the error of the network. Hence, the selection of back-propagation algorithm is one of the challenges when using a neural network [72]. The Levenberg–Marquardt–QNBP algorithm was selected as the back-propagation algorithm, because it is known to work well for non-linear problems, such as those related to meteorological and hydrological data [73]. Also, the number of hidden layers, which can be optimized by a trial and error method, is important for a proper network. The ANN model used in this study has six hidden layers that are optimized. One of the advantages of ANN is that it can be used as an alternative modeling technique when the data show non-linearity, which may cause error with a linear technique [74]. The tree-ring data in California and Oregon have autocorrelation and lagged-correlation characteristics [75], so ANN was employed as an alternative to reconstruct streamflow using tree-ring data. More details on ANN and the back-propagation algorithm can be found in Basheer and Hajmeer [76].

2.4. Drought Frequency Analysis Based on Copula

Unlike precipitation or flood occurrence, drought shows a different statistical behavior for a different duration [62]. Considering drought duration and severity as mutually related variables, the copula method has been employed to capture the dependence between them [47,77]. For a probability distribution $F(x_1, \ldots, x_n)$, which has n-dimensional marginal distributions $F_1(x_1), \ldots, F_n(x_n)$, the copula function C that satisfies the relationship between marginal variables can be expressed as:

$$F(x_1, \ldots, x_n) = C(F_1(x_1), \ldots, F_n(x_n)) = \psi_\alpha^{-1}(\psi_\alpha(F_1(x_1)) + \ldots + \psi_\alpha(F_n(x_n))) \quad (1)$$

$$1/T = E(L) / \{1 - F_1 - F_2 \ldots, F_n + C(F_1, F_2, \ldots, F_n)\} \quad (2)$$

where ψ_α denotes the generating function; ψ_α^{-1} is the pseudo-inverse of that function, which differs with the copula family; T denotes the return period; and $E(L)$ is the interval between events. Unlike univariate frequency analysis, bivariate frequency indicates the probability that the phenomenon under study occurs if and only if a prior condition takes place. There are several types of copula functions, but the Archimedean copula family, which allows for greater flexibility and simplicity of use, is more commonly used in hydrology [78]. From the Archimedean copula family, the Clayton, Gumbel, and Frank copulas were employed and their functions are given in Table 2.

Table 2. Bivariate Archimedean copula family: C is the copula function, t denotes the drought event, α is the copula parameter, and F_1 and F_2 denote cumulative distribution function of each variable [79].

Copula Family	Copula func., $C\left(F_1\left(x_1\right), F_2\left(x_2\right)\right)$	Generator func., $\psi_\alpha\left(t\right)$	Parameter (α)
Clayton	$\left(\max\left\{F_1\left(x_1\right)^{-1} + F_2\left(x_2\right)^{-1} - 1; 0\right\}\right)$	$\frac{1}{\alpha}\left(t^{-\alpha} - 1\right)$	$\alpha \in [-1, \infty]$
Frank	$\frac{-1}{\alpha}\log\left(1 + \frac{\left(\exp\left(-\alpha F_1\left(x_1\right)\right) - 1\right)\left(\exp\left(-\alpha F_2\left(x_2\right)\right) - 1\right)}{\exp\left(-\alpha\right) - 1}\right)$	$-\log\left(\frac{\exp\left(-\alpha t\right) - 1}{\exp\left(-\alpha\right) - 1}\right)$	$\alpha \in [\mathcal{R}]$
Gumbel	$\exp\left(-\left(\left(-\log(F_1\left(x_1\right))\right)^\alpha + \left(-\log(F_1\left(x_1\right))^\alpha\right)^{\frac{1}{\alpha}}\right)\right)$	$-\log\left(t\right)^\alpha$	$\alpha \in [1, \infty]$

3. Results and Discussion

3.1. Tree-Ring Data Screening

Appropriate tree rings, which have correlation with atmospheric-climatic factors, can be used as the predictor for annual streamflow. Therefore, selection of the appropriate input for the ANN model was one of the challenges in this study. Generally, a trial and error method with different input variables is employed, but it can lead to the poor performance of neural networks [80,81]. Alternatively, cross correlation coefficients were employed to select appropriate inputs in this study in the same way as Meko *et al.* [75].

Seven tree-rings indices showed correlation with streamflow (Figure 4), including Antelope Lake (Pinus Jeffreyi and Pinus Ponderosa), Felkner Ridge, Frederick Butte, Lemon Canyon, Piute Mountain, and Sorrel Peak; four tree rings had correlation with a one-year time-lag, including Dalton Reservoir, Hager Basin Reservoir and Antelope Lake (Pinus Jeffreyi and Pinus Ponderosa), as shown in Figures 4 and 5. Hence, nine tree rings were selected as predictors for the ANN model (two tree-rings were overlapped in zero-lagged and one-year lagged). Also, four tree-rings, which had one-year time-lag correlation (Antelope with Pinus Jeffreyi and Pinus Ponderosa, Dalton Reservoir, and Hager Basin Reservoir), are located on the nearby lake or reservoir. Therefore, it seems that the groundwater level or soil moisture influenced tree ring width, but there are no clues to estimate the correlation between them and further studies are thus needed. These nine tree-ring data points composed the input dataset for the ANN model.

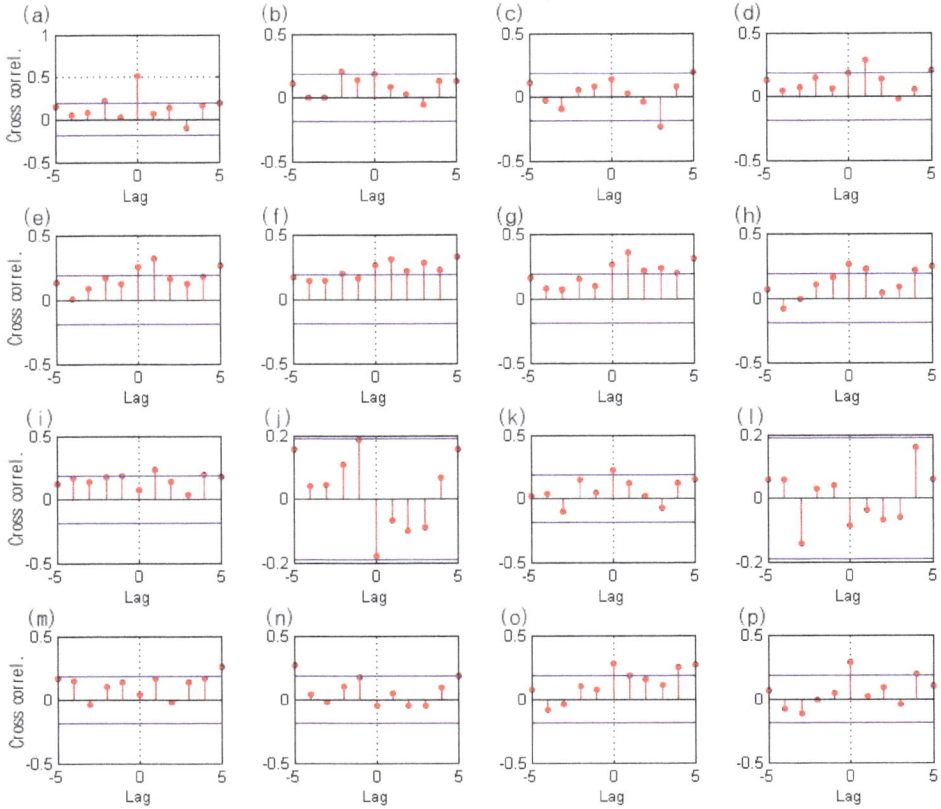

Figure 4. Cross correlation diagram between streamflow and tree ring data at:
(**a**) Felkner Ridge; (**b**) Lemon Canyon; (**c**) Calamity Creek; (**d**) Dalton Reservoir;
(**e**) Hager Basin Reservoir; (**f**) Antelope Lake (Pinis Jeffreyi Balf.); (**g**) Antelope Lake
(Pinus Ponderosa Douglas ex C. Lawson); (**h**) Steens Mountain; (**i**) Hell's Half Acre;
(**j**) Donner Summit; (**k**) Frederick Butte; (**l**) Spring Canyon; (**m**) St. White Mountain;
(**n**) Devil's Dance Floor; (**o**) Sorrel Peak; and (**p**) Piute Mountain.

122

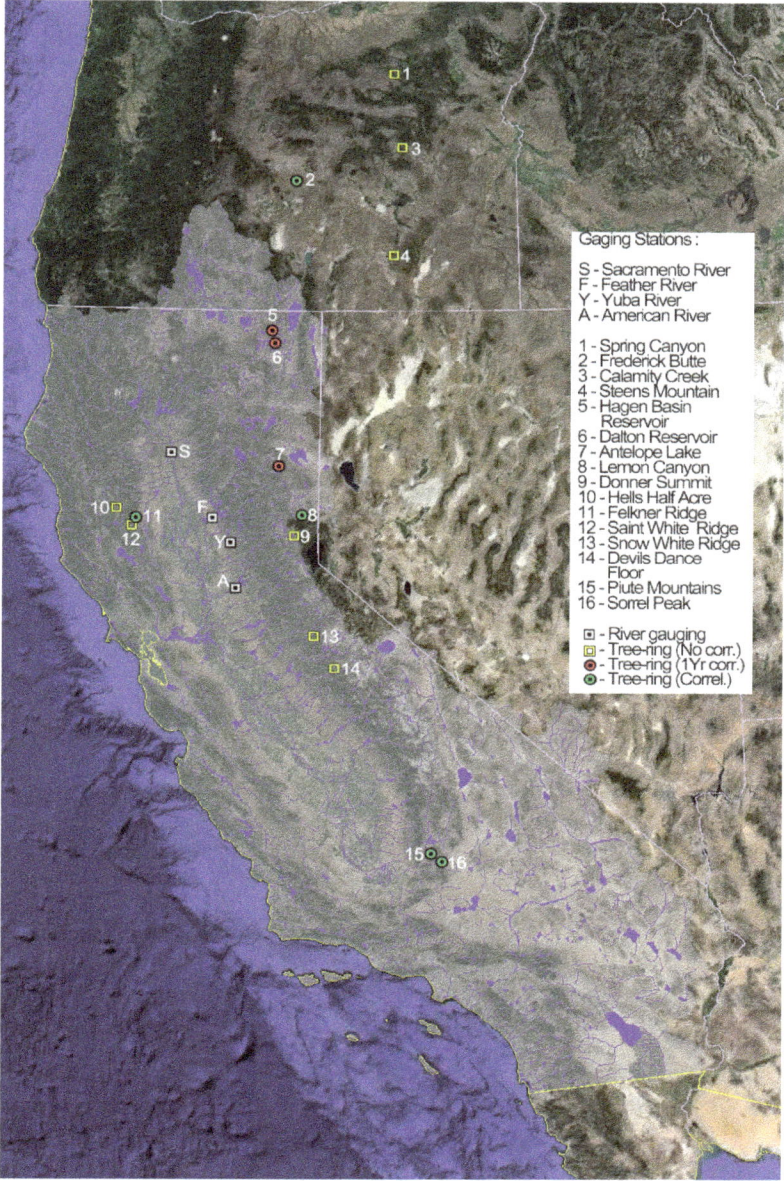

Figure 5. Study area, tree ring sites, and streamflow sites; a green circle indicates that the tree ring has correlation with streamflow, while a red circle indicates correlation with time-lag (1 year) and a yellow box indicates that there is no correlation with streamflow.

3.2. Reconstructed ANN Model Calibration and Validation

The ANN model had six hidden layers that were determined by trial and error and was established with selected predictors. The Sacramento streamflow data were divided into calibration period (1872 to 1957) and validation period (1958 to 1977). To evaluate the results of calibration and validation, R^2 and RMSE [82], and the Nash–Sutcliffe model efficiency coefficient [83] were computed. The RMSE describes a measure of average error in prediction and R^2 and Nash–Sutcliffe model efficiency coefficient have been widely used to assess the predictive performance of models [84]. The calibration and validation results with selected predictors are shown in Figure 6.

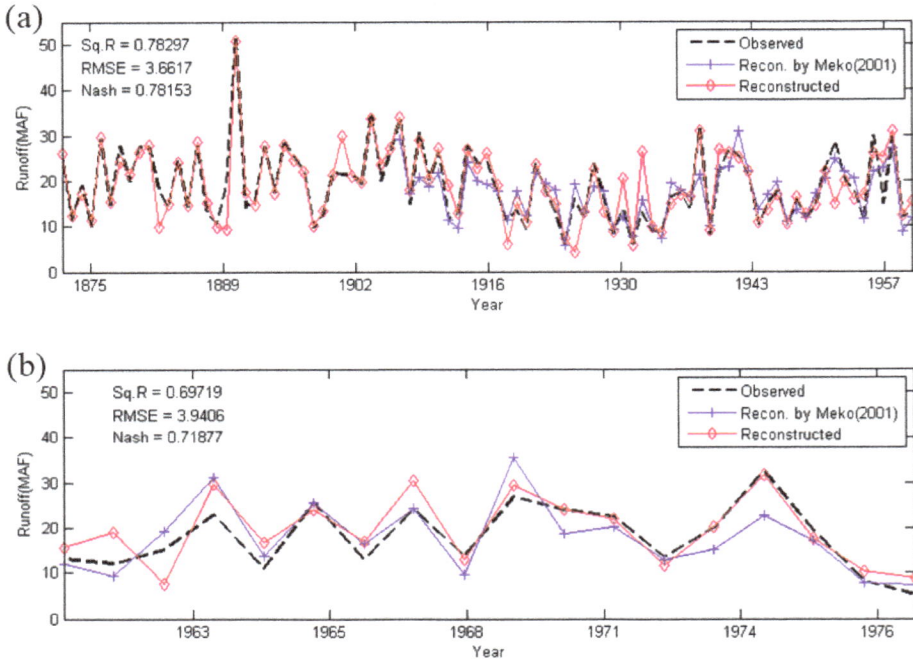

Figure 6. Calibration and validation results with tree ring: (**a**) calibration period (1872 to 1957); and (**b**) validation period (1958 to 1977).

The ANN model showed relatively high values of evaluation measures, with R^2, RMSE, and Nash–Sutcliffe efficiency of 0.78, 3.66, and 0.78 in the calibration and 0.70, 3.94, and 0.72 in the validation period, respectively. Thus, the reconstructed streamflow, based on the ANN model and selected predictor, could be used as the reconstruction model. Also, it could be used for hydro-meteorological simulations. The variability of each period was 5.74 to 51.64 MAF in the calibration period and 5.13 to 32.5 MAF in the validation period.

The reconstructed streamflow using the selected predictor (nine tree rings) is shown in Figure 7 and the basic statistics are shown in Table 3, with an average of 18.9 MAF observed and 20.4 MAF of reconstructed streamflow.

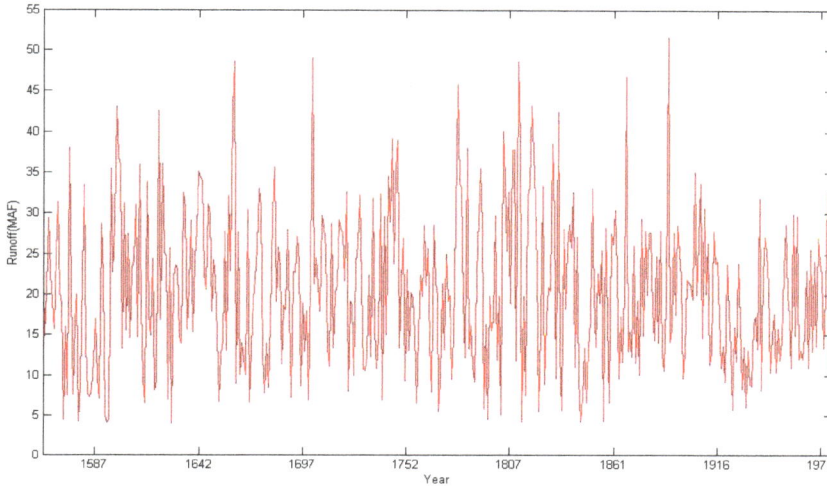

Figure 7. Reconstructed streamflow using the ANN model.

Table 3. Basic statistics of observed and reconstructed streamflow.

Period	Mean (MAF)	Median (MAF)	Standard Deviation (MAF)	Skewness
Observed (1872–1977)	18.9	17.6	7.8	0.8
Reconstructed (1560–1871)	20.4	19.8	9.6	0.6

3.3. Bivariate Drought Analysis and Discussion

Before drought analysis based on reconstructed streamflow, the truncation level that defines the relevant streamflow level was determined to define hydrological drought from the streamflow series. The median value of annual streamflow, which was 19.4 MAF, was employed as the truncation level to define hydrological droughts for the Sacramento River basin and results are shown in Figure 8. In total, 96 droughts occurred during the period from 1560 to 1977. Their statistical characteristics were: median drought duration of about two years, average drought severity of about 15.8 MAF, and average drought arrival time of about 2.1 years during the 15th to 20th centuries. The longest drought duration estimated was 10 years and had 75.31 MAF during 1927 to 1936 (in the observation period), and the severest drought estimated was 76.17 MAF, which had an eight-year drought duration from 1582 to 1589 (in the reconstruction period).

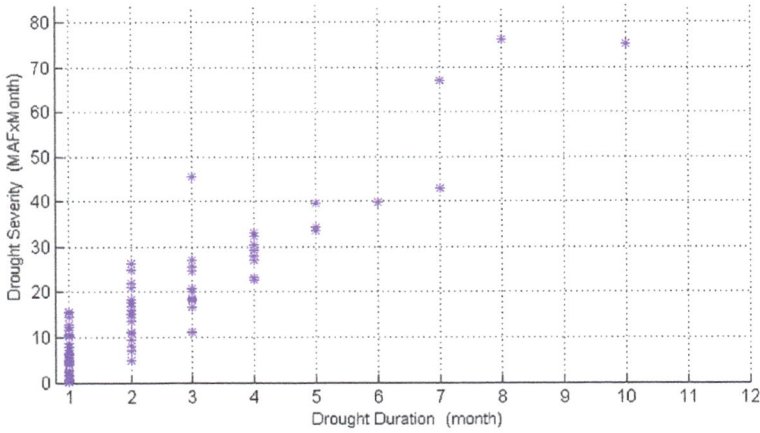

Figure 8. Truncated drought events (1560 to 1977) using observed and reconstructed streamflow.

For bivariate drought frequency analysis with the copula function, marginal distributions of drought variables (duration and severity) were derived. The drought duration was found to have an "exponential" distribution, if treated as a continuous random variable [34]. Also, the drought severity was found to have a "gamma" distribution with 95% confidential level with the PPCC (probability plot correlation coefficient; [85]) distribution goodness of fit test. Then, parameters of the Archimedean family copulas (Frank, Clayton, and Gumbel) were estimated by the method of moments according to their relationship between copula parameter and Kendall's tau [86], which has been found adequate for estimating parameters for small sample sizes [78].

The minimum quadratic distance (L^2) between the empirical and theoretical values of the K criterion, which describes the most appropriate copula [78], was calculated for each copula. As shown in Figure 9, the Frank copula, which generally fitted well throughout ($L^2 = 0.023$), was selected for bivariate drought analysis for the Sacramento River. The Frank copula parameter was estimated as 8.03, and the bivariate joint probability of drought for the Sacramento River basin was described as:

$$F\left(F_d, F_s\right) = -\frac{1}{8.03}\log\left(1 + \frac{\left(\exp\left(-8.03\,F_d\left(t\right)\right) - 1\right)\left(\exp\left(-8.03\,F_s\left(t\right) - 1\right)\right.}{\exp\left(-8.03\right) - 1}\right) \quad (3)$$

where, $F_d\left(t\right)$ and $F_s\left(t\right)$ are the cumulative distribution functions of drought duration and severity. Figure 10 shows the joint CDF of the Sacramento River basin drought.

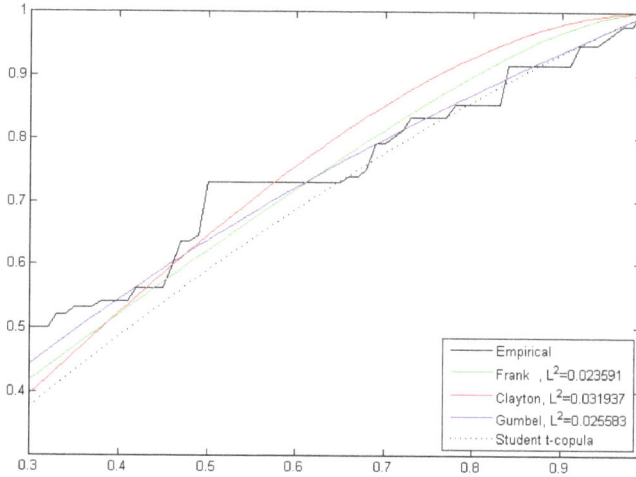

Figure 9. K criterion plot for copula families.

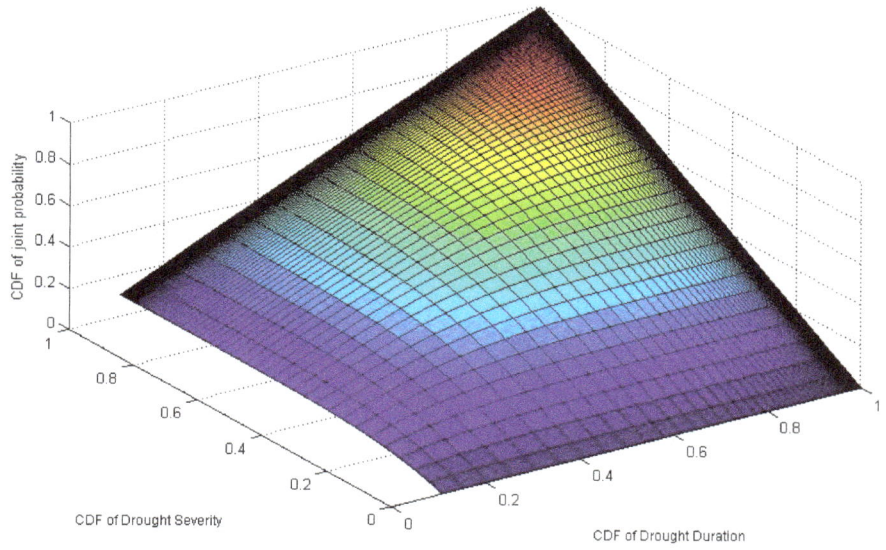

Figure 10. Joint cumulative distribution function (CDF) of the Sacramento River basin drought.

The return period is described as the average time of occurrence of events with the threshold intensity. The bivariate return period can be divided into the exceedance probabilities of both the drought duration and severity [62]. The copula-based return period with the average inter-arrival of occurrences ($E(L)$), which was 2.1 years, can be defined as:

$$T_{return\ period} = \frac{2.1}{P(D > d\ \&\ S > s)} = \frac{2.1}{1 - F_D(d) - F_S(s) + C(F_D(d), F_S(s))} \quad (4)$$

Therefore, the duration and severity of droughts can be expressed in terms of the same return period, which can be illustrated in each "return period plot", as shown in Figure 11.

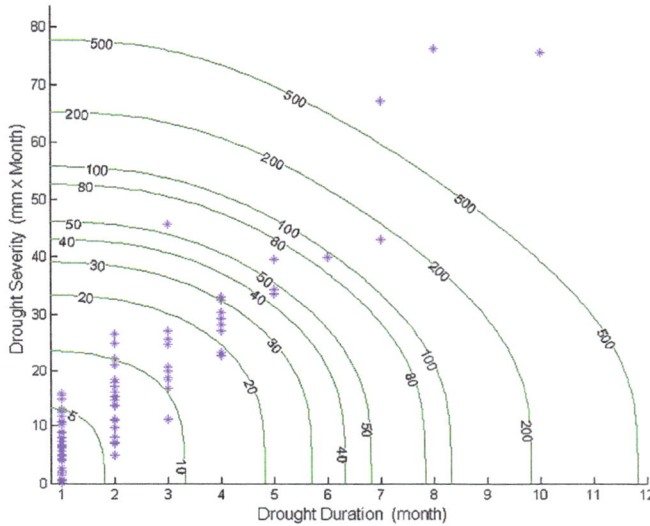

Figure 11. Return period plot for Sacramento River.

The drought event and return period plot in Figure 11 shows the hydrological drought pattern of the Sacramento River basin. Overall, the drought duration and severity seemed to have a positive correlation with each other. To identify the correlation between drought duration and severity, Pearson's linear correlation coefficient, Kendall's rank correlation coefficient (τ), and Spearman's rank correlation coefficient (ρ) were found to be 0.92, 0.73, and 0.85, respectively. Also, the severity of two-year duration drought showed higher variability than did the three-year duration drought. Strong correlation between drought characteristics and higher variability of the two-year drought was the basis for the copula method, especially higher variability of severity or larger statistical irregularity than other durations for the same return period in the univariate frequency analysis [62]. Hence, the copula method would be expected to be more reliable for drought analysis for the Sacramento River basin. Furthermore, California has a 22.4 MAF mean annual runoff and used 5.2 MAF annually to supply the southern area [87]. So, a drought that becomes over 17.2 MAF will cause the shortage of water supply and is equivalent to approximately a 20-year return period with a two-month duration (median value of drought duration) based on the return period plot in Figure 11. Therefore, any return period that causes an actual water shortage could be the appropriate critical level of drought. Also, the return period plot in Figure 11 can be used as elementary data for water resources planning. For instance, if a decision maker or agency determined a three-year design drought for a dam or reservoir to be having a 20-year return period, then its deficiency would be about 11.1×10^9 m^3 (27.0 MAF), and it could be the target storage volume for water resources planning.

Most of the droughts that occurred during the last five centuries did not have more than a 50-year return period, and just six droughts showed 100-year or longer return periods. These high return period events are one of the limitations of the study; for instance, the drought from 1927 to 1936 years, the longest and severest drought in historical data [88], with a 10-year duration and 75.31 MAF severity, which equated to a 9.26×10^9 m^3 streamflow deficiency per year, had approximately a 7500-year return period. That extreme return period was due to the Frank copula and sample size, which shows some bias in high quantile (high return period) events in Figure 9, so it could have overestimated the return period [89], and the number of drought events with high quantiles is also limited. Droughts that have high quantiles or extreme return periods depend significantly on the fitted copula function. Therefore, further studies are needed for generally well-fitted copulas throughout, and also for carefully considering the use of the return period plot in water resources planning.

4. Conclusions

This study reconstructed the past streamflow of the Sacramento River based on the ANN and tree-ring data, and bivariate drought frequency was analyzed using the Archimedean copula. Results of this study can be summarized as follows:

1. The past streamflow for the period from 1560 to 1871 is reconstructed with the ANN model and tree-ring data, which was found to be the appropriate predictor. As shown by calibration and validation results from 1872 to 1977, the R^2 and Nash values are 0.7 or higher. It is therefore concluded that the ANN model reconstructs streamflow of the Sacramento River satisfactorily.
2. Drought characteristics in the Sacramento River basin have strong correlation with each other. The Archimedean copula is found to be appropriate for bivariate drought frequency analysis.
3. It is shown that a drought with a 20-year return period or longer will cause actual water shortages in the perspective of water supply to the southern California area. Hence, it could be considered an appropriate critical level of droughts for actual water shortages.

Acknowledgments: This research was supported by a grant (13SCIPS01) from Smart Civil Infrastructure Research Program funded by Ministry of Land, Infrastructure and Transport (MOLIT) of Korea government and Korea Agency for Infrastructure Technology Advancement (KAIA) and a grant[MPSS-NH-2015-79] through the Natural Hazard Mitigation Research Group funded by Ministry of Public Safety and Security of Korean government.

Author Contributions: The research presented here was carried out in collaboration between all authors. Jaewon Kwak and Soojun Kim had the original idea for the study. Gilho Kim and Jungsool Park designed/conducted the research methods. Vijay P. Singh and Hung Soo Kim contributed to the writing of the paper. All authors discussed the structure and commented on the manuscript at all stages.

Conflicts of Interest: The authors declare no conflict of interest.

References

1. Wang, B.; Jhun, J.G.; Moon, B.K. Variability and singularity of Seoul, South Korea, rainy season (1778–2004). *J. Clim.* **2007**, *20*, 2572–2580.

2. Mun, J.W.; Lee, D.Y. Tree-ring data application for drought mitigation. *J. Korean Soc. Hazard Mitig.* **2011**, *11*, 70–78. (In Korean)

3. Kim, H.S.; Hwang, S.H.; Kim, J.H. Reconstruction of River Flows Using Tree-Ring Series and Neural Network. *J. Korean Soc. Civ. Eng.* **1998**, *18*, 583–589.

4. Ferguson, C.W. Bristlecone Pine: Science and Esthetics A 7100-year tree-ring chronology aids scientists; old trees draw visitors to California mountains. *Science* **1968**, *159*, 839–846.

5. Fritts, H.C. Tree-ring analysis: A tool for water resources research. *Eos Trans. Am. Geophys. Union* **1969**, *50*, 22–29.

6. Fritts, H.C. *Tree Rings and Climate*; Academic Press Inc.: New York, NY, USA, 1976.

7. Hughes, M.K.; Xiangding, W.; Xuemei, S.; Garfin, G.M. A preliminary reconstruction of rainfall in north-central China since AD 1600 from tree-ring density and width. *Quat. Res.* **1994**, *42*, 88–99.

8. Díaz, S.C.; Touchan, R.; Swetnam, T.W. A tree-ring reconstruction of past precipitation for Baja California Sur, Mexico. *Int. J. Climatol.* **2001**, *21*, 1007–1019.

9. Cleaveland, M.K.; Stahle, D.W.; Therrell, M.D.; Villanueva-Diaz, J.; Burns, B.T. Tree-ring reconstructed winter precipitation and tropical teleconnections in Durango, Mexico. *Clim. Chang.* **2003**, *59*, 369–388.

10. Gray, S.T.; Fastie, C.L.; Jackson, S.T.; Betancourt, J.L. Tree-ring-based reconstruction of precipitation in the Bighorn Basin, Wyoming, since 1260 AD. *J. Clim.* **2004**, *17*, 3855–3865.

11. Liu, Y.; Cai, Q.; Shi, J.; Hughes, M.K.; Kutzbach, J.E.; Liu, Z.; An, Z. Seasonal precipitation in the south-central Helan Mountain region, China, reconstructed from tree-ring width for the past 224 years. *Can. J. For. Res.* **2005**, *35*, 2403–2412.

12. Liu, Y.; Ma, L.; Leavitt, S.W.; Cai, Q.; Liu, W. A preliminary seasonal precipitation reconstruction from tree-ring stable carbon isotopes at Mt. Helan, China, since AD 1804. *Glob. Planet. Chang.* **2004**, *41*, 229–239.

13. Schneuwly, D.M.; Stoffel, M. Tree-ring based reconstruction of the seasonal timing, major events and origin of rockfall on a case-study slope in the Swiss Alps. *Nat. Hazards Earth Syst. Sci.* **2008**, *8*, 203–211.

14. Frank, D.; Esper, J. Characterization and climate response patterns of a high-elevation, multi-species tree-ring network in the European Alps. *Dendrochronologia* **2005**, *22*, 107–121.

15. Cook, E.R.; Jacoby, G.C. Tree-ring-drought relationships in the Hudson Valley, New York. *Science* **1977**, *198*, 399–401.

16. Stockton, C.W.; Meko, D.M. Drought Recurrence in the Great Plains as Reconstructed from Long-Term Tree-Ring Records. *J. Clim. Appl. Meteorol.* **1983**, *22*, 17–29.

17. Graumlich, L.J. Precipitation variation in the Pacific Northwest (1675–1975) as reconstructed from tree rings. *Ann. Assoc. Am. Geogr.* **1987**, *77*, 19–29.

18. Till, C.; Guiot, J. Reconstruction of precipitation in Morocco since 1100 AD Based on Cedrus atlantica tree-ring widths. *Quat. Res.* **1990**, *33*, 337–351.

19. Meko, D.; Stockton, C.W.; Boggess, W.R. The Tree-ring Record of Severe Sustained Drought. *J. Am. Water Resour. Assoc.* **1995**, *31*, 789–801.

20. Stahle, D.W.; Cook, E.R.; Cleaveland, M.K.; Therrell, M.D.; Meko, D.M.; Grissino-Mayer, H.D.; Luckman, B.H. Tree-ring data document 16th century mega drought over North America. *EOS Trans. Am. Geophys. Union* **2000**, *81*, 121–125.

21. Raffalli-Delerce, G.; Masson-Delmotte, V.; Dupouey, J.L.; Stievenard, M.; Breda, N.; Moisselin, J.M. Reconstruction of summer droughts using tree-ring cellulose isotopes: A calibration study with living oaks from Brittany (western France). *Tellus B* **2004**, *56*, 160–174.

22. Li, J.; Gou, X.; Cook, E.R.; Chen, F. Tree-ring based drought reconstruction for the central Tien Shan area in northwest China. *Geophys. Res. Lett.* **2006**, *33*.

23. Li, J.; Chen, F.; Cook, E.R.; Gou, X.; Zhang, Y. Drought reconstruction for north central China from tree rings: The value of the Palmer drought severity index. *Int. J. Climatol.* **2007**, *27*, 903–909.

24. Tian, Q.; Gou, X.; Zhang, Y.; Peng, J.; Wang, J.; Chen, T. Tree-ring based drought reconstruction (AD 1855–2001) for the Qilian Mountains, northwestern China. *Tree Ring Res.* **2007**, *63*, 27–36.

25. Mishra, A.K.; Singh, V.P. Drought modeling—A review. *J. Hydrol.* **2011**, *403*, 157–175.

26. Agüero, J.D.L.C.; Rodríguez, F.J.G. Morphometric stock structure of the Pacific sardine Sardinops sagax (Jenyns, 1842) off Baja California, Mexico. In *Morphometrics*; Springer: Berlin, Germany; Heidelberg, Germany, 2004.

27. Stockton, C.W.; Meko, D.M. A long-term history of drought occurrence in western United States as inferred from tree rings. *Weatherwise* **1975**, *28*, 244–249.

28. Gray, S.T.; Betancourt, J.L.; Fastie, C.L.; Jackson, S.T. Patterns and sources of multidecadal oscillations in drought-sensitive tree-ring records from the central and southern Rocky Mountains. *Geophys. Res. Lett.* **2003**, *30*.

29. Helama, S.; Meriläinen, J.; Tuomenvirta, H. Multicentennial megadrought in northern Europe coincided with a global El Niño–Southern Oscillation drought pattern during the Medieval Climate Anomaly. *Geology* **2009**, *37*, 175–178.

30. Ropelewski, C.F.; Halpert, M.S. North American precipitation and temperature patterns associated with the El Niño/Southern Oscillation (ENSO). *Monthly Weather Rev.* **1986**, *114*, 2352–2362.

31. Davi, N.K.; Jacoby, G.C.; D'Arrigo, R.D.; Baatarbileg, N.; Jinbao, L.; Curtis, A.E. A tree-ring-based drought index reconstruction for far-western Mongolia: 1565–2004. *Int. J. Climatol.* **2009**, *29*, 1508–1514.

32. Touchan, R.; Funkhouser, G.; Hughes, M.K.; Erkan, N. Standardized precipitation index reconstructed from Turkish tree-ring widths. *Clim. Chang.* **2005**, *72*, 339–353.

33. Liang, E.; Shao, X.; Liu, H.; Eckstein, D. Tree-ring based PDSI reconstruction since AD 1842 in the Ortindag Sand Land, east Inner Mongolia. *Chin. Sci. Bull.* **2007**, *52*, 2715–2721.

34. Shiau, J.T. Fitting drought duration and severity with two-dimensional copulas. *Water Resour. Manag.* **2006**, *20*, 795–815.

35. Chbouki, N. Spatio-Temporal Characteristics of Drought as Inferred from Tree-Ring Data in Morocco. Ph.D. Thesis, University of Arizona, Tucson, AZ, USA, 1992.

36. Shiau, J.T.; Modarres, R. Copula-based drought severity-duration-frequency analysis in Iran. *Meteorol. Appl.* **2009**, *16*, 481–489.

37. Rajsekhar, D.; Singh, V.P.; Mishra, A.K. Multivariate drought index: An information theory based approach for integrated drought assessment. *J. Hydrol.* **2015**, *526*, 164–182.

38. Hao, Z.; AghaKouchak, A. Multivariate standardized drought index: A parametric multi-index model. *Adv. Water Resour.* **2013**, *57*, 12–18.

39. Brown, J.F.; Wardlow, B.D.; Tadesse, T.; Hayes, M.J.; Reed, B.C. The Vegetation Drought Response Index (VegDRI): A new integrated approach for monitoring drought stress in vegetation. *GISci. Remote Sens.* **2008**, *45*, 16–46.

40. Tadesse, T.; Wardlow, B.D.; Brown, J.F.; Svoboda, M.D.; Hayes, M.J.; Fuchs, B.; Gutzmer, D. Assessing the vegetation condition impacts of the 2011 drought across the US Southern Great Plains using the Vegetation Drought Response Index (VegDRI). *J. Appl. Meteorol. Climatol.* **2015**, *54*, 153–169.

41. Alley, W.M. The Palmer Drought Severity Index: Limitations and assumptions. *J. Clim. Appl. Meteorol.* **1984**, *23*, 1100–1109.

42. Shafer, B.A.; Dezman, L.E. Development of a Surface Water Supply Index (SWSI) to assess the severity of drought conditions in snowpack runoff areas. In Proceedings of the Western Snow Conference, Reno, Nevada, April 1982; pp. 164–175.

43. Valipour, M. Use of surface water supply index to assessing of water resources management in Colorado and Oregon. *Adv. Agric.* **2013**, *3*, 631–640.

44. González, J.; Valdés, J.B. Bivariate drought recurrence analysis using tree ring reconstructions. *J. Hydrol. Eng.* **2003**, *8*, 247–258.

45. Vangelis, H.; Spiliotis, M.; Tsakiris, G. Drought severity assessment based on bivariate probability analysis. *Water Resour. Manag.* **2011**, *25*, 357–371.

46. Murtin, C.M.; Murtin, F. Education Inequalities among World Citizens: 1870–2000. 2006. Working Paper. Available online: http://www.eea-esem.com/files/papers/EEA-ESEM/2006/2780/EducationInequality.pdf (accessed on 21 March 2016).

47. Sklar, A. *Fonctions de Repartition 'a n Dimensions et Leura Marges*; Publication de l'Institut de Statistique de l'Université de Paris: Paris, France, 1959; pp. 229–231. (In French)

48. Shiau, J.T.; Feng, S.; Nadarajah, S. Assessment of hydrological droughts for the Yellow River, China, using copulas. *Hydrol. Process.* **2007**, *21*, 2157–2163.

49. Serinaldi, F.; Bonaccorso, B.; Cancelliere, A.; Grimaldi, S. Probabilistic characterization of drought properties through Copulas. *Phys. Chem. Earth* **2009**, *34*, 596–605.

50. Song, S.; Singh, V.P. Meta-elliptical copulas for drought frequency analysis of periodic hydrologic data. *Stoch Environ. Res. Risk Assess.* **2010**, *24*, 425–444.

51. Mirabbasi, R.; Anagnostou, E.N.; Fakheri-Fard, A.; Dinpashoh, Y.; Eslamian, S. Analysis of meteorological drought in northwest Iran using the Joint Deficit Index. *J. Hydrol.* **2013**, *492*, 35–48.

52. Chen, L.; Singh, V.P.; Guo, S.; Mishra, A.K.; Guo, J. Drought Analysis Using Copulas. *J. Hydrol. Eng.* **2012**, *18*, 797–808.

53. Vergni, L.; Todisco, F.L.; Mannocchi, F. Analysis of agricultural drought characteristics through a two-dimensional copula. *Water Resour. Manag.* **2015**, *29*, 2819–2835.

54. Huang, S.; Huang, Q.; Chang, J.; Chen, Y.; Xing, L.; Xie, Y. Copulas-Based Drought Evolution Characteristics and Risk Evaluation in a Typical Arid and Semi-Arid Region. *Water Resour. Manag.* **2014**, *29*, 1489–1503.

55. Reddy, M.J.; Singh, V.P. Multivariate modeling of droughts using copulas and meta-heuristic methods. *Stoch. Environ. Rese. Risk Assess.* **2014**, *28*, 475–489.

56. Mishra, A.; Singh, V.P.; Desai, V. Drought characterization: A probabilistic approach. *Stoch. Environ. Res. Risk Assess.* **2009**, *23*, 41–55.

57. Jacoby, G.C.; Cook, E.R. Past temperature variations inferred from a 400-year tree-ring chronology from Yukon Territory, Canada. *Arct. Alp. Res.* **1981**, *13*, 409–418.

58. Grissino-Mayer, H.D.; Fritts, H.C. The International Tree-Ring Data Bank: An enhanced global database serving the global scientific community. *Holocene* **1997**, *7*, 235–238.

59. Cook, E.R. A time series analysis approach to tree-ring standardization (Dendrochronology, Forestry, Dendroclimatology, Autoregressive process). Ph.D. Thesis, University of Arizona, Tucson, AZ, USA, 1985.

60. California Data Exchange Center. Available online: http://cdec.water.ca.gov/index.html (accessed on 12 May 2015).

61. Yevjevich, V. An objective approach to definitions and investigations of continental hydrologic droughts. In *Hydrologic Paper*; Colorado State University: Fort Collins, CO, USA, 1967.

62. Kwak, J.; Kim, D.; Kim, S.; Singh, V.P.; Kim, H. Hydrological drought analysis in Namhan river basin, Korea. *J. Hydrol. Eng.* **2014**, *19*.

63. Serinaldi, F.; Grimaldi, S. Fully nested 3-Copula: Procedure and application on hydrological data. *J. Hydrol. Eng.* **2007**, *12*, 420–430.

64. Yu, K.X.; Xiong, L.; Gottschalk, L. Derivation of low flow distribution functions using copulas. *J. Hydrol.* **2014**, *508*, 273–288.

65. Wong, G.; Lambert, M.F.; Metcalfe, A.V. Trivariate copulas for characterization of droughts. *ANZIAM J.* **2008**, *49*, 306–315.

66. Sadri, S.; Burn, D.H. Copula-based pooled frequency analysis of droughts in the Canadian Prairies. *J. Hydrol. Eng.* **2012**, *19*, 277–289.

67. Chen, Y.D.; Zhang, Q.; Xiao, M.; Singh, V.P. Evaluation of risk of hydrological droughts by the trivariate Plackett copula in the East River basin (China). *Nat. Hazards* **2013**, *68*, 529–547.

68. Saghafian, B.; Mehdikhani, H. Drought characterization using a new copula-based trivariate approach. *Nat. Hazards* **2014**, *72*, 1391–1407.

69. Black, P.E. *Watershed Hydrology*; John Wiley & Sons: Hoboken, NJ, USA, 1991.

70. Hopfield, J.J. Neural networks and physical systems with emergent collective computational abilities. *Proc. Natl. Acad. Sci. USA* **1982**, *79*, 2554–2558.

71. Rosenblatt, F. The perceptron: A probabilistic model for information storage and organization in the brain. *Psychol. Rev.* **1958**, *65*, 386–408.

72. Battiti, R. Accelerated backpropagation learning: Two optimization methods. *Complex. Syst.* **1989**, *3*, 331–342.

73. Gill, P.E.; Murray, W.; Wright, M.H. *Practical Optimization*; Academic Press: New York, NY, USA, 1981.

74. Bourquin, J.; Schmidli, H.; van Hoogevest, P.; Leuenberger, H. Advantages of Artificial Neural Networks (ANNs) as alternative modelling technique for data sets showing non-linear relationships using data from a galenical study on a solid dosage form. *Eur. J. Pharm. Sci.* **1998**, *7*, 5–16.

75. Meko, D.M.; Therrell, M.D.; Baisan, C.H.; Hughes, M.K. Sacramento river flow reconstructed to AD 869 from tree rings. *J. Am. Water Resour. Assoc.* **2001**, *37*, 1029–1039.

76. Basheer, I.A.; Hajmeer, M. Artificial neural networks: Fundamentals, computing, design, and application. *J. Microbiol. Methods* **2000**, *43*, 3–31.

77. AghaKouchak, A. Entropy–Copula in Hydrology and Climatology. *J. Hydrometeorol.* **2014**, *15*, 2176–2189.

78. Saad, C.; El Adlouni, S.; St-Hilaire, A.; Gachon, P. A nested multivariate copula approach to hydrometeorological simulations of spring floods: The case of the Richelieu River (Québec, Canada) record flood. *Stoch. Environ. Res. Risk Assess.* **2015**, *29*, 275–294.

79. Rodriguez, J.C. Measuring financial contagion: A copula approach. *J. Empir. Financ.* **2007**, *14*, 401–423.

80. Haykin, S. *Neural Networks: A Comprehensive Foundation*, 2nd ed.; Prentice Hall: Upper Saddle River, NJ, USA, 1999.

81. Maier, H.R.; Dandy, G.C. Neural networks for the prediction and forecasting of water resources variables: A review of modelling issues and applications. *Environ. Model. Softw.* **2000**, *15*, 101–124.

82. Hyndman, R.J.; Khandakar, Y. *Automatic Time Series for Forecasting: The Forecast Package for R*; Department of Econometrics and Business Statistics, Monash University: Melbourne, Australia, 2007.

83. Nash, J.E.; Sutcliffe, J.V. River flow forecasting through conceptual models part I—A discussion of principles. *J. Hydrol.* **1970**, *10*, 282–290.

84. Moriasi, D.N.; Arnold, J.G.; Van Liew, M.W.; Bingner, R.L.; Harmel, R.D.; Veith, T.L. Model Evaluation Guidelines for Systematic Quantification of Accuracy in Watershed Simulations. *Trans. ASABE* **2007**, *50*, 885–900.

85. Vogel, R.M.; Hosking, J.R.; Elphick, C.S.; Roberts, D.L.; Reed, J.M. Goodness of fit of probability distributions for sightings as species approach extinction. *Bull. Math. Biol.* **2009**, *71*, 701–719.

86. El Adlouni, S.; Ouarda, T.B. Joint Bayesian model selection and parameter estimation of the generalized extreme value model with covariates using birth-death Markov chain Monte Carlo. *Water Resour. Res.* **2009**, *45*.

87. Carle, D. *Introduction to Water in California*; University of California Press: Berkeley, CA, USA, 2004.

88. Paulson, R.W.; Chase, E.B.; Roberts, R.S.; Moody, D.W. *National Water Summary 1988–89: Hydrologic Events and Floods and Droughts (No. 2375)*; US Government Printing Office: Washington, DC, USA, 1991.

89. Poulin, A.; Huard, D.; Favre, A.C.; Pugin, S. Importance of tail dependence in bivariate frequency analysis. *J. Hydrol. Eng.* **2007**, *12*, 394–403.

Multi-Basin Modelling of Future Hydrological Fluxes in the Indian Subcontinent

Ilias G. Pechlivanidis, Jonas Olsson, Thomas Bosshard, Devesh Sharma and K.C. Sharma

Abstract: The impact of climate change on the hydro-climatology of the Indian subcontinent is investigated by comparing statistics of current and projected future fluxes resulting from three RCP scenarios (RCP2.6, RCP4.5, and RCP8.5). Climate projections from the CORDEX-South Asia framework have been bias-corrected using the Distribution-Based Scaling (DBS) method and used to force the HYPE hydrological model to generate projections of evapotranspiration, runoff, soil moisture deficit, snow depth, and applied irrigation water to soil. We also assess the changes in the annual cycles in three major rivers located in different hydro-climatic regions. Results show that conclusions can be influenced by uncertainty in the RCP scenarios. Future scenarios project a gradual increase in temperature (up to 7 °C on average), whilst changes (both increase and decrease) in the long-term average precipitation and evapotranspiration are more severe at the end of the century. The potential change (increase and decrease) in runoff could reach 100% depending on the region and time horizon. Analysis of annual cycles for three selected regions showed that changes in discharge and evapotranspiration due to climate change vary between seasons, whereas the magnitude of change is dependent on the region's hydro-climatic gradient. Irrigation needs and the snow depth in the Himalayas are also affected.

Reprinted from *Water*. Cite as: Pechlivanidis, I.G.; Olsson, J.; Bosshard, T.; Sharm, D.; Sharma, K.C. Multi-Basin Modelling of Future Hydrological Fluxes in the Indian Subcontinent. *Water* **2016**, *8*, 177.

1. Introduction

Climate change impacts can be particularly complex in regions which are additionally subject to other environmental and socio-economic changes, *i.e.*, population growth, urbanization, land use change, and change in industrial and hydropower sectors [1]. India is a developing country with nearly two-thirds of the population depending directly on the climate- and water-sensitive sectors. The country already faces high risks of water shortages due to population growth, urbanization, and increasing demands in the agricultural, industrial, and hydropower sectors; hence, India offers a unique opportunity to examine the impacts of climate change, which in some areas have already been observed [2–4]. The region is characterised by a strong hydro-climatic gradient due to the monsoon (tropical climatic regions in the south; temperate and alpine regions in the Himalayan north, where elevated areas receive sustained winter snowfall) and the geographic features; hence, posing extraordinary scientific challenges to understand, quantify, and predict future availability of water resources. Of particular interest are the Northern Indian Himalayan plains given the sensitivity of snow and glacier melt processes to climate variability and change [5–7]. Arid and semi-arid regions might also experience changes in their hydrological cycle [8,9].

Assessment of future climate change impacts on water resources commonly involves climate variables (*i.e.* precipitation, temperature) from global circulation models (GCMs) in combination with hydrological models [10,11]. GCMs demonstrate significant skill at the continental and hemispheric spatial scales and incorporate a large proportion of the complexity of the global system; however, they are inherently unable to represent local basin-scale features and dynamics [12]. To narrow the gap between GCMs' abilities and hydrological needs, regional climate models (RCMs) have been developed to downscale the GCM output and, thus, provide high-resolution meteorological inputs to hydrological models. To improve the confidence in regional trends of hydro-climatic key variables and increase robustness in hydrological long term predictions, the World Climate Research Programme (WCRP) has recently launched a framework, called COordinated Regional climate Downscaling EXperiment (CORDEX), to generate and evaluate fine-scale ensembles of regional climate projections for all continents globally [13]. CORDEX has several domains that are defined as regions for which the regional downscaling is taking place. In particular, the efforts in the South Asia (SA) domain aim to translate regionally-downscaled climate data into meaningful sustainable development information in the monsoon South Asia area [14]. CORDEX-SA was initiated in 2012 and the RCM outputs have only recently become available.

While RCMs transfer the large-scale information from GCMs to scales which are closer to the basin scale (10–50 km), the output often shows large bias in the magnitude and spatial distribution of precipitation and, to a lesser extent, temperature [15]. RCM data are, therefore, not considered to be directly useful for assessing hydrological impacts at the regional and/or local scale [16]. A way to tackle the problem of RCM misrepresentation is to bias correct the RCM data to make them reproduce historical observed statistics to the degree possible [17]. Different approaches to bias correction have been made, with various complexity [18]. Simpler methods include shifting long-term annual or seasonal means to agree with observations whereas more advanced methods include adjustment of the full frequency distribution. A distribution-based approach is attractive not least for precipitation, for which both bias and future change are generally found to depend on the intensity level [19]. Bias correction often includes an implicit downscaling component, in that higher-resolution reference observations are used when fitting the RCM mapping functions. Bias correction generally preserves the variability described by different climatic conditions generated by RCM projections [20]; however, the RCM may perform differently depending on the season or governing atmospheric circulation. For instance, a typically wet weather regime (e.g., pattern or season) can have a different precipitation distribution in time and space than a dry regime.

Projected hydrologic information is prone to considerable uncertainty/errors at various steps of the modelling chain, *i.e.*, climate projection, bias correction and downscaling techniques, and hydrological simulation [21–23]. These errors can propagate in a very complex way (e.g., magnitude of error could vary both in space and time) which could be misinformative for management decisions [24,25]. A major source of uncertainty, among others, concerns the future emission scenarios, described by the representative concentration pathways (RCP), which further results in different climate projections. [26] showed that towards the end of the 21st century, the emission scenarios (here RCPs) are the dominant source of uncertainty in climate projections. The spatiotemporal variability of water fluxes differs between RCPs, particularly in areas with unique weather systems, *i.e.*, monsoon [27,28]. However, the choice of GCM and RCM may also have a large impact on the results and generally an ensemble of projections—encompassing different GCMs, RCMs, and emission scenarios—is recommended in hydrological climate change impact assessments [29,30].

Conventionally, hydrologic impacts are investigated on small (~0.1–10^2 km^2) or medium-sized basins (~10^2–10^3 km^2); however, current needs require assessment on larger areas and river basins, which requires the use of large scale hydrological models [31,32]. This type of modelling has the potential to encompass many river basins, cross-regional, and international boundaries and represents a number of different geophysical and climatic zones [33]. In addition, according to [34], large scale modelling can balance "depth with breadth", enhance process understanding, increase robustness of generalizations, facilitate catchment classification and regionalization schemes, and support better understanding of prediction uncertainty.

This paper contributes to ongoing efforts on assessing the potential impacts of climate change on water availability in the Indian subcontinent. In particular, we aim to answer the following questions: (i) what is the quantified impact of climate change on India's water resources? (ii) how is the uncertainty due to RCP propagated in hydrological impact modelling? and (iii) how does the potential impact vary in different climatic regions (*i.e.*, tropical, humid subtropical, and montane)? Although previous investigations have, at least to some degree, addressed similar questions, our contribution is associated with three novel features in that we apply: (1) three recently-generated high-resolution CORDEX-SA projections (RCP2.6, RCP4.5, and RCP8.5), (2) the DBS (distribution-based scaling; [35]) method to correct biases in the climate projections, and (3) the large-scale multi-basin HYPE (HYdrological Predictions for the Environment; [36]) hydrological model, to quantify climatic as well as anthropogenic impacts of climate change on hydrology and water availability over the entire Indian subcontinent. Section 2 introduces the study area, whereas the hydrological model and methodology are presented in Section 3. Section 4 presents the results of climate change impacts, followed by a discussion in Section 5 and conclusions in Section 6.

2. Study Area and Data

2.1. Study Area

India is the seventh-largest country by area and the second-most populated country with over 1.2 billion people. The country covers an area of about 3.3 million km^2 and some of its river basins extend into several neighboring countries (*i.e.*, China, Nepal, Pakistan, and Bangladesh; see Figure 1). Major rivers of Himalayan origin that are mainly located in India include the Ganga and the Brahmaputra, both of which drain into the Bay of Bengal. Major peninsular rivers, whose steeper gradients prevent them from flooding, include the Godavari, the Mahanadi, and the Krishna, which also drain into the Bay of Bengal; and the Narmada and the Tapi, which drain into the Arabian Sea. Coastal features include the marshy Rann of Kutch of Western India and the alluvial Sundarbans delta of Eastern India; the latter is shared with Bangladesh.

The spatiotemporal variation in climate is perhaps greater than in any other area of similar size in the world. The climate is strongly influenced by the Himalayas and the Thar Desert in the northwest, both of which drive the summer and winter monsoons [37]. Four seasons can be distinguished: winter (January–February), pre-monsoon (March–May), monsoon (June–September), and post-monsoon (October–December). In terms of spatial variability, the rainfall pattern roughly reflects the different climate regimes of the country, which vary from humid in the northeast (precipitation is 2068 mm/year and occurs about 180 days/year), to arid in Rajasthan (precipitation is 313 mm/year and occurs about 20 days/year). Moreover, India is characterized by strong temperature variations in different seasons ranging from a mean temperature of about 10 °C in winter to about 32 °C in pre-monsoon season.

The monsoon season is very important for water resources (and in turn their use for, e.g., power generation and agriculture) in the country since 75% of the annual rainfall (877 out of 1182 mm) is received in this period [38]. In particular, India's mean monthly rainfall during July (286.5 mm) is highest and constitutes about 24% of the annual total. The contribution in August is slightly lower (~21%) and in June and September ~14% [39]. The contribution of pre-monsoon and post-monsoon

139

rainfall to the annual total is roughly the same (11%). Higher variation is observed during the end of post-monsoon and winter (*i.e.*, November–February). To provide a better understanding of the system behavior, Figure 1b–d shows the annual cycles of the hydro-climatic fluxes in three different climatic regions, *i.e.*, humid subtropical (Ganga), montane (Indus), and tropical (Godavari).

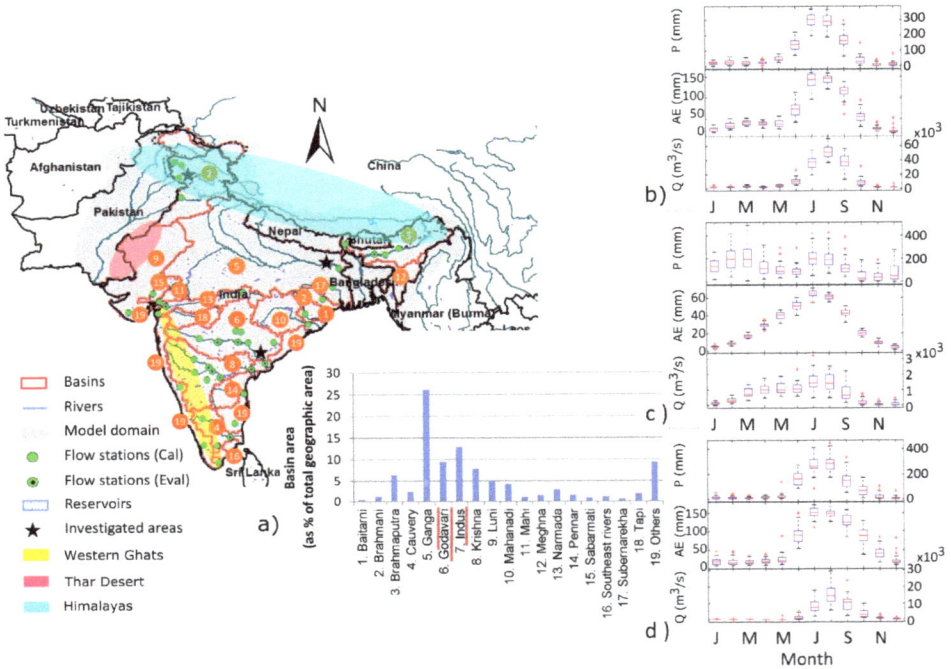

Figure 1. (**a**) The Indian subcontinent (model domain)—the numbers correspond to the basins in which regional analysis is conducted. Mean annual hydro-climatic cycles (precipitation (P), actual evapotranspiration (AE), and discharge (Q)) during the period 1976–2005 for the rivers: (**b**) Ganga (at Farakka station), (**c**) Indus (Chenab at Akhnoor station), and (**d**) Godavari (at Polavaram station). These results are shown from investigation areas in (**a**) with a star.

2.2. Spatial Input Data

Data availability is usually a severe constraint in the analysis of large-scale domains. To overcome such a problem, we use global datasets to extract the information required for hydrological applications (see Table 1 in [40]).

140

2.3. Meteorological Reference Data

Daily precipitation inputs for the period 1971–2005 are obtained from the Asian Precipitation—Highly-Resolved Observational Data Integration Towards Evaluation of Water Resources (APHRODITE) project [41,42] at 0.25° resolution (Table 1). Similarly, AphroTEMP [43] provides daily temperature inputs for the same period at 0.5° resolution. APHRODITE and AphroTEMP (in the following jointly denoted APHRODITE) are the only long-term continental-scale gridded datasets that are based on a dense network of daily data for Asia including the Himalayas. Therefore, the datasets have contributed to studies including among others water resources, climate change analysis, and statistical downscaling [28]. In this study, as a reference period we have chosen the 30-year period 1976–2005. The APHRODITE dataset fully covers this reference period and the period does not overlap with the future climate projections that start in 2006.

Table 1. The three CORDEX-SA climate projections used in this study.

RCP	GCM	RCM	Reference Data
2.6 4.5 8.5	EC-EARTH	RCA4 (0.44 × 0.44 deg)	APHRODITE (0.25 × 0.25 deg) AphroTEMP (0.5 × 0.5 deg)

2.4. Climate Projections

Our ensemble of three climate projections consists of modelling chains that use the same GCM (EC-EARTH; [44]) and RCM (RCA4; [45]), but three different representative concentration pathways, RCPs (see). RCPs are numbered after their increased radiative forcing until year 2100 (+2.6, +4.5, and +8.5 W/m^2, respectively; [46]). Note that more climate projections are becoming available over the South Asian domain through the CORDEX initiative (e.g., [47]).

A total of 129 years of hydrological simulations have been conducted for each climate scenario (1971–2099). However, the analysis is based on three 30-year periods: reference period (1976–2005), mid-century period (2021–2050), and end-century period (2070–2099). Note that in here we only analyze three projections from the CORDEX-SA ensemble; the CORDEX experiment is ongoing and more projections are being generated but these were the only ones available at the time of the study.

3. Methodology

3.1. India-HYPE: Description, Setup, and Calibration

The Hydrological Predictions for the Environment, HYPE, model [36] is a semi-distributed rainfall-runoff model capable of describing the hydrological processes at the basin scale. The model represents processes for snow and ice accumulation and melting, evapotranspiration, soil moisture, discharge generation, groundwater recharge, and routing through rivers and lakes. HYPE simulates the water flow paths in soil, which is divided into three layers with a fluctuating groundwater table. Parameters are linked to physiographical characteristics in the landscape, such as hydrological response units (HRUs) linked to soil type and depths and vegetation. Elevation is used to get temperature variations within a sub-basin for estimating the snow cover dynamics.

Lakes are defined as classes with specified areas and receive the local runoff and the river flow from upstream sub-basins. Precipitation falls directly on lake surfaces and lake water evaporates at the potential rate until the lake is dry. Each lake has a defined depth below an outflow threshold. The outflow from lakes is determined by a general rating curve, unless a specific one is given or if the lake is regulated. Lakes and man-made reservoirs are not separated in the simulation. A simple regulation rule can be used, in which the outflow is constant or follows a seasonal function for water levels above the threshold. A rating curve for the spillways can be used when the reservoir is full.

Irrigation in HYPE is simulated based on crop water demands calculated either with the FAO-56 crop coefficient method [48] or relative to a reference flooding level for submerged crops (e.g., rice). The demands are withdrawn from rivers, lakes, reservoirs, and/or groundwater within and/or external to the sub-basin where the demands originated. The demands are constrained by the water availability at these sources. After subtraction of conveyance losses, the withdrawn water is applied as additional infiltration to the irrigated soils from which the demands originated (in here named as applied irrigation water to soil (AIW)).

The HYPE model is set up for the entire Indian subcontinent (4.9 million km^2) divided into 6010 sub-basins, *i.e.*, with an average size of 810 km^2, and is referred to as India-HYPE. The model runs at a daily time step using APHRODITE as input data, but due to lack of daily discharge observations it was calibrated and evaluated (both in space and time) against monthly observations from 42 stations in the GRDC (Global Runoff Data Centre) database. For the Indian subcontinent, GRDC data are limited to monthly discharge for chosen river basins in the period 1971–1979. More discharge data are held in the Indian government agencies, but are not released to the public domain due to confidentiality; this generally sets a constraining factor for a model setup. Many of the parameters in the model are coupled to soil type or land use, while others are assumed to be general to a larger region. This approach fosters the potential of parameter transferability within reasonably homogeneous regions. In applications of HYPE, we generally consider the parameter identifiability and their regionalization to ungauged regions to be acceptable if the model performs adequately in the gauged basins over the entire model domain.

The HYPE model was spatiotemporally calibrated and evaluated in a multi-basin approach by considering the median performance in selected stations; 30 stations were selected for model calibration and 12 "blind" stations for spatial evaluation. The years 1969–1970 were used as a model warm-up period, the next five years for model calibration (1971–1975) and the final four years for temporal performance evaluation (1976–1979). The model's predictability was tested using various performance measures (*i.e.*, objective functions and flow signatures) and additional data sources (*i.e.*, remote sensing potential evapotranspiration records); see details in [40]. Here, we assess the model's predictability based on the Nash-Sutcliffe Efficiency, NSE [49] and relative error, RE (defined as the difference between the mean modelled value and the mean observed value divided by the mean observed value). The former has been widely applied in hydrology as a benchmark measure of fit. It can also be interpreted as a classic skill score, where skill is interpreted as the comparative ability with regards to a baseline model (here taken to be the "mean of the observations"; *i.e.*, NSE < 0 indicates that the mean of the observed time series provides, on average, a better prediction than the model). NSE ranges between 1 (perfect fit) and $-\infty$ whereas RE ranges between $-\infty$ and $+\infty$, with the "ideal" value being 0.

3.2. Bias Correction of RCM Data

The RCM projections (mean daily precipitation and temperature) were bias corrected against the APHRODITE dataset using the distribution based scaling, DBS, statistical method [35]. In brief, DBS aims to map the quantile distributions of precipitation and temperature in the RCM data to those of the reference data. For precipitation, a two-step procedure is applied: 1) correction of the wet-day frequency by applying a wet-day cut-off threshold (in case of wet frequency bias) or by adding wet-days to pre-existing wet-spells (if dry frequency bias), and 2) quantile-mapping of the precipitation data using a double-gamma distribution to accurately represent both normal and extreme precipitation intensity ranges. For temperature, a quantile-mapping correction based on a Gaussian distribution is used. The temperature correction model is dependent on the wet/dry state of the corresponding precipitation. This means that DBS takes into account different biases on wet and dry days (see details in [35]). The bias-corrected projections were used to force the hydrological model for the assessment of climate change impacts on water resources. DBS was used for bias-correcting GCM projections over India (Mumbai region) by [50].

3.3. Climate Change Impact Assessment

For the present climate, water availability in space and time was simulated using the reference APHRODITE dataset as input. The same framework is then used to project the impact of climate change on the water resources with the assumption that the land use shall not change over time.

3.3.1. Long-Term Averages

Firstly, we assess the impact of climate change on the hydro-climatic variables, *i.e.*, long-term means of precipitation (P, in mm), temperature (T, in C), actual evapotranspiration (AE, in mm), runoff (R, in mm), soil moisture deficit (SMD, in mm), snow depth (SD, in cm), and applied irrigation water to soil (AIW, in Mm^3). The daily series for each 30-year period (reference, mid- and end-century) is used to extract the statistics. The relative future change in the long-term average (%) between two periods (mid- or end-century *versus* reference period) due to climate change is estimated for each sub-basin. Positive (negative) change indicates increase (decrease) from the average value in the reference period. Note that for T we calculate absolute differences between future and reference periods; hence, we can express properly also changes in the sign of T.

The spatial variability of change at the basin scale is further summarized (here presented as a boxplot) allowing comparison of the overall change between basins; note that this analysis is only presented for P and R.

3.3.2. Annual Cycles

To complement the above assessment, we investigate the changes in the annual cycles in different locations and governing climatic conditions. We, therefore, compare the annual cycles of P, E, and discharge (Q, in m^3/s) for the Ganga (at Farakka station), Indus (Chenab at Akhnoor station), and Godavari (at Polavaram station) basins (see annual cycles of the present climate in Figure 1b–d). The climate in these basins is characterized as humid subtropical, montane, and tropical, respectively; hence, the analysis can capture the sensitivity to the dominant climatic gradients in the subcontinent.

4. Results

4.1. Model Evaluation

Here, we investigate the model's reliability (which involves measures sensitive to high flows, timing, variability, and volume), and present performances for the benchmark objective functions, *i.e.*, NSE and RE. The NSE and RE for all calibration and evaluation stations and periods are presented in Table 2. Overall, the India-HYPE model achieved an acceptable performance for our purpose and is, therefore, considered adequate to describe the dominant hydrological processes in the region (median NSE and RE for the calibration period is 0.76 and −5.26%, respectively, while for the evaluation period the values are 0.68 and +8.01%, respectively). However, as expected, the performance decreased when the model is validated both in space and time (NSE and RE equal 0.40 and 16.81%, respectively); the model is not "trained" to capture the flow conditions at independent areas and time periods. A more detailed analysis during the evaluation period showed that the model could not fully capture the variability (e.g., standard deviation) of the observed data in the validation stations. The ratio of the standard deviation of modelled over the standard deviation of observed data decreases during the evaluation period at the evaluation stations from 0.78 to 0.58 which consequently affects the NSE values; see the discussion on the decomposition of the NSE and the importance of its decomposed terms (dealing with timing, variability, and volume) on the overall NSE in [51]. However other flow characteristics, *i.e.* timing and volume, are better represented than variability during the evaluation period. Further analysis and discussion on model performance and consistency can be found in [40].

Table 2. Median model performance for calibration and evaluation stations and periods.

Space	Time/Periods	NSE	RE (%)
Cal. (30 stations)	1971–1975	0.76	−5.26
	1976–1979	0.63	−5.43
Eval. (12 stations)	1971–1975	0.68	8.01
	1976–1979	0.40	16.81

4.2. Bias Correction

The effect of bias-correction using the DBS methodology is illustrated here for the case of mean P (Figure 2). The highest P amounts are measured along the mountain ridges, e.g., on the western side of the Western Ghats, Southwest India, and at the foothills of the Himalaya in the northern part of the study region. Those wet regions are strongly associated with atmospheric flows during the monsoon season. Dry regions are located in the northwestern part and on the eastern side of the Western Ghats. The projections employed in this study all share the same historical run, thus Figure 2 is representative for all three projections. The historical projection shows large positive and negative biases in mean P, with sharp gradients between regions of positive and negative biases. The regions with sharp gradients coincide very well with regions of complex topography (e.g., in the Himalaya or on the lee and luv side of the Western Ghats). DBS is able to correct for most of the bias although some remaining dry bias is apparent in regions where the historical projection strongly underestimates mean precipitation. This is related to a dry frequency bias, *i.e.*, too few wet days in the projection. DBS is not able to fully correct for the pronounced deficits in rain days, but can only correct moderate dry frequency bias by adding wet-days to existing wet-spells. Such a limitation of bias-correction methods in case of dry frequency biases is a general limitation of current quantile-mapping bias-correction methods [35,52].

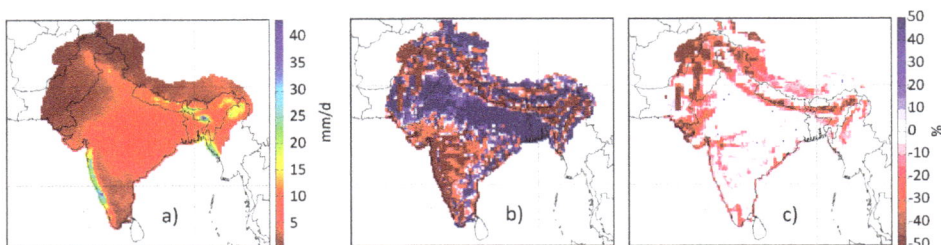

Figure 2. Observed precipitation and RCM bias in the period 1976–2005: (**a**) mean daily precipitation in APHRODITE; (**b**) relative bias in uncorrected historical GCM-RCM projection, and (**c**) relative bias in bias-corrected historical GCM-RCM projection.

Concerning T, in most parts of India it was underestimated by 1–2 C in the historical GCM-RCM projection (not shown). After the DBS bias-correction, only a negligible bias remained and the spatial pattern became essentially identical to the observations (see Figure 3c).

4.3. Reference Data Analysis

To infer a quantitative understanding of the magnitude of the climate change impact on the hydro-climatic components (P was analyzed in Section 4.2), it is necessary to estimate the long-term averages of India-HYPE driven by APHRODITE for the entire reference period (Figure 3). Long-term average R is controlled by the spatial patterns of P and AE; consequently high R is generated at the southwestern rivers and at the Himalayan mountain range. AE, which in this model is controlled by water availability and T, shows similar spatial patterns to P but lower than the latter by almost 25%; hence, indicating that most precipitated water is evaporated back to the atmosphere. Note, also, the high temperatures which are almost homogeneously distributed over the region (average T is about 25 °C); however, the T gradient is very strong at the Himalayan mountain range (average T varies between −5 and 2 °C).

The AIW varies between 100 and 900 Mm^3 per year with high spatial variability in the region (note that the spatial pattern of this variable is subject to the GMIA irrigation map). The SMD is between 0 and 140 mm/year, with the Himalayas exhibiting very low values and the Thar Desert exhibiting high values; note the presence of wetlands covering this northwest region. Finally, SD can reach 100 cm in the Himalayas with large local variability.

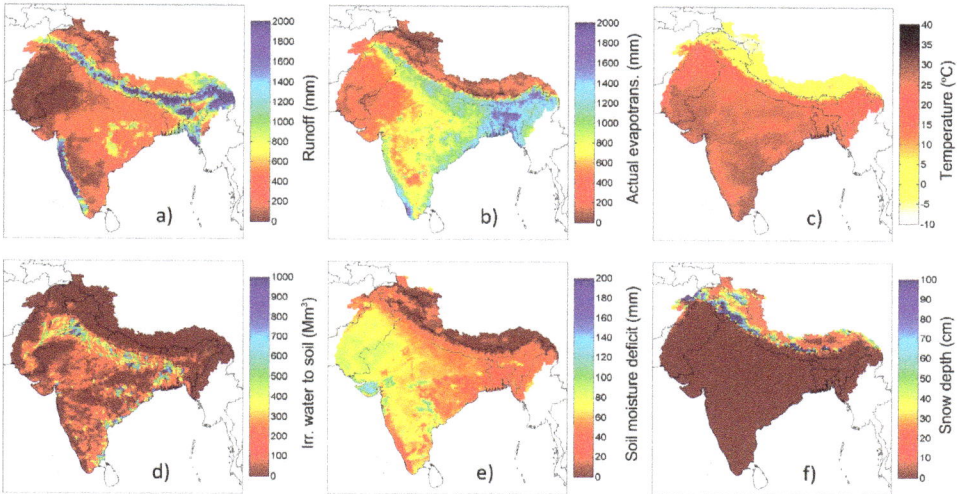

Figure 3. Annual averages (period 1976–2005) for the variables: (**a**) runoff, R; (**b**) actual evapotranspiration, AE; (**c**) temperature, T; (**d**) applied irrigation water to soil, AIW; (**e**) soil moisture deficit, SMD; and (**f**) snow depth, SD.

4.4. Climate Change Impacts

As outlined in Section 3.3, we analyze the effects of climate change for various variables in two different scenario periods. We firstly present changes in the variables which control the hydro-climatic conditions and continue with the applied irrigation water to soil which is more "end-user related". We then investigate the changes in the annual cycles of three regions.

4.4.1. Long-Term Averages

All projections show an increase of T for both scenario periods and in the whole modelling domain (Figure 4). In RCP2.6, the increase in T is limited to ranges below +1.5 °C until the end of the 21st century. For the same scenario period and RCP4.5, the increase in T is projected to fall into the range of +1.0 to +2.0 °C over most of the domain, but is higher—up to +3.0 °C—in the mountainous parts in the north. The same spatial pattern is seen in RCP8.5 in which T is projected to increase between

1.5 and 6.0 °C with the highest increase also occurring in the mountainous regions. This elevation dependency could be due to the snow albedo effect which leads to higher increases in T when the snow cover duration decreases [53].

Regarding P, the projections show strongly increasing mean P over most of the region (Figure 5). This also includes areas where present day P is already very high (e.g. along the Western Ghats and the southern side of the Himalayan mountain range). The signal is stronger towards the end of the 21st century than in mid-century and the higher the RCP is, with the exception of RCP2.6 for which there is no intensification of the signal from the mid- to the end-century period. The clear dependency of the change signal on the period and the RCP is true for all variables discussed below. The dry regions in the northwestern and central parts, including e.g., the Indus, Luni, Sabarmati, and parts of the Ganga River, are projected to receive less P. Additionally, the spatial variability of the changes within the major river basins (see boxplots in Figure 5) is generally increasing towards the end of the century. The above-mentioned river basins include sub-basins with a very heterogeneous pattern of P changes, ranging from decreases to increases for all projections except for the RCP8.5 end-century evaluation. Those basins are located in a zone where slight changes in the spatial patterns of P decrease or increase might lead to a pronounced effect on the basin-wide qualitative changes (*i.e.*, decrease or increase of mean P).

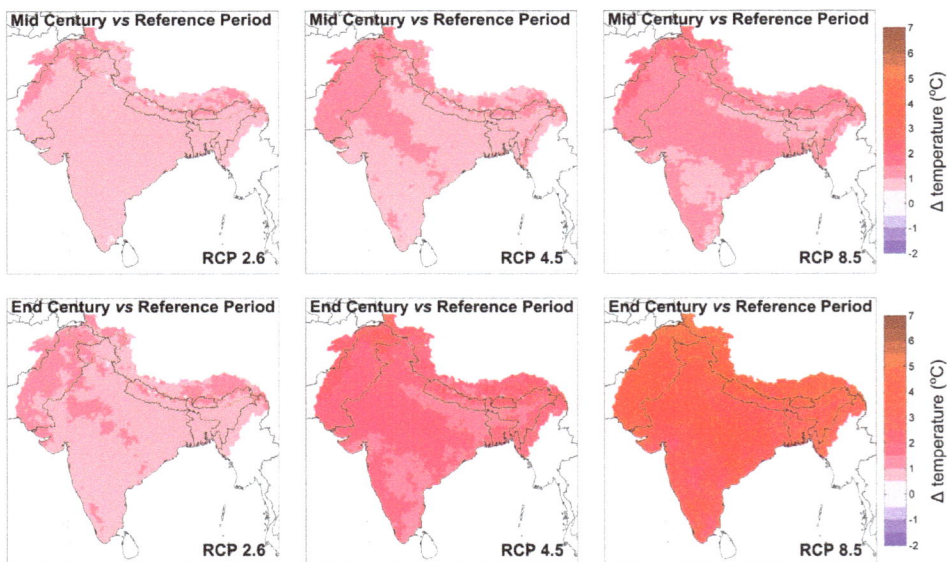

Figure 4. Change in temperature T for each climate projection (columns 1–3) and period (top and bottom row for the mid- and end-century, respectively).

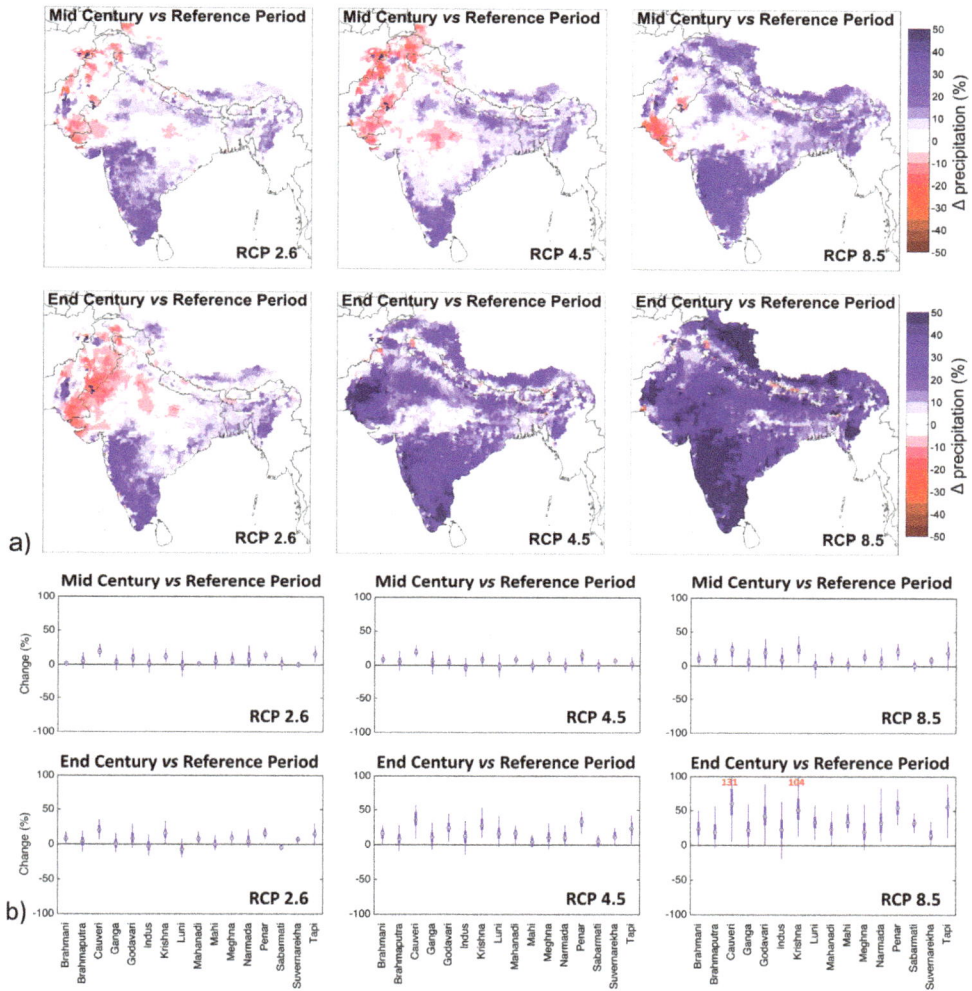

Figure 5. (a) Relative change in precipitation P for each climate projection (columns 1–3) and period (top and bottom row for the mid- and end-century, respectively); and (b) variability of the change at the basin scale for each climate projection (columns 1–3) and period (rows 1–2).

AE generally either remains essentially unchanged or increases. In areas with a pronounced decrease in P, such as in the northwest, a slight decrease of AE may, however, occur (see Figure 6). The increase of evaporation in the south is due to decrease in SMD (see Figure 8). Over the Himalayan mountain range, strong increases are projected in all RCPs and both scenario periods. In that area, the increase of potential evapotranspiration (not shown) due to increased temperature and a shortened period of snow cover (see Figure 8 for snow depth changes as an indicator of snow cover changes) both contribute to higher AE.

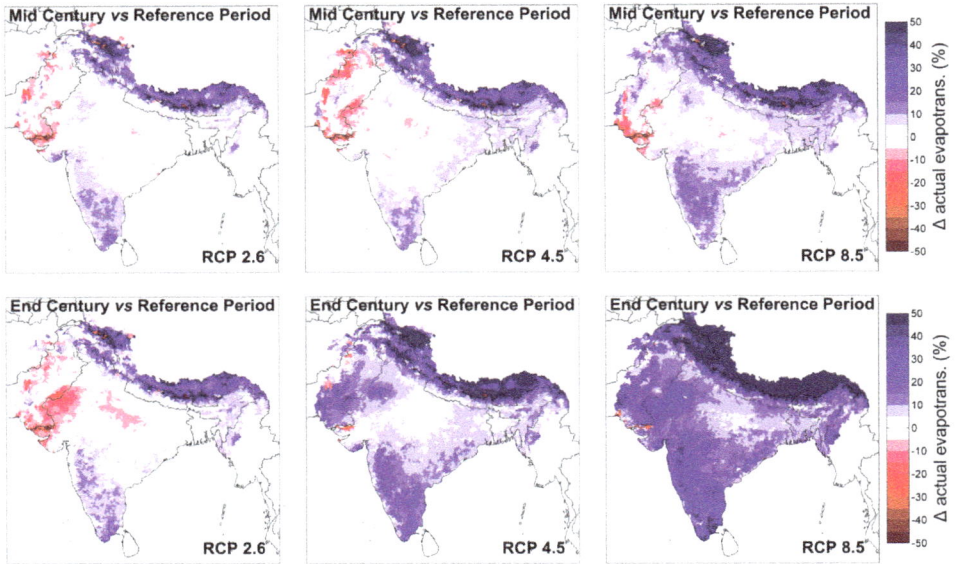

Figure 6. Same as Figure 5a for actual evapotranspiration AE.

The pattern of R changes follows the one of P but the magnitudes of the relative changes are larger than for P (Figure 7). Consequently, the spatial variability within the major river basins is also much larger than for P, shown by the wider boxplots in the Figure 7. Thus, the basins, which are susceptible to small spatial changes of areas with increasing and decreasing P, are even more sensitive to spatial changes when it comes to R.

Within the model domain, snow only occurs in the Himalayas (Figure 3). For a large fraction of that area, SD is projected to decrease strongly, a change which is already apparent in the mid-century period (Figure 8). Only in some high-elevated areas is T cold enough to make SD increase due to increased mean P.

The changes in irrigated water to soil (Figure 9) should be interpreted in a way that they represent the changes in future demand for irrigation. Wherever irrigation is projected to increase, a higher demand for irrigation is implied. In that sense, the results indicate that the demand for irrigation will increase in the Northwestern and central regions in all RCPs and both scenario periods. Towards the end of the century in RCP8.5, the region with increased irrigation demand is enlarged and encompasses the whole Indus and Ganga river basins. Decreases in irrigation needs are projected for the southern part of the model domain.

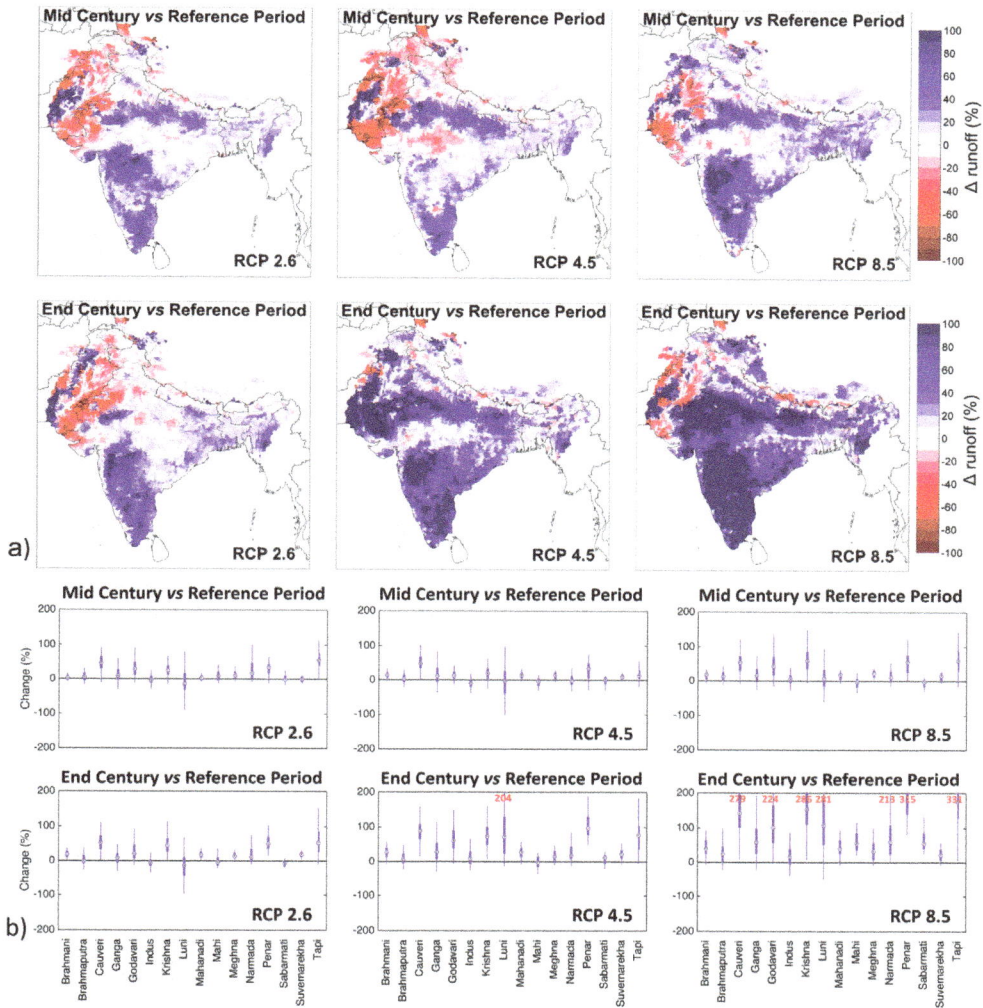

Figure 7. (**a**) Same as Figure 5a for runoff R. Maximum values for the regions in which change is greater than 200% are presented in red and (**b**) variability of the change at the basin scale for each climate projection (columns 1–3) and period (rows 1–2).

4.4.2. Annual Cycles

Whereas in Section 4.4.1 we presented spatially detailed maps of changes in annual mean quantities, we show here the results of changes in the annual cycle at three selected gauges (see stars in Figure 1) in order to provide more details about the seasonal dynamics.

151

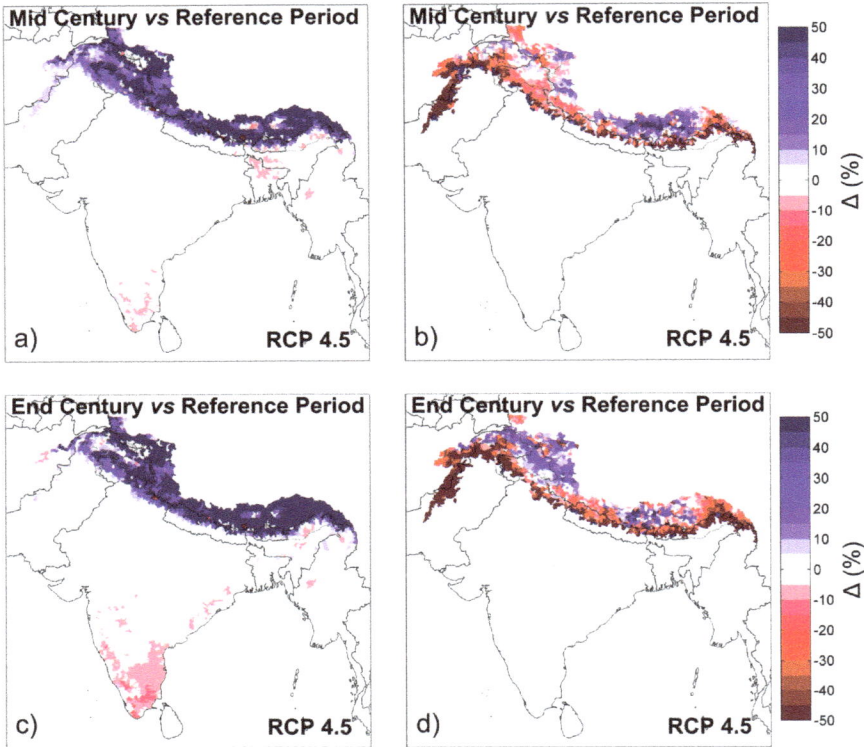

Figure 8. Relative change in soil moisture deficit SMD (column 1) and snow depth SD (column 2) for climate projection RCP4.5 until mid-century (row 1) and end-century (row 2). Only RCP4.5 is shown; the spatial pattern of changes is very similar for the other RCPs. (**a**) SMD change for mid-century; (**b**) SD change for mid-century; (**c**) SMD change for end-century, and (**d**) SD change for end-century. See the results for all RCPs and future periods in Figures S1 and S2 in the Supplement.

The Ganga River has clear Q seasonality with peak Q in the monsoon season and very little Q in the remaining months (Figure 1b). P is projected to increase during the monsoon period, and little change happens in the rest of the year (Figure 10a). This trend is stronger for the end-century than for the mid-century changes in RCP4.5 and RCP8.5, while RCP2.6's changes remain on a low level even at the end of the century. For the end-century period, RCP8.5 also shows a longer season of increasing P, lasting until the start of the post-monsoon season in October. The seasonal pattern of changes in AE closely follows the one of P, but the magnitudes of the changes are smaller. Thus, the increases in P partly become runoff effective, which is seen as increasing Q in the monsoon season. The largest increases in Q towards the end of the century are projected to range from 5000 to 20,000 m^3/s depending on the chosen RCP, corresponding to a 12 to 50 % increase compared to the reference period.

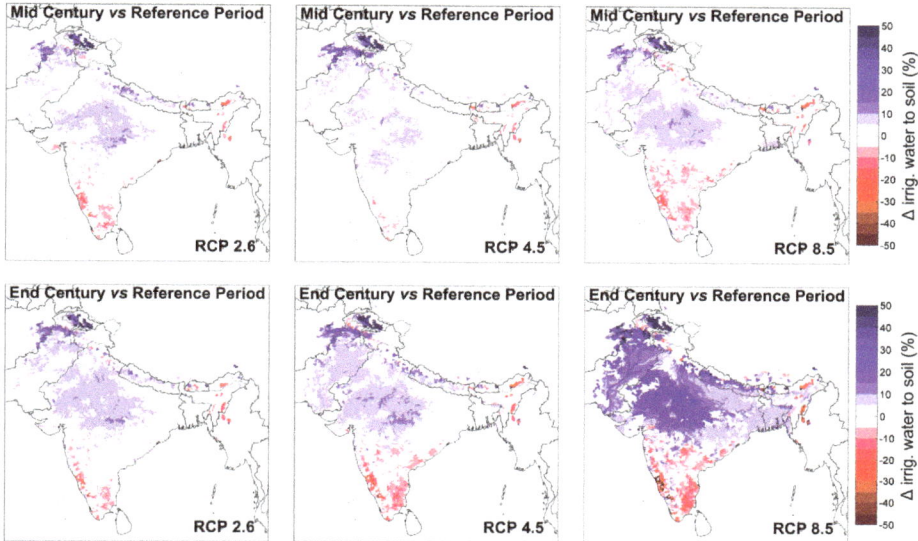

Figure 9. Same as Figure 5a for applied irrigation water to soil AIW.

The gauge in the Indus catchment lies close to the Himalayas. In present day conditions, the seasonal cycle of P shows a two-peak pattern with one peak in March and the second one in July/August (Figure 1c). The three projections do not agree on the change signal (Figure 10b). For the end-century period, a slight tendency towards increasing P in summer and decreasing P in March is noticeable. This would mean that the first P peak weakens, whereas the second one in summer gets stronger. AE increases throughout the year, with the largest increases being projected for the summer. There is no clear link between the P and the AE change signal, indicating that the soil moisture is not the limiting factor for AE in that area. The change signal in Q is associated with more uncertainty and small projected changes switch sign several times in the course of the annual cycle. This is also reflected in the boxplots in Figure 7 where the Indus River shows a large spatial variability of runoff changes around the no-change level. Small changes in some sub-basins might lead to a switch of the sign of change in the response on the aggregated basin level. Overall, it seems that the climate change signal of discharge is small.

Under the current conditions, the Godavari River has a clear seasonal discharge regime with a peak in the monsoon season and low discharge in the rest of the year (Figure 1d). Compared to the Ganga River, the seasonality is even more pronounced due to the lack of snow-melt from the Himalayas during the pre- and post-monsoon

153

season. Additionally, the seasonal pattern of changes in all variables looks similar to the ones in the Ganga River; however, the magnitudes are larger (Figure 10c). For total Q in the monsoon and post-monsoon period, the changes amount to +10% to +80%.

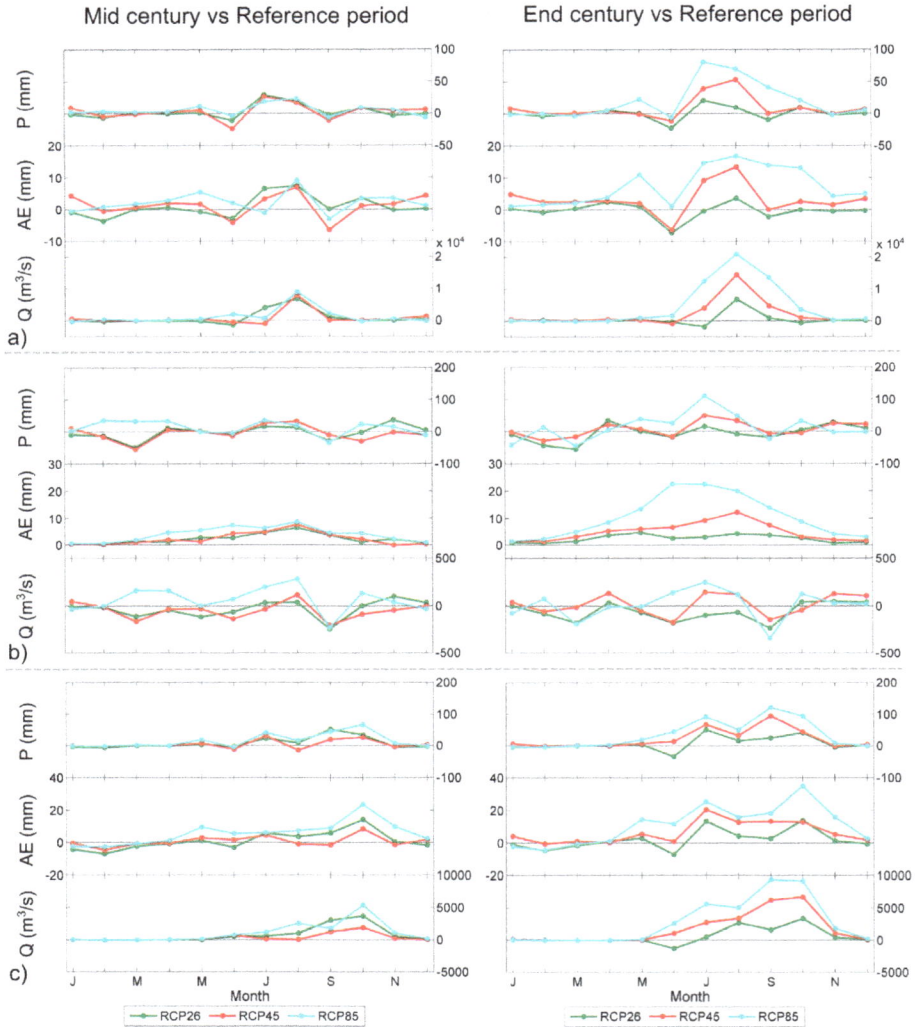

Figure 10. Absolute changes in the annual cycles of precipitation P, actual evapotranspiration AE and discharge Q for each climate scenario and period (column 1 and 2 for the mid- and end-century, respectively) for the rivers: (**a**) Ganga (at Farakka station); (**b**) Indus (Chenab at Akhnoor station), and (**c**) Godavari (at Polavaram station).

5. Discussion

5.1. Enhancing Understanding of Future Climate Change Impacts

Our study contributes to the previously reported assessments of future climatic patterns (which in most cases are driven by projections from different GCMs and RCMs) in India and their impact on water resources. Overall our results support previous findings that climate projections reveal an increase in monsoon precipitation in the mid-century [54,55] and a possible extension of the monsoon period [56–58]. Our results over Northern and Western India and the southeastern coastline differ from previous projections based on the CGCM3.1-PRECIS model chain and the A1B emission scenario [31,59]. According to [59], in those regions precipitation is expected to decrease both in the mid- and end-century. Moreover, analysis (not shown here) on the number of wet spells was consistent with the general trends of previous finding, showing a higher frequency of wet spells from the 2060s and beyond, mostly in northern and coastal regions [14,60]. [61] define the river basins of Kutch, Saurashtra, Luni, and Indus in the northwest and Pennar in the southeast as physically water-scarce, most of the rest of India as economically water-scarce, and only Brahmaputra, Meghna, Brahmani, Mahanadi, and some smaller basins along the east coast as non-water-scarce (Figure 1). Our results support this regional classification, however attention is needed in results for R in the northwestern basins due to their pronounced spatial variability (although overall results point at decreasing runoff). The overall decreasing future trend of water availability in the northwestern basins with a high spatial variability agrees with the results in [31]. In our results river Pennar will clearly become less water-scarce by the end of the century, which is in contrast to [31] who found a reduced water yield in this basin on this time horizon.

A key component for assessing water scarcity is the evapotranspiration. Similarly to P, AE in our results exhibit a pronounced spatial variability in the northwest and overall an increase is projected for the Indian subcontinent. This can be contrasted with recent findings of observed decreasing trends in temperature (and consequently evaporation and water scarcity) over Northeastern India [62], the Godavari basin [63], as well as many regions world-wide [64,65]. For India, this finding has been attributed to mainly a reduction of wind speeds ('stilling') and an increase of atmospheric humidity [63]. These impacts cannot be simulated with the purely T-based relationship for AE used here.

In the Himalayan region, we found a generally consistent increase in runoff for different projections and future periods, which agrees with the general trend in the literature [5,6,66–68]; however this increase varies significantly in space (following the topographic pattern). In addition, it is generally expected that the changes in the Ganga and Meghna rivers will be larger than in the Brahmaputra river, probably due to the impact of melted water from snow and glaciers [2]. This conclusion is consistent with our results. Change in runoff in the Krishna river is ambiguous with studies showing both a future increase [32,69], as here, and a decrease [31].

5.2. Limitations of This Study

Reliability of the input data used to drive large-scale multi-basin hydrological models, particularly those derived from global datasets, has been questioned [70]. Although a preliminary comparison of the applied climatic and physiographic data against national data did not show significant discrepancies (e.g., when comparing the APHRODITE data against P and T observations from the Indian Meteorological Department), some inconsistencies were generally observed at the local scale.

Here, we have only focused on the uncertainty from the RCP scenario. This was motivated by previous findings indicating that towards the end of the 21st century, generally the emission scenario (here, RCP) is the dominant source of uncertainty in climate projections (e.g., [26]). However, it is clear that also the other sources may substantially contribute to the uncertainty, e.g., the choice of GCMs, RCMs, bias-correction method, hydrological model structure, and parameterization can have a substantial impact [71,72].

Climate change impact results are subject to the impact model's predictability and consistency, here assessed by the model's performance in the present climate. The India-HYPE model can adequately represent the long-term average fluxes and their seasonal variation (see details in [40]); however, it is expected that a larger number of discharge gauges (representing different hydro-climatic systems) and/or temporally extended time series (in terms of length and resolution) and/or additional variables (e.g., evapotranspiration, snow cover area *etc.*) would provide additional information to drive the parameterization of the model and guide towards model structural improvements.

Finally, our study is limited by the assumed stationarity in the investigated hydrological systems. It is recognized that non-stationarity exists as a characteristic of the natural world due to various environmental changes (land use and other man-made alterations) [73].

6. Conclusions

We have explored the potential impact of climate change on the hydrology and water resources of the Indian subcontinent, based on an RCP-ensemble of regional climate projections from the CORDEX-SA framework. Climate projections with GCM EC-EARTH and RCM RCA4 based on three RCP scenarios (RCP2.6, RCP4.5, and RCP8.5) are bias-corrected and introduced into the HYPE hydrological model to assess average changes in various hydro-climatic fluxes for two future periods (mid- and end-century). Additionally, the intra-annual variability of key fluxes (precipitation, evapotranspiration, and discharge) in three river systems.

Overall, the distribution of change in runoff varies considerably with both hydro-climatic region and climate projection. In particular, the following future changes were indicated:

- Temperature will increase in the entire subcontinent, with the highest increase in the mountainous regions.
- An increase in long-term average precipitation and evapotranspiration in wet regions; however, less precipitation and evapotranspiration are expected at the dry regions.
- Average snow depth in the Himalayan region will be reduced; this is consistent in all projections and time horizons.
- A general increase in the need for irrigation; however, the need is reduced in the south.
- Large relative changes in runoff, and large spatial variability at the basin scale, particularly towards the end of the century.
- Changed seasonality in discharge, with more pronounced changes in the tropical and subtropical zones than in the mountainous regions.

Overall, the conclusions support previous findings in most parts of India (including the Himalayas), with respect to future trends in runoff. This study, however, indicates larger changes under the severe emission scenario RCP8.5. Regionally, some differences are found in comparison with existing results, e.g., for precipitation and evapotranspiration in northern and western India and the southeastern coastline and for runoff changes in the Krishna river. These differences highlight the need for further studies focusing on climate change impacts on hydrology in India. In addition to pure climate impact modelling, as performed here, we believe more efforts are needed to better understand and quantify the sensitivity of the river systems to climate changes, as well as other changes (e.g., land-use and population). This work should preferably include multiple scenarios of all considered changes used in combination with multiple hydrological models. Work in this direction is ongoing and will be reported elsewhere.

Supplementary Materials: The following are available online at www.mdpi.com/2073-4441/8/5/177/s1. Figure S1: Relative change in soil moisture deficit (SMD) for each climate projection (columns 1–3) and period (top and bottom row for mid- and end-century respectively), Figure S2: Relative change in snow depth (SD) for each climate projection (columns 1–3) and period (top and bottom row for mid- and end-century, respectively).

Acknowledgments: This work was funded mainly by the Swedish International Development Cooperation Agency (SIDA; grant no. AKT-2012-022) with additional support from the Swedish Research Council Formas (grant No. 2010-121) and the Swedish Secretariat for Environmental Earth System Sciences (SSEESS; grant No. KVA/2015/111/75). We would also like to acknowledge David Gustafsson, Kean Foster, Kristina Isberg, Jörgen Rosberg, Jafet Andersson and Sian de Koster for assistance with background material for this study and manuscript checks.

Author Contributions: Ilias G. Pechlivanidis led the project-related activities on setting up the hydrological model, conducting the analyses, and writing the manuscript. Jonas Olsson assisted on the manuscript writing and supervised the efforts on bias-correction of the climate projections. Thomas Bosshard led the activities on bias-correcting the climate projections and assisted on writing the manuscript. Devesh Sharma and K.C. Sharma provided local knowledge during the hydrological model setup and made improvements to the manuscript.

Conflicts of Interest: The authors declare no conflict of interest.

References

1. Buytaert, W.; De Bièvre, B. Water for cities: The impact of climate change and demographic growth in the tropical Andes. *Water Resour. Res.* **2012**, *48*.

2. Mirza, M.; Warrick, R.; Ericksen, N. The implications of climate change on floods of the Ganges, Brahmaputra and Meghna rivers in Bangladesh. *Clim. Chang.* **2003**, *57*, 287–318.

3. Neupane, R.P.; Yao, J.; White, J.D. Estimating the effects of climate change on the intensification of monsoonal-driven stream discharge in a Himalayan watershed. *Hydrol. Process.* **2014**, *28*, 6236–6250.

4. Singh, P.; Kumar, N. Impact assessment of climate change on the hydrological response of a snow and glacier melt runoff dominated Himalayan river. *J. Hydrol.* **1997**, *193*, 316–350.

5. Pervez, M.S.; Henebry, G.M. Assessing the impacts of climate and land use and land cover change on the freshwater availability in the Brahmaputra River basin. *J. Hydrol. Reg. Stud.* **2014**, *3*, 285–311.

6. Singh, P.; Bengtsson, L. Impact of warmer climate on melt and evaporation for the rainfed, snowfed and glacierfed basins in the Himalayan region. *J. Hydrol.* **2005**, *300*, 140–154.

7. Singh, P.; Arora, M.; Goel, N.K. Effect of climate change on runoff of a glacierized Himalayan basin. *Hydrol. Process.* **2006**, *20*, 1979–1992.

8. Gosain, A.; Rao, S.; Basuray, D. Climate change impact assessment on hydrology of Indian river basins. *Curr. Sci.* **2006**, *90*, 346–353.

9. Pechlivanidis, I.G.; Olsson, J.; Sharma, D.; Bosshard, T.; Sharma, K.C. Assessment of the climate change impacts on the water resources of the Luni region, India. *Glob. NEST J.* **2015**, *17*, 29–40.

10. Graham, L.P.; Hagemann, S.; Jaun, S.; Beniston, M. On interpreting hydrological change from regional climate models. *Clim. Chang.* **2007**, *81*, 97–122.

11. Pechlivanidis, I.G.; Jackson, B.; McIntyre, N.; Wheater, H.S. Catchment scale hydrological modelling: A review of model types, calibration approaches and uncertainty analysis methods in the context of recent developments in technology and applications. *Glob. NEST J.* **2011**, *13*, 193–214.

12. Fowler, H.J.; Blenkinsop, S.; Tebaldi, C. Linking climate change modelling to impacts studies: Recent advances in downscaling techniques for hydrological. *Int. J. Climatol.* **2007**, *27*, 1547–1578.

13. Giorgi, F.; Jones, C.; Asrar, G.R. Addressing climate information needs at the regional level: The CORDEX framework. *WMO Bull.* **2009**, *58*, 175–183.

14. Chaturvedi, R.; Joshi, J.; Jayaraman, M.; Bala, G.; Ravindranath, N. Multi-model climate change projections for India under representative concentration pathways. *Curr. Sci.* **2012**, *103*, 1–12.

15. Foley, A. Uncertainty in regional climate modelling: A review. *Prog. Phys. Geogr.* **2010**, *34*, 647–670.

16. Haylock, M.R.; Cawley, G.C.; Harpham, C.; Wilby, R.L.; Goodess, C.M. Downscaling heavy precipitation over the United Kingdom: A comparison of dynamical and statistical methods and their future scenarios. *Int. J. Climatol.* **2006**, *26*, 1397–1415.

17. Wetterhall, F.; Pappenberger, F.; He, Y.; Freer, J.; Cloke, H.L. Conditioning model output statistics of regional climate model precipitation on circulation patterns. *Nonlinear Process. Geophys.* **2012**, *19*, 623–633.

18. Teutschbein, C.; Seibert, J. Bias correction of regional climate model simulations for hydrological climate-change impact studies: Review and evaluation of different methods. *J. Hydrol.* **2012**, *456–457*, 12–29.

19. Déqué, M. Frequency of precipitation and temperature extremes over France in an anthropogenic scenario: Model results and statistical correction according to observed values. *Glob. Planet. Change* **2007**, *57*, 16–26.

20. Lenderink, G.; Buishand, A.; van Deursen, W. Estimates of future discharges of the river Rhine using two scenario methodologies: Direct *versus* delta approach. *Hydrol. Earth Syst. Sci.* **2007**, *11*, 1145–1159.

21. Chen, H.; Xu, C.-Y.; Guo, S. Comparison and evaluation of multiple GCMs, statistical downscaling and hydrological models in the study of climate change impacts on runoff. *J. Hydrol.* **2012**, *434–435*, 36–45.

22. Minville, M.; Brissette, F.; Leconte, R. Uncertainty of the impact of climate change on the hydrology of a nordic watershed. *J. Hydrol.* **2008**, *358*, 70–83.

23. Preston, B.L.; Jones, R.N. Evaluating sources of uncertainty in Australian runoff projections. *Adv. Water Resour.* **2008**, *31*, 758–775.

24. Chiew, F.H.S.; Teng, J.; Vaze, J.; Kirono, D.G.C. Influence of global climate model selection on runoff impact assessment. *J. Hydrol.* **2009**, *379*, 172–180.

25. Jones, R.N.; Chiew, F.H.S.; Boughton, W.C.; Zhang, L. Estimating the sensitivity of mean annual runoff to climate change using selected hydrological models. *Adv. Water Resour.* **2006**, *29*, 1419–1429.

26. Hawkins, E.; Sutton, R. The potential to narrow uncertainty in regional climate predictions. *Bull. Am. Meteorol. Soc.* **2009**, *90*, 1095–1107.

27. Bueh, C.; Cubasch, U.; Hagemann, S. Impacts of global warming on changes in the East Asian monsoon and the related river discharge in a global time-slice experiment. *Clim. Res.* **2003**, *24*, 47–57.

28. Lucas-Picher, P.; Christensen, J.H.; Saeed, F.; Kumar, P.; Asharaf, S.; Ahrens, B.; Wiltshire, A.J.; Jacob, D.; Hagemann, S. Can regional climate models represent the Indian monsoon? *J. Hydrometeorol.* **2011**, *12*, 849–868.

29. Aich, V.; Liersch, S.; Vetter, T.; Huang, S.; Tecklenburg, J.; Hoffmann, P.; Koch, H.; Fournet, S.; Krysanova, V.; Müller, E.N.; Hattermann, F.F. Comparing impacts of climate change on streamflow in four large African river basins. *Hydrol. Earth Syst. Sci.* **2014**, *18*, 1305–1321.

30. Huang, S.; Krysanova, V.; Hattermann, F. Projections of climate change impacts on floods and droughts in Germany using an ensemble of climate change scenarios. *Reg. Environ. Chang.* **2014**, *15*, 461–473.

31. Gosain, A.; Rao, S.; Arora, A. Climate change impact assessment of water resources of India. *Curr. Sci.* **2011**, *101*, 356–371.

32. Raje, D.; Priya, P.; Krishnan, R. Macroscale hydrological modelling approach for study of large scale hydrologic impacts under climate change in Indian river basins. *Hydrol. Process.* **2013**, *28*, 1874–1889.

33. Blöschl, G.; Sivapalan, M.; Wagener, T.; Viglione, A.; Savenije, H. *Runoff Prediction in Ungauged Basins. Synthesis across Processes, Places and Scales*; Cambridge University Press: Cambridge, UK, 2013.

34. Gupta, H.V.; Perrin, C.; Blöschl, G.; Montanari, A.; Kumar, R.; Clark, M.; Andréassian, V. Large-sample hydrology: A need to balance depth with breadth. *Hydrol. Earth Syst. Sci.* **2014**, *18*, 463–477.

35. Yang, W.; Andréasson, J.; Graham, P.L.; Olsson, J.; Rosberg, J.; Wetterhall, F. Distribution-based scaling to improve usability of regional climate model projections for hydrological climate change impacts studies. *Hydrol. Res.* **2010**, *41*, 211–229.

36. Lindström, G.; Pers, C.; Rosberg, J.; Strömqvist, J.; Arheimer, B. Development and testing of the HYPE (Hydrological Predictions for the Environment) water quality model for different spatial scales. *Hydrol. Res.* **2010**, *41*, 295–319.

37. Attri, S.D.; Tyagi, A. *Climate Profile of India*; Government of India Ministry of Earth Sciences: New Delhi, India, 2010; p. 129.

38. Mall, R.K.; Singh, R.; Gupta, A.; Srinivasan, G.; Rathore, L.S. Impact of Climate Change on Indian Agriculture: A Review. *Clim. Change* **2006**, *78*, 445–478.

39. Li, L.; Xu, C.-Y.; Zhang, Z.; Jain, S.K. Validation of a new meteorological forcing data in analysis of spatial and temporal variability of precipitation in India. *Stoch. Environ. Res. Risk Assess.* **2013**, *28*, 239–252.

40. Pechlivanidis, I.G.; Arheimer, B. Large-scale hydrological modelling by using modified PUB recommendations: the India-HYPE case. *Hydrol. Earth Syst. Sci.* **2015**, *19*, 4559–4579.

41. Yatagai, A.; Kamiguchi, K.; Arakawa, O.; Hamada, A.; Yasutomi, N.; Kitoh, A. APHRODITE: Constructing a long-term daily gridded precipitation dataset for Asia based on a dense network of rain gauges. *Bull. Am. Meteorol. Soc.* **2012**, *93*, 1401–1415.

42. Yatagai, A.; Arakawa, O.; Kamiguchi, K. A 44-year daily gridded precipitation dataset for Asia based on a dense network of rain gauges. *Sola* **2009**, *5*, 137–140.

43. Yasutomi, N.; Hamada, A.; Yatagai, A. Development of a long-term daily gridded temperature dataset and its application to rain/snow discrimination of daily precipitation. *Glob. Environ. Res.* **2011**, *3*, 165–172.

44. Hazeleger, W.; Wang, X.; Severijns, C.; Ştefănescu, S.; Bintanja, R.; Sterl, A.; Wyser, K.; Semmler, T.; Yang, S.; van den Hurk, B.; van Noije, T.; van der Linden, E.; van der Wiel, K. EC-Earth V2.2: Description and validation of a new seamless earth system prediction model. *Clim. Dyn.* **2012**, *39*, 2611–2629.

45. Samuelsson, P.; Collin, G.J.; Willén, U.; Ullerstig, A.; Gollvik, S.; Hansson, U.; Jansson, C.; Kjellström, E.; Nikulin, G.; Wyser, K. The Rossby Centre Regional Climate model RCA3: Model description and performance. *Tellus* **2011**, *63A*, 4–23.

46. Moss, R.H.; Edmonds, J.A.; Hibbard, K.A.; Manning, M.R.; Rose, S.K.; van Vuuren, D.; Carter, T.R.; Emori, S.; Kainuma, M.; Kram, T.; *et al.* The next generation of scenarios for climate change research and assessment. *Nature* **2010**, *463*, 747–756.

47. Ghimire, S.; Choudhary, A.; Dimri, A.P. Assessment of the performance of CORDEX-South Asia experiments for monsoonal precipitation over the Himalayan region during present climate: Part I. *Clim. Dyn.* **2015**.

48. Allen, R.G.; Pereira, L.S.; Raes, D.; Smith, M. *Crop Evapotranspiration, Guidelines for Computing Crop Water Requirements-FAO Irrigation and Drainage Paper 56*; Food and Agriculture Organization of the United Nations (FAO): Rome, Italy, 1998.

49. Nash, J.E.; Sutcliffe, J.V. River flow forecasting through conceptual models. *J. Hydrol.* **1970**, *10*, 282–290.

50. Rana, A.; Foster, K.; Bosshard, T.; Olsson, J.; Bengtsson, L. Impact of climate change on rainfall over Mumbai using Distribution-based Scaling of Global Climate Model projections. *J. Hydrol. Reg. Stud.* **2014**, *1*, 107–128.

51. Gupta, H.V.; Kling, H.; Yilmaz, K.K.; Martinez, G.F. Decomposition of the mean squared error and NSE performance criteria: Implications for improving hydrological modelling. *J. Hydrol.* **2009**, *377*, 80–91.

52. Themeßl, M.J.; Gobiet, A.; Heinrich, G. Empirical-statistical downscaling and error correction of regional climate models and its impact on the climate change signal. *Clim. Change* **2012**, *112*, 449–468.

53. Kotlarski, S.; Bosshard, T.; Lüthi, D.; Pall, P.; Schär, C. Elevation gradients of European climate change in the regional climate model COSMO-CLM. *Clim. Change* **2012**, *112*, 189–215.

54. Cherchi, A.; Alessandri, A.; Masina, S.; Navarra, A. Effects of increased CO_2 levels on monsoons. *Clim. Dyn.* **2010**, *37*, 83–101.

55. Kripalani, R.H.; Oh, J.H.; Kulkarni, A.; Sabade, S.S.; Chaudhari, H.S. South Asian summer monsoon precipitation variability: Coupled climate model simulations and projections under IPCC AR4. *Theor. Appl. Climatol.* **2007**, *90*, 133–159.

56. Goswami, B.N.; Venugopal, V.; Sengupta, D.; Madhusoodanan, M.S.; Xavier, P.K. Increasing trend of extreme rain events over India in a warming environment. *Science* **2006**, *314*, 1442–1445.

57. Kumar, K.K.; Patwardhan, S.K.; Kulkarni, A.; Kamala, K.; Rao, K.K.; Jones, R. Future projection of Indian summer monsoon variability under climate change scenario: An assessment from CMIP5 climate models. *Curr. Sci.* **2011**, *101*, 312–326.

58. Sharmila, S.; Joseph, S.; Sahai, A.K.; Abhilash, S.; Chattopadhyay, R. Future projection of Indian summer monsoon variability under climate change scenario: An assessment from CMIP5 climate models. *Glob. Planet. Chang.* **2014**, *124*, 62–78.

59. Salvi, K.; Kannan, S.; Ghosh, S. High-resolution multisite daily rainfall projections in India with statistical downscaling for climate change impacts assessment. *J. Geophys. Res. Atmos.* **2013**, *118*, 3557–3578.

60. Ojha, R.; Kumar, D.N.; Sharma, A.; Mehrotra, R. Assessing severe drought and wet events over India in a future climate using a nested bias-correction approach. *J. Hydrol. Eng.* **2013**, *18*, 760–772.

61. Amarasinghe, U.A.; Sharma, B.R.; Aloysius, N.; Scott, C.; Smakhtin, V.; de Fraiture, C. *Spatial Variation in Water Supply and Demand across River Basins of India*; Research Report 83; International Water Management Institute: Colombo, Sri Lanka, 2004.

62. Jhajharia, D.; Shrivastava, S.K.; Sarkar, D.; Sarkar, S. Temporal characteristics of pan evaporation trends under the humid conditions of northeast India. *Agric. For. Meteorol.* **2009**, *149*, 763–770.

63. Jhajharia, D.; Dinpashoh, Y.; Kahya, E.; Choudhary, R.R.; Singh, V.P. Trends in temperature over Godavari River basin in Southern Peninsular India. *Int. J. Climatol.* **2014**, *34*, 1369–1384.

64. McVicar, T.R.; Roderick, M.L.; Donohue, R.J.; Li, L.T.; van Niel, T.G.; Thomas, A.; Grieser, J.; Jhajharia, D.; Himri, Y.; Mahowald, N.M.; *et al.* Global review and synthesis of trends in observed terrestrial near-surface wind speeds: Implications for evaporation. *J. Hydrol.* **2012**, *416–417*, 182–205.

65. Schewe, J.; Heinke, J.; Gerten, D.; Haddeland, I.; Arnell, N.W.; Clark, D.B.; Dankers, R.; Eisner, S.; Fekete, B.M.; Colón-González, F.J.; *et al.* Multimodel assessment of water scarcity under climate change. *PNAS* **2014**, *111*, 3245–3250.

66. Immerzeel, W.W.; Droogers, P.; de Jong, S.M.; Bierkens, M.F.P. Large-scale monitoring of snow cover and runoff simulation in Himalayan river basins using remote sensing. *Remote Sens. Environ.* **2009**, *113*, 40–49.

67. Masood, M.; Yeh, P.J.-F.; Hanasaki, N.; Takeuchi, K. Model study of the impacts of future climate change on the hydrology of Ganges–Brahmaputra–Meghna (GBM) basin. *Hydrol. Earth Syst. Sci. Discuss.* **2014**, *11*, 5747–5791.

68. Mukhopadhyay, B. Signature and hydrologic consequences of climate change within the upper-middle Brahmaputra Basin. *Hydrol. Process.* **2013**, *27*, 2126–2143.

69. Meenu, R.; Rehana, S.; Mujumdar, P. Assessment of hydrologic impacts of climate change in Tunga–Bhadra river basin, India with HEC-HMS and SDSM. *Hydrol. Process.* **2013**, *27*, 1572–1589.

70. Kauffeldt, A.; Halldin, S.; Rodhe, A.; Xu, C.-Y.; Westerberg, I.K. Disinformative data in large-scale hydrological modelling. *Hydrol. Earth Syst. Sci.* **2013**, *17*, 2845–2857.

71. Hagemann, S.; Chen, C.; Haerter, J.O.; Heinke, J.; Gerten, D.; Piani, C. Impact of a statistical bias correction on the projected hydrological changes obtained from three GCMs and two hydrology models. *J. Hydrometeorol.* **2011**, *12*, 556–578.

72. Pechlivanidis, I.G.; Arheimer, B.; Donnelly, C.; Hundecha, Y.; Huang, S.; Aich, V.; Samaniego, L.; Eisner, S.; Shi, P. Analysis of hydrological extremes at different hydro-climatic regimes under present and future conditions. *Clim. Chang.* **2016**. submitted.

73. Wagner, P.D.; Kumar, S.; Schneider, K. An assessment of land use change impacts on the water resources of the Mula and Mutha Rivers catchment upstream of Pune, India. *Hydrol. Earth Syst. Sci.* **2013**, *17*, 2233–2246.

Runoff and Sediment Yield Variations in Response to Precipitation Changes: A Case Study of Xichuan Watershed in the Loess Plateau, China

Tianhong Li and Yuan Gao

Abstract: The impacts of climate change on hydrological cycles and water resource distribution is particularly concerned with environmentally vulnerable areas, such as the Loess Plateau, where precipitation scarcity leads to or intensifies serious water related problems including water resource shortages, land degradation, and serious soil erosion. Based on a geographical information system (GIS), and using gauged hydrological data from 2001 to 2010, digital land-use and soil maps from 2005, a Soil and Water Assessment Tool (SWAT) model was applied to the Xichuan Watershed, a typical hilly-gullied area in the Loess Plateau, China. The relative error, coefficient of determination, and Nash-Sutcliffe coefficient were used to analyze the accuracy of runoffs and sediment yields simulated by the model. Runoff and sediment yield variations were analyzed under different precipitation scenarios. The increases in runoff and sediment with increased precipitation were greater than their decreases with reduced precipitation, and runoff was more sensitive to the variations of precipitation than was sediment yield. The coefficients of variation (C_Vs) of the runoff and sediment yield increased with increasing precipitation, and the C_V of the sediment yield was more sensitive to small rainfall. The annual runoff and sediment yield fluctuated greatly, and their variation ranges and C_Vs were large when precipitation increased by 20%. The results provide local decision makers with scientific references for water resource utilization and soil and water conservation.

Reprinted from *Water*. Cite as: Li, T.; Gao, Y. Runoff and Sediment Yield Variations in Response to Precipitation Changes: A Case Study of Xichuan Watershed in the Loess Plateau, China. *Water* **2015**, *7*, 5638–5656.

1. Introduction

Climate change as a result of both natural factors and human activities is altering the earth's hydrologic cycles to various degrees [1,2]. Climate change affects hydrology mainly through changes in precipitation, temperature, and evaporation [3,4], and it subsequently influences the temporal-spatial distributions of runoff and sediment, as well as the patterns of runoff and sediment transport [5]. The impacts of climate change on water resources and the hydrologic cycles have long been a focus of the international community [6,7]. Research on this issue began as early as the 1980s. In 1985, the World Meteorological Organization (WMO) published a summary report of their study of the impacts of climate change on hydrology and water resources and proposed several evaluation and test methods. In 1987, the WMO proposed analyzing the sensitivity of hydrology and water resources to climate change. This issue was also discussed in the 2007 international conference of the International Union of Geodesy and Geophysics (IUGG). The Intergovernmental Panel on Climate Change (IPCC) of the United Nations analyzed the impacts of climate change on hydrology and water resources from 1990 to 2007. In its technical report [8], the IPCC highlighted that the global and regional water resource problems caused by climate change are crucial issues. Changes in precipitation and temperature have significant effects on runoff and water availability, particularly in semiarid and arid regions [9].

China has always considered the impacts of climate change on water resources to be important and has actively carried out a series of scientific studies to support research on the impact of the changing environment (due to global changes and human activities) on water cycles [5,10]. For instance, the *National Planning Outline for Mid- and Long-term Scientific and Technological Development (2006–2020)* issued by the State Council of China in 2006 pointed out that research on the impacts of global climate change in China is a focus, with special emphasis on the impacts of climate change on hydrologic cycles and regional water resources, especially in arid regions with fragile ecological environments [11].

Currently, studies of the impacts of climate change on runoff and sediment mainly focus on two aspects. Some studies analyze the changes in the temporal-spatial distributions of runoff and sediment and the patterns of runoff and sediment transport that are caused by changes in climate factors, such as precipitation and temperature, whereas other studies analyze the trends of the changes in runoff and sediment under future climate change scenarios. The main method for quantitatively evaluating and studying the impacts of climate change on runoff and sediment is to use watershed hydrological models. The most commonly used models are statistical regression models, water balance models, and distributed physical models [9,12]. Of these models, the Soil and Water Assessment Tool (SWAT) [13], which was developed by the US Department of Agriculture in the 1990s, has been widely applied to watersheds around the world [14–20]. There are two types of predicted future climate change scenarios. First, changes in temperature, precipitation, and evaporation are hypothesized based on the trends and ranges of the meteorological changes in the study area, as well as specialized knowledge, experience, and the time-series statistical analysis method, which is easy to design and apply [21–23]; Second, different climate change scenarios can be simulated using models, such as the General Circulation Models [19,24].

Previous studies have shown that the precipitation in the Yellow River Basin has decreased significantly since the 1970s [10,25] although variation trends may differ in sub-basins. Precipitation is the main source of runoff and one of the driving factors of soil and water losses in the Loess Plateau [26,27], where water resources are scarce. Therefore, in the context of global climate change, studying the impacts of changes in precipitation on the runoff and sediment production in the Yellow River Basin is important for the sustainable utilization of water resources. Most previous studies of the impacts of changes in precipitation on water resources in the Yellow River Basin have focused on the entire basin [28,29] or the basin at relatively large scales [30–32]. Relatively few studies have focused on small watersheds. In addition, most studies have focused on the impact of precipitation on the runoff and have rarely investigated the impact of precipitation on the sediment yield. In fact, high sediment content is an important and unique characteristic of the rivers in the Loess Plateau, China. Sediment transport requires a considerable amount of water [33] and competes with other water uses. Thus, it is imperative to consider sediment when studying the water resource problems of these rivers.

This study used the Xichuan Watershed, a typical small basin in the hilly and gully area in the Loess Plateau, as a target area, and used ArcGIS, MATLAB, and SPAW [34,35] to process observed meteorological and hydrological data. Then, a localized SWAT model was constructed, calibrated, and validated. Using the SWAT model under different precipitation scenarios, the study quantitatively predicted the impacts of changes in precipitation on the runoff and sediment yield in this small watershed and analyzed the characteristics of the changes in runoff and sediment production with the aim of providing a scientific basis for the management and sustainable utilization of water resources in basins that are similar to the study area.

2. Methodology

Based on the spatial and attribute data, including meteorological data, hydrological data, soil map, land use map, and a digital elevation model (DEM), this study investigated the characteristics of the changes in the precipitation, runoff, and sediment yield in the study area. Spatial and attribute databases of the SWAT model were developed. After determining the parameters of the SWAT model and verifying the predicted results from the model, we quantitatively analyzed the impacts of changes in precipitation on the runoff and sediment production in the study area using precipitation change scenarios. Figure 1 shows the technical workflow of the study.

This study used AVSWAT, developed by integrating the SWAT into ArcView for the analysis. AVSWAT has powerful spatial analysis and processing functions and is convenient to use. The SWAT model consists of three sub-models: the hydrological process sub-model, the soil erosion sub-model and the pollution load sub-model. This study mainly uses the hydrological process and the soil erosion sub-models.

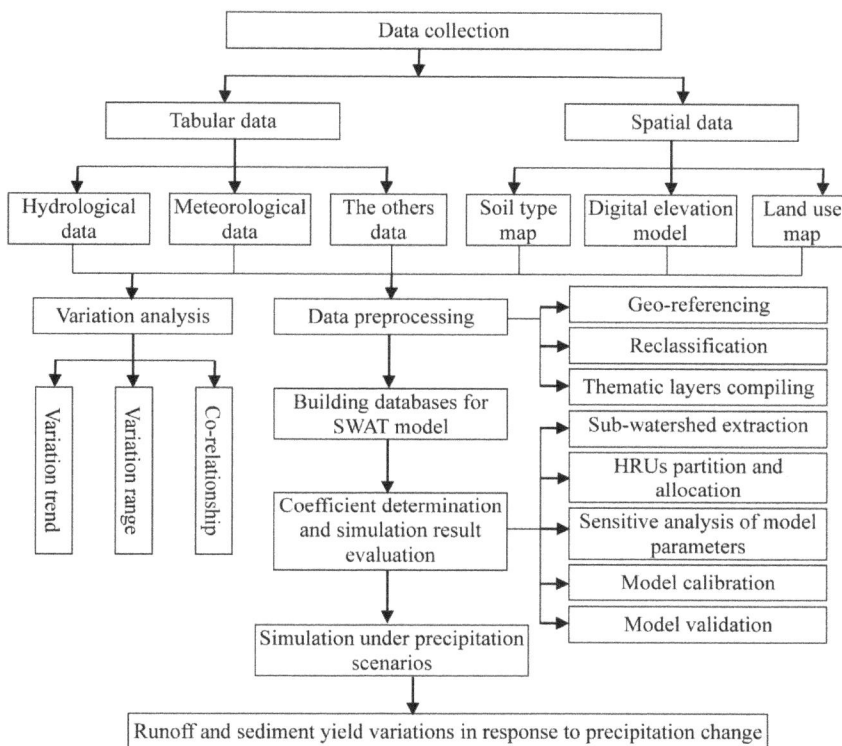

Figure 1. Technical workflow of this study.

2.1. Study Area

The Xichuan River is a tributary of the Yanhe River (a tributary of the Yellow River) with a total length of 61.5 km. The Xichuan Watershed is located west of the Yanhe River Basin between 108°50′ E and 109°20′ E and between 36°30′ N and 36°45′ N, covering an area of 801 km^2 [36]. The mean runoff of the watershed was 169.04 × 10^6 m^3 from 2001 to 2010. The river originates in Caofeng Village, Zhidan Town in Zhidan County and flows past Xihekou Village, Zhuanyaowan Town and Gaoqiao Village in Ansai County and Zaoyuan Village in the Baota District and eventually flows into the Yanhe River near Shifogou in the Baota District. The Zaoyuan Hydrological Station (ZHS) is located 13 km upstream from the mouth of the Xichuan River, and it controls 90% area (719 km^2) of the whole watershed (Figure 2).

167

The Xichuan Watershed has a continental monsoon climate where winters are cold and dry with little precipitation, whereas summers are warm with abundant precipitation. Precipitation is unevenly distributed and mainly concentrated in the summer and fall, accounting for 54.3% and 27.7% of the total annual precipitation. Floods in this watershed have relatively short durations, rising and falling suddenly with high sediment concentrations [37].

The soil types in this watershed include yellow loessial soil, red clay soil, alluvial soil, and dark loessial soil. The yellow loessial soil, developed from the parent loess, is the main soil type, covering more than 80% of the total basin area. The vegetation coverage of the watershed is very low, belonging to the forest steppe zone. Both natural and artificial vegetation types are present, mainly consisting of crops, evergreen coniferous forests, deciduous coniferous forests, deciduous broad-leaved forests, shrubs, and grasslands.

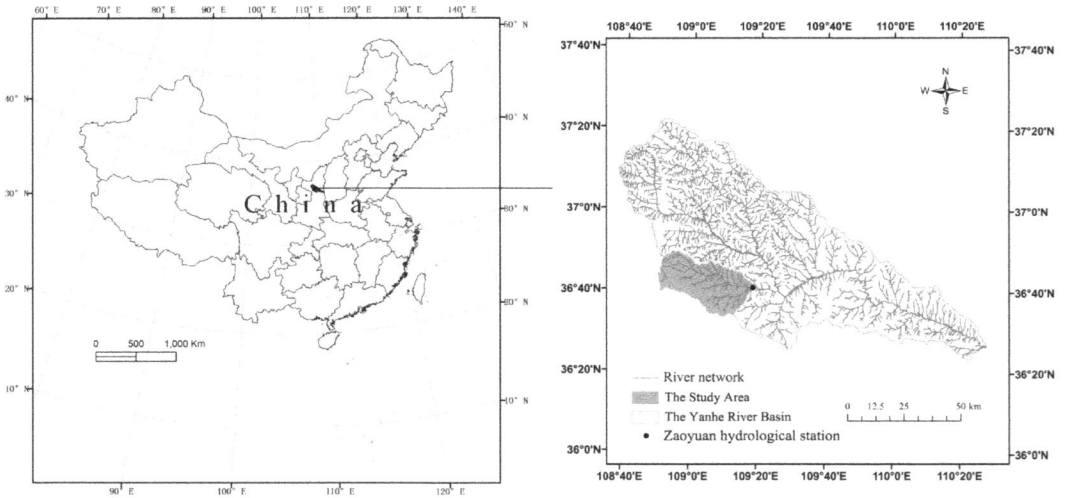

Figure 2. Location of the study area.

More than 80% of the basin area suffers soil erosion by water. The multi-year mean sediment discharge is 1330.2×10^4 t/a. Since the 1970s, a series of water and soil conservation and ecological construction projects have been implemented and have substantially improved the ecological environment in the watershed [38].

2.2. Data and Data Preprocessing

The main data used in this study included Digital Elevation Model (DEM) data, a land use map, a soil type map, precipitation data, temperature data, and the boundary of the Yanhe River basin. Table 1 lists the data descriptions and sources.

168

Table 1. List of data that were used in this study.

Data Type	Temporal/Spatial Resolution	Source
DEM data	Grid format, 30 m/grid	Data Application Environment Sharing Platform of the Chinese Academy of Sciences
Land use map	At the scale of 1:100,000, compiled at 2005	Data Application Environment Sharing Platform of the Chinese Academy of Sciences
Soil type map	At the scale of 1:1,000,000, compiled at 2005	Data Application Environment Sharing Platform of the Chinese Academy of Sciences
Meteorological data	Daily precipitation, daily maximum temperature and daily minimum temperature between 1990 and 2010	China Meteorological Data Sharing Service Website
Runoff and sediment yield	Monthly runoff and sediment yield between 2001 and 2010	The Zaoyuan Meteorological Station in the Yan'an City

The data required for the SWAT model included geospatial data, a non-spatial attribute database, meteorological data, and hydrological and sediment data for model calibration and verification. Spatial data processing, grid calculations, and interpolations were conducted using ArcGIS, and the statistical analysis was performed using Excel. All spatial data used in the SWAT were converted to the Albers equal-area conic projection. The whole area was divided into 31 sub-basins using the DEM and each sub-basin contained 5–16 hydrological response units (HRUs), which is the basic unit in SWAT model.

2.3. SWAT Model Development

2.3.1. Model Construction

The SWAT model requires the land use classification scheme developed by the US Geological Survey (USGS). It also requires the auxiliary land use attributes with parameters provided by the USGS. The land use map used in this study had to be reclassified to meet these requirements. After reclassification, the main land use types included farmland (39.30%), typical grassland (32.60%), meadow grassland (13.09%), deciduous coniferous forest and deciduous broad-leaved forest (8.90%), bush wood (5.05%), evergreen forest (0.86%), barren land (0.09%), water body (0.05%), and rural villages (0.05%) [35].

The soil data included the spatial distribution and physical and chemical attributes of the soils. The soil map was produced based on the 2005 soil survey of the study area. The physical attributes of the soils mainly included the thicknesses, silt contents, clay contents, bulk densities, organic carbon contents, effective water contents, saturated hydraulic conductivities, and the available field capacities. These attributes control the movement of the water and air in the soil and have an important role in the water cycles. This study established a users' soil parameter database based on these characteristics.

The soils were divided into four hydrologic groups. For the same precipitation and surface conditions, soils with similar runoff production capacities were classified into a single hydrologic group [39].

The wet density of a soil, the available effective water in a soil layer and the saturated hydraulic conductivity coefficient can be calculated using the SPAW model [34,35]. The SPAW model is a daily hydrologic budget model for agricultural fields. It also includes a routine for the daily water budgets of inundated ponds and wetlands that utilizes the field hydrology of the watershed.

The soil erosion factor (K) is often used to evaluate soil erodibility. K is calculated based on the organic carbon and particle compositions of the soil using the method proposed by Williams [40].

The observed meteorological input data mainly included the precipitation, daily maximum and minimum temperature data from 2001 to 2010. The SWAT model includes a built-in weather generator. If some data were not available, the weather generator simulated daily meteorological data based on multi-year monthly mean data that were provided in advance. The "pcpSTAT" and "dew02" procedures were used to calculate daily precipitation and temperature to obtain the related parameters and generate the weather data that were needed for simulations.

The measured runoff and sediment data were collected at ZHS (Figure 2) from 2001 to 2010. They were used in sensitivity analysis and parameters calibration. The measured daily precipitation data were used to simulate daily runoff using the Soil Conservation Service (SCS) curve method [39]. The potential evaporation was derived using the Penman-Monteith method [41]. The variable storage coefficient method [42] was used in the river channel routing simulation.

2.3.2. Sensitivity Analysis, Validation, and Testing of the SWAT Parameters

The parameter sensitivity analysis module was used to analyze the sensitivities of the parameters in the runoff and sediment simulations. This module uses the Latin hypercube one-factor-at-a-time (LH-OAT) method [43]. The objective of this analysis is to analyze and determine which input parameters have the most significant impacts on the output when their values are changed. Important parameters are selected to highlight their impacts on the simulation and to reduce the time that is needed for parameter adjustment. In this study, the simulated runoff and sediment yield were compared with actual gauged data at ZHS (Figure 2). The important factors affecting the precision of the simulation in the watershed were determined after analyzing the sensitivity of each parameter.

The runoff and sediment yield parameters were calibrated in sequence. Three indexes, including the relative error (Re) [44], the coefficient of determination (R^2) [45] and the Nash-Sutcliffe coefficient (Ens) [46] were chosen to statistically test the accuracy of the calibrated and validated runoff and sediment yield outputs. If $Re = 0$, the model prediction is the same as the observed data. If $Re > 0$, the model prediction is larger than the observation. If $Re < 0$, the model prediction is less than the observed value. R^2 was obtained from the linear regression in Microsoft Excel. The larger the R^2 value, the better simulation of the model. If the value of Ens is greater than 0.75, the simulation is excellent. If Ens is between 0.36 and 0.75, the simulation is satisfactory, and if Ens is less than 0.36, the simulation is unsatisfactory.

2.4. Precipitation Scenarios

Based on land use maps in 2005 and the climate conditions from 2001 to 2010, the calibrated and verified SWAT model was used to simulate the impacts of precipitation on the runoff and sediment yield by altering the input climate conditions (precipitation). Spatial variability in precipitation was not considered in the simulations because the study area is a small watershed with limited precipitation stations.

Precipitation scenarios were determined based on the variation characteristics of the precipitation in this area. During the study period, the annual precipitation did not show significant increasing or decreasing trend in this area. A previous study [47] also showed no significant increasing or decreasing trend in the Yanhe River basin. The mean annual precipitation in the study area from 2001 to 2010 was 514 mm. Thus, we considered four precipitation change scenarios:

(1)   the annual mean precipitation increases by 20%, *i.e.*, 617 mm;
(2)   the annual mean precipitation increases by 10%, *i.e.*, 565 mm;
(3)   the annual mean precipitation decreases by 10%, *i.e.*, 462 mm; and
(4)   the annual mean precipitation decreases by 20%, *i.e.*, 411 mm.

2.5. Variance Analysis

The coefficient of variation (Cv) was used to reflect the inter-annual changes in the precipitation and runoff. It is calculated using the following equation:

$$Cv = \frac{SD}{M} \tag{1}$$

where SD is the standard deviation of a variable and M is the average value of the variable. The greater the value of Cv of precipitation or runoff, the greater the extent of the inter-annual change in the precipitation or runoff is, and the possibility of occurrences of floods or droughts increases. The smaller the value of Cv of the inter-annual precipitation or runoff is, the smaller the extent of the inter-annual change in the precipitation or runoff is, which is more beneficial to the utilization of water resources. Cv also reflects the characteristics of the inter-annual change in the sediment yield. The greater the value of Cv of the sediment yield indicates that the sediment yield changed greatly, and disasters such as soil erosion are more common. On the other hand, the smaller the value of Cv of sediment yield is, the smaller changing extent of the sediment yield is, which is more beneficial to water and soil conservation.

The changing rate (CR) is another parameter that expresses the changes in the runoff and sediment yield:

$$CR = \frac{X_i - X_0}{X_0} \times 100\% \tag{2}$$

where X_i represents the simulated annual mean runoff or sediment yield under the ith precipitation scheme, and X_0 represents the simulated annual mean runoff or sediment yield under the actual conditions.

3. Results

3.1. Characteristics of the Variations of Runoff and Sediment Yield

The Xichuan Watershed is in a continental monsoon climate region. The runoff mainly originates from precipitation and is thus significantly affected by precipitation. With the data measured at the ZHS between 2001 and 2010, the changes in the runoff and sediment yield in the watershed during this 10-year period were analyzed and compared with precipitation changes.

Figure 3 shows the distributions of the monthly mean precipitation and runoff between 2001 and 2010. The precipitation and runoff are both concentrated in the flood season (from June to September). The most intense monthly precipitation occurs in July, which accounts for 20.4% of the annual total, and the maximum monthly runoff occurs in August, accounting for 20.1% of the total annual runoff. The minimum monthly precipitation occurs in December (0.75% of the total annual precipitation), and the minimum monthly runoff occurs in January (3.38% of the total annual runoff). The seasonal precipitation and runoff distributions in the watershed are extremely uneven. Precipitation and runoff are mainly concentrated in the summer and fall. Summer precipitation accounts for 54.3% of the total annual precipitation, and summer runoff accounts for 45.7% of the total annual runoff. Fall precipitation accounts for 27.7% of the total annual precipitation, and fall runoff accounts for 23.3% of the total annual runoff. Figure 3 also shows that, during the spring and winter, the changes in the precipitation and runoff are gentle, and the precipitation and runoff are relatively stable. The runoff is slightly affected by precipitation during this period. Generally speaking, the runoff in the Xichuan Watershed is mainly generated by the base flows in the spring and winter and by precipitation in the summer and spring.

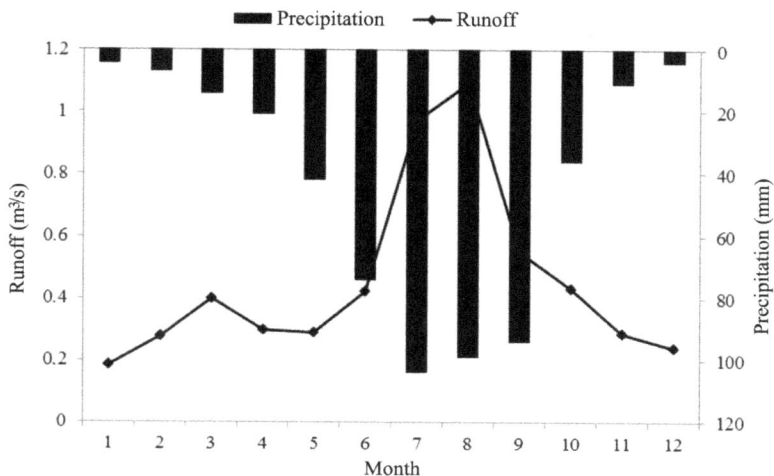

Figure 3. Comparison of the monthly measured runoff and precipitation.

Figure 4 shows the distributions of the monthly mean precipitation and sediment yields between 2001 and 2010. The annual sediment yield is generally consistent with the runoff pattern. The sediment yield is mainly concentrated in flood seasons. The maximum sediment output (16,406 t/m) occurs in August. Both the maximum runoff and the maximum sediment yield occur one month after the maximum precipitation. In the Loess Plateau, sediment production is usually accompanied by runoff with eroding capability. That the maximum peaks of runoff and sediment yield occurred within the same month is understandable. Besides precipitation, runoff and sediment generation processes are related to many other factors, including soil properties, topography, and land cover change. These factors are heavily influenced by human activities. In this watershed, a series of infrastructure reforms, water and soil conservation, and ecological projects, especially the well-known Grain-to-Green Program started in 1998 [48], were undertaken before and during the study period. Specific practices, including changing slopes into terrace fields, afforestation in barren land, constructing silt dams and reservoirs, returning farmland to forest or grassland *etc.*, could retard the processes of runoff generation and soil erosion [49].

In general, high-intensity precipitation will affect the runoff and sediment yield in a basin only after time has passed since the precipitation. The sediment yield in the watershed is mainly concentrated in the summer because high-intensity precipitation frequently occurs in that season, and precipitation is the main source of the runoff,

173

which is main driving force of sediment output; therefore, the runoff and sediment exhibit significant seasonal variations [9].

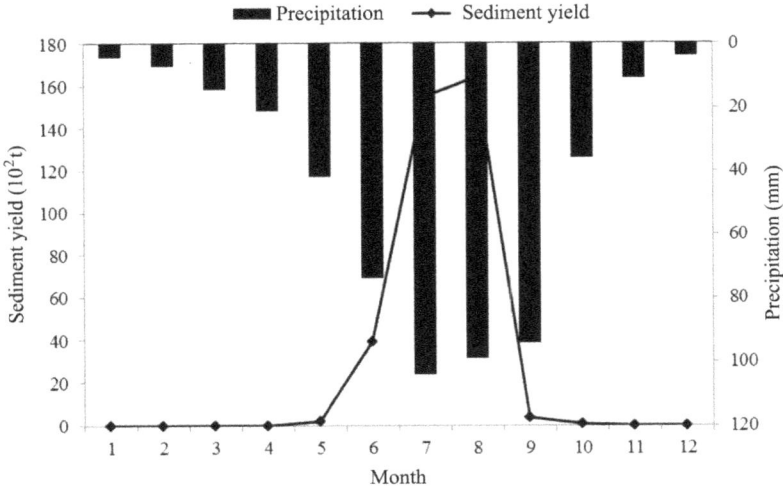

Figure 4. Comparison of the monthly measured sediment yield and precipitation.

Inter-annual runoff changes are often affected by factors such as climate change, human activities, and changes in the underlying surface conditions. Table 2 shows the statistical characteristics of the precipitation and runoff at ZHS. The annual mean precipitation and runoff at the ZHS from 2001 to 2010 are 513.38 mm and 169.04×10^6 m^3, respectively. The maximum precipitation at ZHS (634.3 mm) occurred in 2007, and the maximum runoff (235.89×10^6 m^3) occurred in 2002. The minimum precipitation (441.6 mm) and minimum runoff (95.08×10^6 m^3) at ZHS both occurred in 2008. The Cv values of the inter-annual precipitation and runoff between 2001 and 2010 are both relatively small (0.13 and 0.29, respectively). Therefore, in this 10-year period, the variation in precipitation in the watershed was insignificant, and the water resources were relatively stable, favoring the use of water resources.

The sediment discharge is related to many factors, such as the topography and landforms of the basin, vegetation cover, precipitation, and precipitation intensity. In recent decades, the changes in the sediment yield in the Xichuan Watershed were relatively complicated due to the impacts of climate change and human activities. The mean sediment yield at ZHS between 2001 and 2010 is 1330.2×10^4 t (Table 2). During 2001 to 2010, the sediment discharge changed significantly, and soil erosion was relatively severe. The Cv value of the inter-annual sediment discharge at ZHS between 2001 and 2010 is relatively large (0.84). The maximum sediment yield at ZHS between 2001 and 2010 (3161.4×10^4 t) occurred in the same year as the maximum runoff (2002), and the minimum sediment yield (122.7×10^4 t) occurred in 2006.

174

Table 2. Annual characteristics of precipitation, runoff and sediment yield during 2001–2010.

Variable	Mean Value	Cv	Maximum Value		Minimum Value	
			Value	Year of Occurrence	Value	Year of Occurrence
Precipitation	513.38 mm	0.13	634.30 mm	2007	441.60 mm	2008
Runoff	169.04×10^6 m^3	0.29	235.89×10^6 m^3	2002	95.08×10^6 m^3	2008
Sediment	$1,330.20 \times 10^4$ t	0.84	$3,161.40 \times 10^4$ t	2002	122.70×10^4 t	2006

3.2. SWAT Calibration and Validation Results

In this study, the year of 2001 was used as the warming period, data in 2002–2006 was used for model calibration, and data in 2007–2010 was used for model validation. The calibration procedure follows the method introduced by [36] with the following steps:

Step 1: model initiation—run the model and read the required data and parameters;

Step 2: runoff simulation—produce the simulated monthly runoffs and compare them to the actual observed values;

Step 3: if runoff simulation reaches the condition of $-20\% < Re < 20\%$, $R^2 > 0.6$, and $Ens > 0.5$, then go to the next step for sediment yield simulation; otherwise, the adjust the parameters "Base flow recession constant," "Snow pack temperature lag factor," "Soil evaporation compensation factor," "Available water capacity," "Threshold depth of water in the shallow aquifer required for return flow to occur," and "Groundwater 'revap' coefficient" and go back to step 2;

Step 4: sediment yield simulation—produce the simulated monthly sediment yields and compare them to the actual observed values;

Step 5: if sediment yield simulation reaches the condition of $-20\% < Re < 20\%$, $R^2 > 0.6$, and $Ens > 0.5$, then the simulation is successfully ended; otherwise, adjust the parameters "USLE equation support practice factor," "Linear parameters for calculating the channel sediment rooting," and "Peak rate adjustment factor for sediment routing in the main channel" and go back to step 4.

Figures 5 and 6 show comparisons between the observed and simulated monthly runoffs and sediment yield at ZHS during the model calibration period. Figures 7 and 8 show the results during the model validation period. Table 3 compares the parameters of the simulated monthly runoff and sediment yield in the model calibration and validation periods.

175

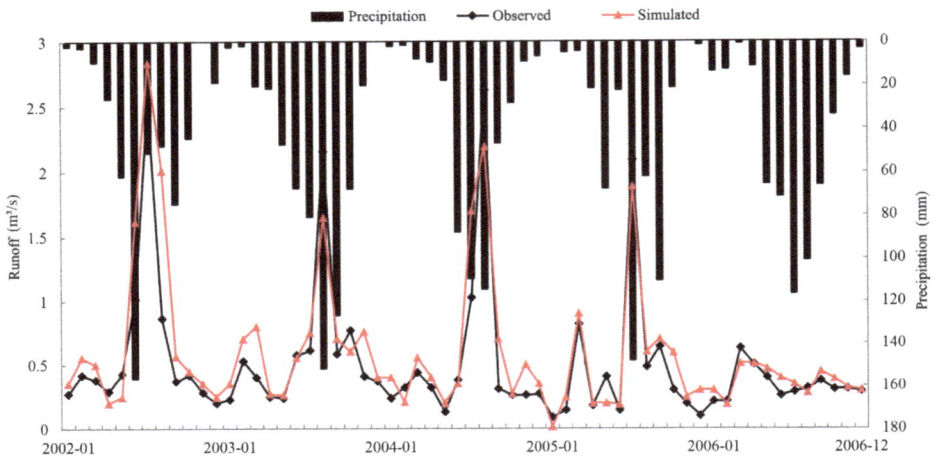

Figure 5. Comparison between the simulated and observed monthly runoff in model calibration.

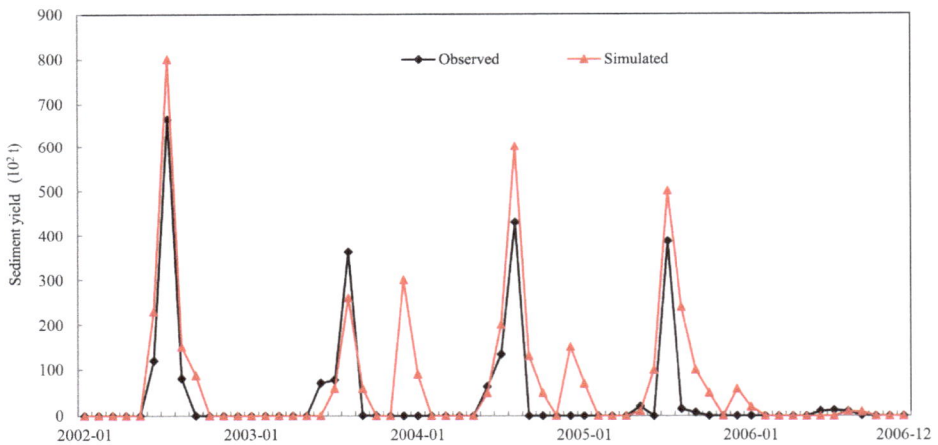

Figure 6. Comparison between the simulated and observed monthly sediment yield in model calibration.

Table 3. Evaluation results of the SWAT model performance.

Simulation Period	Runoff			Sediment Yield		
	Re	R^2	Ens	Re	R^2	Ens
Calibration (2002–2006)	9.10%	0.79	0.73	14.20%	0.78	0.67
Verification (2007–2010)	11.20%	0.88	0.82	17.50%	0.83	0.71

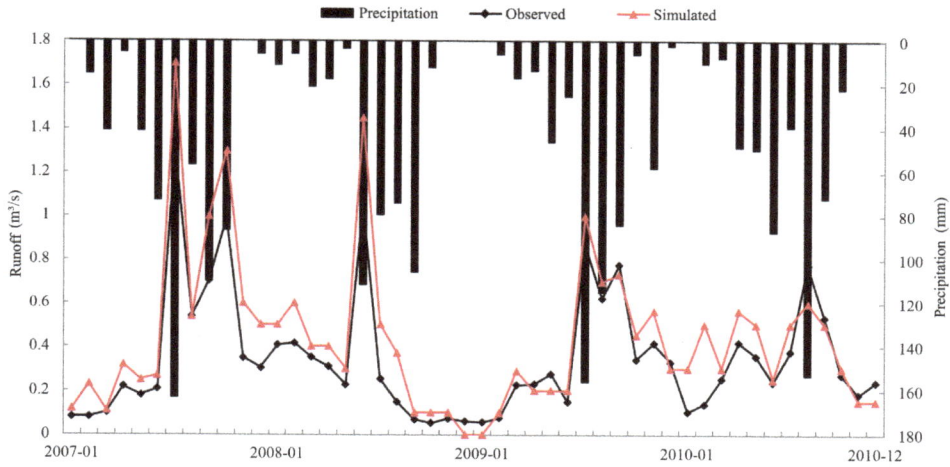

Figure 7. Comparison between the simulated and observed monthly runoff in model validation.

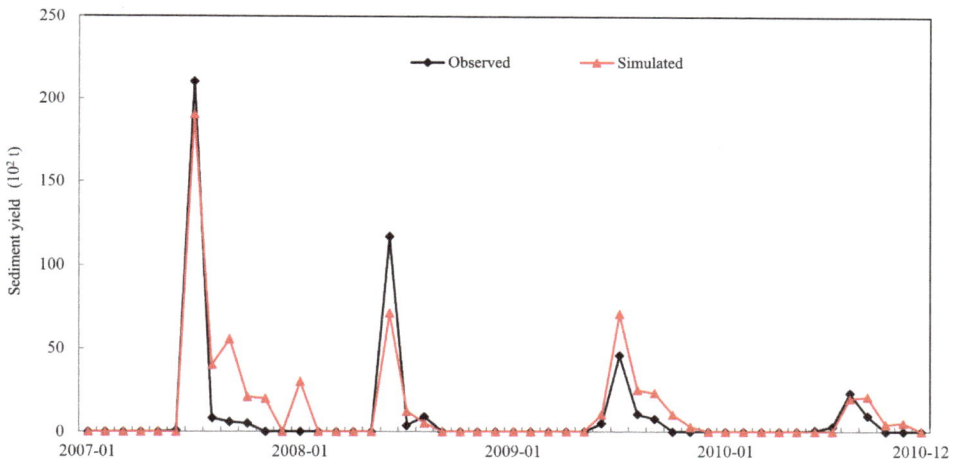

Figure 8. Comparison between the simulated and observed monthly sediment yield in model validation.

As illustrated in Figures 5–8, the difference between simulated runoff/sediment yield and observed values is smaller during the calibration period than during the validation period, while the variation trend of runoff/sediment yield is more consistent with the trend of observed data during the validation period than in the calibration period. These differences can also be supported by the statistics in Table 3. Table 3 also shows that in the model validation period, the values of *Re* between the simulated and observed monthly runoffs and between the simulated and observed

177

monthly sediment yields are 11.2% and 17.5%, respectively, and the values of R^2 are 0.88 and 0.83, respectively. The Ens values of runoffs and sediment yields are 0.82 and 0.71, respectively. The SWAT simulated values generally reflect the actual changes in the runoff and sediment yield, and the SWAT model can be used for the subsequent scenario analysis.

3.3. Responses of Runoff and Sediment Yield to Precipitation Changes

We also simulated the runoff and sediment yields under the four precipitation scenarios described in Section 2.4. Keeping the other inputs the same, the four precipitation scenarios were input to the validated SWAT model, and the daily runoff and sediment yield were simulated for the year of 2002 to 2010. Table 4 shows the nine-year (2002–2010) mean values. Figure 9 shows the trends and changes in the runoff and sediment yield under these precipitation scenarios.

Table 4. Responses of annual runoff and sediment yield to precipitation changes.

Simulated Item	Compared Value	P	P (1% + 20%)	P (1% + 10%)	P (1% − 10%)	P (1% − 20%)
Runoff	Simulated value (m^3)	156.14	207.20	184.81	135.28	112.58
Sediment yield	Simulated value (10^4 t)	101.09	120.49	112.87	90.93	85.04

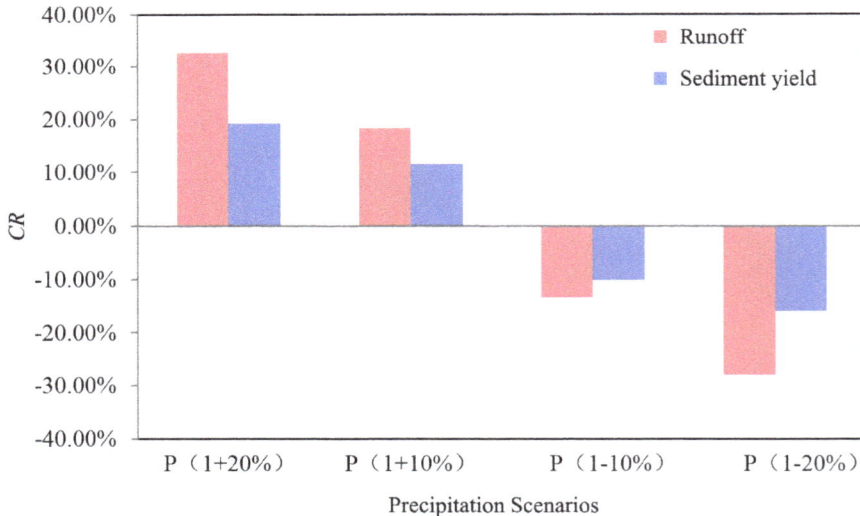

Figure 9. Changing rate of annual runoff and sediment yield change to precipitation change.

178

The results show that the runoff and sediment yield in this watershed under the four precipitation scenarios have the following characteristics:

(1)   The runoff and sediment yield increase with increasing precipitation and decrease with decreasing precipitation, which is consistent with the actual situation. Precipitation has a direct impact on runoff, and sediment is transported by runoff. Therefore, the trends of the changes in the precipitation, runoff, and sediment are similar.

(2)   When precipitation increases by 10%, the runoff and sediment yield increase by 18.36% and 11.54%, respectively. When precipitation decreases by 10%, the runoff and sediment yield decrease by 13.36% and 10.05%, respectively. The increases in the runoff and sediment yield are greater than the decreases in the runoff and sediment yield. The change in the runoff with precipitation is greater than the change in the sediment yield with precipitation. Therefore, precipitation has a more significant impact on the runoff than the sediment yield. The runoff generated by precipitation is only one of several factors that affect sediment production and sediment production may also be affected by other factors, such as vegetation cover, soil bulk density and land use changes.

(3)   When precipitation increases by 20%, the runoff and sediment yield increase by 32.7% and 19.20%, respectively. Thus, water resources will become relatively abundant when the annual precipitation intensity is relatively high, so it will be necessary to focus on preventing floods and sediment loss. When precipitation decreases by 20%, the runoff and sediment yield decrease by 27.9% and 15.88%, respectively. In these cases, water resources will be relatively scarce and it is necessary to take measures to prevent and combat droughts.

Based on the simulation results, the Cv values of the annual runoff and sediment yield are statistically calculated for the four precipitation scenarios. Figure 10 shows the trends and changes in the Cv values of the annual runoff and sediment yield for the precipitation scenarios.

The Cv values of the annual runoff and sediment yield both decrease with decreasing precipitation. The Cv values vary between 0.18 and 0.46 for the annual runoff, and vary between 0.51 and 0.98 for the sediment yield. Therefore, the Cv values of the annual runoff are smaller than those of the sediment yield. The decreased Cv value of the sediment yield with decreasing precipitation is more apparent than its increase with increasing precipitation, indicating that the Cv of the sediment yield is more sensitive to a decrease in precipitation. When precipitation increases by 20%, the Cv value of the annual runoff is 0.46 and that of the annual sediment yield is 0.98, which also demonstrates that when the precipitation intensity is relatively high, the annual runoff and sediment yield fluctuate significantly, and floods and soil erosion will occur frequently, which will be detrimental to the utilization and management of the water resources in the watershed.

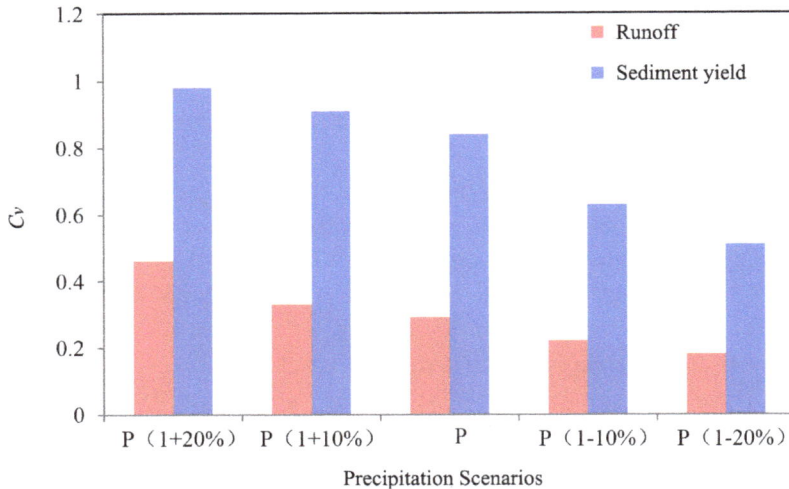

Figure 10. Coefficients of variation of annual runoff and sediment yield to precipitation changes.

4. Discussion

By studying the impacts of different precipitation scenarios on runoff and sediment yield in the Xichuan Watershed, we found that the runoff and sediment yield in the basin increase with increasing precipitation and decrease with decreasing precipitation and that the increase is more significant than the decrease. Precipitation has a more significant impact on runoff than on sediment yield. However, researchers have not yet reached a consensus about whether runoff and sediment production is more sensitive to a decrease or an increase in precipitation, as well as whether runoff or sediment production is more sensitive to changes in precipitation.

In a study of the response of runoff production in the Fox Basin in Illinois (US) to changes in precipitation, Elias [19] found that runoff production is more sensitive to increases in precipitation. In an investigation of the impact of climate change on runoff and sediment production in the purple, hilly area of Sichuan Province, China, Zeng et al. [50] found that the annual mean runoff and sediment yield increased with increasing precipitation. For the same change in precipitation, the percentage change in the sediment yield was almost twice the percentage change in the runoff. The runoff and sediment yield were more sensitive to decreases in precipitation than to increases, and the changes in the runoff and sediment yield with decreasing precipitation were more significant. The surface conditions of an area can thus affect the impact of precipitation on the runoff and sediment production in the area. In addition, another important parameter of climate change, namely temperature, will affect runoff variations. In the parameter sensitive analysis, the snow pack temperature factor is the second most sensitive parameter in the SMAT model [36], which means that melting snow is an important source of precipitation in the study area. Xia et al. [51] also discovered that when the temperature varied between 0 °C and 1 °C, the monthly runoff in the Hanjiang River Basin increased with increasing precipitation more than it decreased with decreasing precipitation. However, with increasing temperature, the extent to which the monthly mean runoff increased with increasing precipitation gradually decreased, whereas the extent to which it decreased with decreasing precipitation gradually increased. This might be attributed to the fact that an increase in temperature had a greater impact than precipitation on evapotranspiration in the basin.

This study uses the SWAT model to simulate changes in the runoff and sediment yield under several precipitation scenarios and preliminarily reveals the characteristics of the changes in the runoff and sediment yield in a small basin typical of the hill and gully area of the Loess Plateau. However, several problems in this study merit further research and investigation. (1) Several climate factors have impacts on runoff and sediment production. However, this study only considers the impact of precipitation and does not consider the impact of changes in temperature on the runoff and sediment production. Evapotranspiration is an important factor that affects surface runoff. Due to the lack of evapotranspiration data for the basin, it is not possible to calibrate and verify the evapotranspiration in the basin. Therefore, this study does not investigate the changes in evapotranspiration in the basin that are caused by changes in precipitation. Additional studies should be conducted to investigate the impact of other factors (e.g., temperature, evapotranspiration) on the runoff and sediment production; (2) This study did not consider the spatial distribution of precipitation that will bring uncertainty to the results of model simulation. Because the study area only covers 719 km^2, the assumption of a uniform precipitation is acceptable. When the method is extended to large river basins, the spatial distribution of precipitation should be considered. Since the HRU is the basic spatial unit in the simulation, the spatial distribution of precipitation can be considered by inputting different precipitation data for different HRUs if more precipitation stations are available. (3) Due to the data availability, the precipitation scenarios were set using the empirical method, a method used by other researchers in this situation [21,23]. In future studies, efforts should be made to collect basic data and use climate output models to predict precipitation to conduct in-depth investigations on the trends of future climate change.

5. Conclusions

A SWAT model was parameterized in the Xichuan Watershed, a typical hilly and gully loess area in the Loess Plateau, China. The variations of runoff and sediment yield were simulated using the calibrated SWAT model and scenario analyses. Based on the simulation, we found that the increases of runoff and sediment yield with increasing precipitation are more apparent than their decreases with decreasing precipitation. Precipitation has a more significant impact on runoff than on sediment yield. The Cv values of the annual runoff and sediment yield both increase with increasing precipitation. However, the Cv value of the annual runoff is relatively smaller than the Cv value of the sediment yield when precipitation increases, and the Cv value of sediment yield is more sensitive to a decrease in precipitation than is that of runoff. The different characteristics of variations in runoff and sediment yield suggests proper strategies in the utilization and management of the water resources in this watershed.

Acknowledgments: This work was supported by the National Natural Science Foundation of China with Grant No. 50979003. Support from the Collaborative Innovation Center for Regional Environmental Quality is also acknowledged. Special thanks also extend to the editors and all of the anonymous reviewers.

Author Contributions: Tianhong Li conceived and designed the research, and wrote the paper; Yuan Gao analyzed the data, localized SWAT model and performed scenario analysis.

Conflicts of Interest: The authors declare no conflict of interest.

References

1. Allen, M.R.; Ingram, W.J. Constrains on future changes in climate and the hydrologic cycle. *Nature* **2002**, *419*, 224–232.
2. Oki, T.; Kanae, S. Global hydrological cycles and world water resources. *Science* **2006**, *313*, 1068–1072.
3. Dong, L.H.; Xiong, L.H.; Yu, K.X.; Li, S. Research advances in effects of climate change and human activities on hydrology. *Adv. Water Sci.* **2012**, *2*, 278–285. (In Chinese)
4. Trenberth, K.E. Conceptual framework for changes of extremes of the hydrological cycle with climate change. In *Weather and Climate Extremes—Changes, Variations and a Perspective from the Insurance Industry*; Springer Netherlands: Dordrecht, The Netherlands, 1999; pp. 327–339.
5. Zhang, J.Y.; Wang, G.Q. *Research on Impacts of Climate Change on Hydrology and Water Resources*; Science Press: Beijing, China, 2007. (In Chinese)
6. Piao, S.L.; Ciais, P.; Huang, Y.; Shen, Z.; Peng, S.; Li, J.; Zhou, L.; Liu, H.; Ma, Y.; Ding, Y.; *et al.* The impacts of climate change on water resources and agriculture in China. *Nature* **2010**, *467*, 43–51.
7. Barnett, T.P.; Adam, J.C.; Lettenmaier, D.P. Potential impacts of a warming climate on water availability in snow-dominated regions. *Nature* **2005**, *438*, 303–309.

8. IPCC. *Climate Change 2007: Impacts, Adaptation and Vulnerability Contribution of Working Group 11 to the Fourth Assessment Report of the Intergovernmental Panel on Climate Change*; Cambridge University: Cambridge, UK; New York, NY, USA, 2007.

9. Li, F.P.; Zhang, G.X.; Dong, L.Q. Studies for impact of climate change on hydrology and water resources. *Sci. Geogr. Sin.* **2013**, *4*, 457–464.

10. Song, X.M.; Zhang, J.Y.; Zhan, C.S.; Liu, C.Z. Review for impacts of climate change and human activities on water cycle. *China J. Hydrol.* **2013**, *44*, 779–790. (In Chinese)

11. Xia, J.; Liu, C.Q.; Ren, G.Y. Opportunity and challenge of the climate change impact on the water resources of China. *Adv. Earth Sci.* **2011**, *1*, 1–12.

12. Wang, G.Q.; Zhang, J.Y.; Liu, J.F.; Jin, J.L.; Liu, C.S. The sensitivity of runoff to climate change in different climatic regions in China. *Adv. Water Sci.* **2011**, *3*, 307–314. (In Chinese)

13. Arnold, J.G.; Williams, J.R.; Srinivasan, R.; King, K.W. *The Soil and Water Assessment Tool (SWAT) User's Manual*; USDA-ARS: Temple, TX, USA, 1995.

14. Eckhardt, K.; Ulbrich, U. Potential impacts of climate change on groundwater recharge and streamflow in a central European low mountain range. *J. Hydrol.* **2003**, *284*, 244–252.

15. Ficklin, D.L.; Luo, Y.Z.; Luedeling, E.; Zhang, M. Climate change sensitivity assessment of a highly agricultural watershed using SWAT. *J. Hydrol.* **2009**, *374*, 16–29.

16. Xu, Z.X.; Zuo, D.P.; Tang, F.F. *Response of Water Cycle to Future Climate Change in Typical Watershed in the Yellow River Basin*; Press of Hehai University: Nanjing, China, 2012; pp. 37–49. (In Chinese)

17. Githui, F.; Gitau, W.; Mutuab, F.; Bauwens, W. Climate change impact on SWAT simulated streamflow in Western Kenya. *Int. J. Climatol.* **2009**, *29*, 1823–1834.

18. Yu, L.; Gu, J.; Li, J.X.; Zhu, X.J. A Study of hydrologic responses to climate change in medium scale basin based on SWAT. *Bull. Water Soil Conserv.* **2008**, *28*, 152–154. (In Chinese)

19. Bekele, E.G.; Knapp, H.V. Watershed Modeling to Assessing Impacts of Potential Climate Change on Water Supply Availability. *Water Resour. Manag.* **2010**, *24*, 3299–3320.

20. Zhu, C.H.; Li, Y.K. Long-term hydrological impacts of land use/land cover change from 1984 to 2010 in the Little River Watershed, Tennessee. *Int. Soil Water Conserv. Res.* **2014**, *2*, 11–22.

21. Nowak, K.; Hoerling, M.; Rajagopalan, B.; Zagona, E. Colorado River Basin Hydro-climatic Variability. *J. Clim.* **2012**, *25*, 4389–4403.

22. Fan, Y.T.; Chen, Y.N.; Li, W.H.; Wang, H.J.; Li, X.G. Impacts of temperature and precipitation on runoff in the Tarim River during the past 50 years. *J. Arid Land* **2011**, *3*, 220–230.

23. Cayan, D.R.; Dettinger, M.D.; Kammerdiener, S.A.; Caprio, J.M.; Peterson, D.H. Changes in the onset of spring in the western United States. *Bull. Am. Meteorol. Soc.* **2001**, *82*, 399–416.

24. Colman, R. A comparison of climate feedbacks in general circulation models. *Clim. Dynam.* **2003**, *20*, 865–873.

25. Liu, Q.; Yang, Z.; Cui, B. Spatial and temporal variability of annual precipitation during 1961–2006 in Yellow River Basin, China. *J. Hydrol.* **2008**, *361*, 330–338.

26. Lu, X.X. Vulnerability of water discharge of large Chinese rivers to environmental changes: An overview. *Reg. Environ. Chang.* **2004**, *4*, 182–191.

27. Wei, W.; Chen, L.; Fu, B.; Chen, J. Water erosion response to rainfall and land use in different drought-level years in a loess hilly area of China. *Catena* **2010**, *81*, 24–31.

28. Miao, C.Y.; Ni, J.R.; Borthwick, A.G.L.; Yang, L. A preliminary estimate of human and natural contributions to the changes in water discharge and sediment load in the Yellow River. *Glob. Plenary Chang.* **2011**, *76*, 196–205.

29. Miao, C.Y.; Ni, J.R.; Borthwick, A.G.L. Recent changes of water discharge and sediment load in the Yellow River basin, China. *Phys. Geogr.* **2010**, *34*, 541–561.

30. Ren, Z.P.; Zhang, G.H.; Yang, Q.K. Characteristics of runoff and sediment variation in the Yanhe River Basin in last 50 years. *J. China Hydrol.* **2012**, *32*, 81–86. (In Chinese)

31. Sui, J.; He, Y.; Liu, C. Changes in sediment transport in the Kuye River in the Loess Plateau in China. *Int. J. Sediment Res.* **2009**, *24*, 201–213.

32. Li, Z.; Liu, W.Z.; Zhang, X.C.; Zheng, F.L. Impacts of land use change and climate variability on hydrology in an agricultural catchment on the Loess Plateau of China. *J. Hydrol.* **2009**, *377*, 35–42.

33. Luo, H.M.; Li, T.H.; Ni, J.R.; Wang, Y.D. Water Demand for Ecosystem Protection in River with Hyper-concentrated Sediment-laden Flow. *Sci. China Ser. E* **2004**, *47*, 186–198.

34. Saxton, K.E. Soil Water Characteristics Hydraulic Properties Calculator [EB/OL]. Available online: http://www.bsyse.wsu.edu/saxton/soilwater (accessed on 20 March 2004).

35. Saxton, K.E.; Willey, P.H.; Rawls, W.J. Field and pond hydrologic analyses with the SPAW model. In Proceedings of the Annual International Meeting of American Society of Agricultural and Biological Engineers, Portland, OR, USA, 9–12 July 2006; pp. 1–13.

36. Gao, Y.; Li, T.H. Responses of runoff and sediment yield to LUCC with SWAT model: A case study in the Xichuan River Basin, China. *Sustain. Environ. Res.* **2015**, *25*, 27–35.

37. Bai, Y.M. Effect of returning farmland to forests to benefit of soil and water conservation in Xichuanhe River basin. *J. Water Resour. Water Eng.* **2011**, *22*, 176–178. (In Chinese)

38. Liu, K.W. *River Systems in Yanan City*; Press of Yan'An Education College: Yan'an, China, 2000. (In Chinese)

39. USDA (United States Department of Agriculture). *Urban Hydrology for Small Watersheds: TR-55*; No. 210-VI-TR-55; Government Printing Office: Washington, DC, USA, 1986.

40. Williams, J.; Nearing, M.; Nicks, A.; Skidmore, E.; Valentin, C.; King, K.; Savabi, R. Using soil erosion models for global change studies. *J. Soil Water Conserv.* **1996**, *51*, 381–385.

41. Chiew, F.H.S.; McMahon, T.A. The applicability of Morton's and Penman's evapotranspiration estimates in rainfall-runoff modelling. *Water Resour. Bull.* **1991**, *27*, 611–620.

42. Williams, J.R. Flood routing with variable travel time or variable storage coefficients. *Trans. ASAE* **1969**, *12*, 100–103.

43. Van Griensven, A.; Meixner, T.; Grunwald, S.; Bishop, T.; Diluzio, M.; Srinivasan, R. A global sensitivity analysis tool for the parameters of multi-variable catchment models. *J. Hydrol.* **2006**, *324*, 10–23.

44. Nagelkerke, N.J.D. A Note on a General Definition of the Coefficient of determination. *Biometrika* **1991**, *78*, 691–692.

45. Quinlan, J.R. Learning with continuous classes. In Proceedings of the AI'92, the 5th Australian Joint Conference on Artificial Intelligence, Tasmania, Australia, 16–18 November 1992; Adams, A., Sterling, L., Eds.; World Scientific: Singapore, 1992; pp. 343–348.

46. Nash, J.E.; Sutcliffe, J.V. River flow forecasting through conceptual models. Part I—A discussion of principles. *J. Hydrol.* **1970**, *10*, 282–290.

47. Yue, B.J.; Shi, Z.H.; Fang, N.F. Evaluation of rainfall erosivity and its temporal variation in the Yanhe River catchment of the Chinese Loess Plateau. *Nat. Hazards* **2014**, *74*, 585–602.

48. Fu, B.J.; Wang, Y.F.; Lu, Y.H.; He, C.H.; Chen, L.D.; Song, C.J. The effects of land-use combinations on soil erosion: A case study in the Loess Plateau of China. *Prog. Phys. Geogr.* **2009**, *33*, 793–804.

49. Zhu, L.Q.; Zhu, W.B. Research on effects of land use/cover change on soil erosion. *Adv. Mater. Res.* **2012**, *433*, 1038–1043.

50. Zeng, Y.; Wei, L. Impacts of Climate and Land Use Changes on Runoff and Sediment Yield in Sichuan Purple Hilly Area. *Bull. Water Soil Conserv.* **2013**, *33*, 1–6. (In Chinese)

51. Xia, Z.H.; Zhou, Y.H.; Xu, H.M. Water resources responses to climate changes in Hanjiang River basin based on SWAT model. *Resour. Environ. Yangtze Basin* **2010**, *19*, 158–163. (In Chinese)

Assessment on Hydrologic Response by Climate Change in the Chao Phraya River Basin, Thailand

Mayzonee Ligaray, Hanna Kim, Suthipong Sthiannopkao, Seungwon Lee, Kyung Hwa Cho and Joon Ha Kim

Abstract: The Chao Phraya River in Thailand has been greatly affected by climate change and the occurrence of extreme flood events, hindering its economic development. This study assessed the hydrological responses of the Chao Phraya River basin under several climate sensitivity and greenhouse gas emission scenarios. The Soil and Water Assessment Tool (SWAT) model was applied to simulate the streamflow using meteorological and observed data over a nine-year period from 2003 to 2011. The SWAT model produced an acceptable performance for calibration and validation, yielding Nash-Sutcliffe efficiency (NSE) values greater than 0.5. Precipitation scenarios yielded streamflow variations that corresponded to the change of rainfall intensity and amount of rainfall, while scenarios with increased air temperatures predicted future water shortages. High CO_2 concentration scenarios incorporated plant responses that led to a dramatic increase in streamflow. The greenhouse gas emission scenarios increased the streamflow variations to 6.8%, 41.9%, and 38.4% from the reference period (2003–2011). This study also provided a framework upon which the peak flow can be managed to control the nonpoint sources during wet season. We hope that the future climate scenarios presented in this study could provide predictive information for the river basin.

Reprinted from *Water*. Cite as: Ligaray, M.; Kim, H.; Sthiannopkao, S.; Lee, S.; Cho, K.H.; Kim, J.H. Assessment on Hydrologic Response by Climate Change in the Chao Phraya River Basin, Thailand. *Water* **2015**, *7*, 6892–6909.

1. Introduction

Current environmental changes due to either natural or anthropogenic influences are creating a significant impact on natural resources and the living conditions of humans [1]. In particular, as a critical natural resource, water bodies have been subjected to pollution and are reaching scarcity levels around the globe [2,3]. Climate change is a key factor that has greatly affected water resources, due to its uncertainty and variability [4–6]; the intensities and frequencies of rainfall have been fluctuating over the years, thereby changing the spatiotemporal distributions of water resources [7]. Furthermore, it is apparent that climate change influences a change in water quality by modifying the surface and groundwater components [8,9].

186

Floods and droughts brought about by climate change may also lead to a change in water quality by increasing the effects of erosion or dilution [8,10,11].

There are several signs that the varying rainfall intensity has caused a change in climate change factors. These change factors include the increasing global surface temperature and significant local impacts such as high-magnitude floods, prolonged droughts, flow variability, temperature rise, and decreased rainfall [12–14]. As such, there is a need to change the perception of climate change and its uncertainty and vulnerability since its weaknesses should be prioritized. Though a certain level of uncertainty has always existed in water resource management and planning, alternative approaches to manage water resources have been proposed, such as creating climate change scenarios with respect to speed and intensity of changes in baseline conditions [15,16]. In addition, though it is difficult to mitigate the effect of climate on water quantity since policy makers should obtain scientific and predictive information, for effective water resources management, the accurate prediction of water quantity and quality is a necessary response to the climate change scenario [17–19].

The Chao Phraya River, a major river basin in the Mekong subregion and the largest watershed in Thailand, serves mostly as a source for irrigation water and a transportation route in Central Thailand [20]. Over the years, climate change has greatly affected the river, which may subsequently hinder the economic development and the eradication of poverty in adjacent countries [21]. The Chao Phraya River basin has experienced extreme floods, which has brought about the subsequent contamination water used for human consumption [22], health problems [23,24], and economic loss on Thai Rice exports [25]. For example, the flood event that occurred in 2011 caused 813 fatalities nationwide and $45.7 billion in economic damages and losses [26,27]. This event prompted several researchers to conduct further case studies in different fields, including disaster management [28] and medicine [29,30].

Water management of irrigation water is very important for the Chao Phraya River basin because it is one of the world's major agricultural producers, having a cultivated area of approximately 51% (cropland) [31–33]. However, climate change has also played a significant role in the rate of landuse changes in the basin [34]. Farmers respond to weather conditions by changing crops seasonally, the timing of planting and harvesting, and other daily activities [35]. In cases such as this, there remains a need to understand the characteristics of the basin and its hydrological response to climate change.

To address the issues presented by extreme flood events and climate change, this study investigates the impact of climate change scenarios on the basin. The aims of this investigation are thus to: (1) calibrate and validate the water quantity in the Chao Phraya River basin using the Soil and Water Assessment Tool (SWAT) model; and (2) assess hydrological responses under hypothetical climate sensitivity scenarios and greenhouse gas emission scenarios. The results of this research will both help improve the foundation of water resource management based on these climate change scenarios and also provide an evaluation of highly variable climate changes on the basin.

2. Methodology

2.1. Site Description

The Chao Phraya River is an important water resource that supplies water to irrigated, urban, and domestic areas of the central part of Thailand [36]. Figure 1 presents the land use classes used in the SWAT model. It also includes the weather and outlet stations of the Chao Phraya River basin. The basin has 51.66% agriculture area, which depends on the main river and its tributaries to cultivate crops, such as rice, sugarcane, and corn. The river has four main tributaries: the Wang, Yom, Ping, and Nan Rivers and a significant lateral tributary, the Pasak River. The river drains an area of approximately 160,000 km^2 (98° E–103° E, 13° N–20° N), covering 30% of the country, and receives annual precipitation of about 1,179 mm/year and an average discharge of 196 m^3/s at Chai Nat Station [37]. The basin is in a tropical climate, and is influenced by northeast and southwest monsoons. The northeast monsoon brings in cool and dry air from November to February, whereas the southwest monsoon brings very humid air, thus causing heavy rains from May to October [38].

2.2. SWAT Model

The Soil and Water Assessment Tool (SWAT) is watershed model that can be applied to simple and complex watersheds. It is a continuous-time model that operates on a daily time step and was developed for the USDA Agricultural Research Service (ARS) to predict the impact of management on water, sediment, and agricultural chemical yields in large complex watersheds [39]. The model is physically-based, uses readily available inputs, is computationally efficient, and is able to continuously simulate long-term impacts. Major model components include hydrology, weather, soil, temperature, plant growth, nutrients, pesticides, and land management. The SWAT model divides the watershed into multiple subwatersheds, which are then further subdivided into hydrologic response units (HRUs) that consist of homogenous land use, management, and soil characteristics. Models calibrated using watershed and water quality data have been used to forecast water quantity/quality in response to climate change scenarios; the SWAT model has been widely used to predict water quantity and quality in response to various management and climate scenarios [40–51].

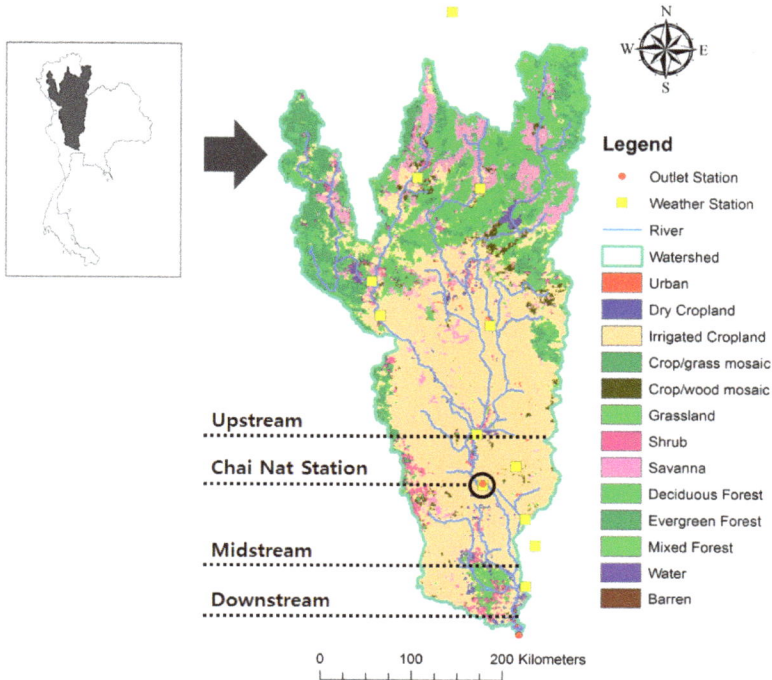

Figure 1. Chao Phraya River basin is shown in the map. It includes the locations of the twelve weather stations (yellow squares) of the basin as well as the outlet stations (red circles) of the river, which also served as monitoring stations. Chai Nat Station has been chosen for the SWAT model simulation of the streamflow.

2.3. Model Application

To construct the Chao Phraya River basin model, a model database was compiled using topographical data, consisting of a digital elevation map, landuse, soil and river basin, meteorological data (e.g., precipitation, and maximum and minimum temperature), and observed monitoring data (e.g., flow discharge). The topographical data included a digital elevation model (DEM) in GeoTiff grid tiles (5° × 5° tiles) created from a water database [52]; the data was derived from USGS/NASA Shuttle Radar Topography Mission (SRTM) data with a 90 m resolution. The land use data has a spatial resolution of 1 km that included 14 classes of landuse representations from the USGS Global Land Cover Characterization (GLCC) database [53]. There were also 14 soil types in the soil grid provided by the Food and Agriculture Organization of the United Nations [54]. We also obtained nine-year meteorological data (2003–2011) from the Thai Meteorological Department, which included daily precipitation and maximum/minimum temperatures from 12 stations within the Chao Phraya River basin.

189

The SWAT model was able to delineate an area of 119,663 km^2 of the Chao Phraya River basin. Figure 1 illustrates the Chao Phraya River basin built by SWAT. Landuse classes of the basin are listed in Table 1, showing that the basin is mainly comprised of agriculture area and broadleaf forest. The table also indicates that half of the area used for agriculture was irrigated cropland and pasture (CRIR). Streamflow was monitored at the Chai Nat Station located in Chai Nat province. The observed flow data of the said station was stable and were acquired from the Royal Irrigation Department Computer Center in Sanphaya, Chai Nat.

Table 1. Area and percentage of land cover in the Chao Phraya River basin.

Landuse	Definition	Area (ha)	Percentage (%)
CRIR	Irrigated cropland and pasture	6,181,831	51.66
FODB	Deciduous broadleaf forest	1,947,509	16.27
FOEB	Evergreen broadleaf forest	1,503,653	12.57
SAVA	Savanna	1,038,567	8.68
FOMI	Mixed forest	495,301	4.14
CRWO	Cropland/woodland mosaic	294,706	2.46
SHRB	Shrubland	270,081	2.26
WATB	Water bodies	113,731	0.95
CRDY	Dryland cropland and pasture	79,365	0.66
URMD	Urban residential medium density	37,820	0.32
GRAS	Grassland	3060	0.03
CRGR	Cropland/grassland mosaic	381	0
BSVG	Barren or sparsely vegetated	249	0
	Watershed	11,966,254	100

2.4. Sensitivity Analysis

The Latin Hypercube—One-factor-At-a-Time (LH-OAT) method is a sensitivity analysis technique that combined the robustness of the Latin Hypercube (LH) sampling method and a one-factor-at-a-time (OAT) design [55]. It searches for good performance using a limited parameter number of important factors for model calibration and model output for a particular basin. LH is a stratified sampling method developed by McKay *et al.* [56] and is based on the Monte Carlo simulation; the OAT design was developed by Morris [57] and is used to observe the changes in the output by changing a particular input. LH-OAT method firstly divides the range of the parameters into segments then takes LH samples from each parameter to create parameter sets. The OAT design can then achieve global sensitivity by changing an entire parameter range using the LH samples [58]. Table 2 shows the 15 SWAT model parameters that were subjected to sensitivity analysis.

190

Table 2. Soil and water assessment tool (SWAT) model parameters used for the sensitivity analysis.

Name	Definition	Range	Process
Cn2	Soil Conversion Service (SCS) runoff curve number for moisture condition 2	35–98	Runoff
Alpha_Bf	Baseflow alpha factor (days)	0.00–1.00	Groundwater
Rchrg_Dp	Deep aquifer percolation fraction	0.00–1.00	Groundwater
Esco	Soil evaporation compensation factor	0.00–1.00	Evaporation
Revapmn	Threshold depth of water in the shallow aquifer for percolation to the deep aquifer (mmH_2O)	0–500	Groundwater
Ch_K2	Effective hydraulic conductivity in main channel alluvium (mm/h)	−0.01–150	Channel
Gwqmn	Threshold depth of water in the shallow aquifer required for return flow to occur (mm)	0–5000	Soil
Sol_Awc	Available water capacity of the soil layer (mm/mm soil)	0–100	Soil
Sol_Z	Maximum canopy index Soil depth	0–3000	Soil
Gw_Revap	Groundwater "revap" coefficient	0.02–0.2	Groundwater
Surlag	Surface runoff lag coefficient	0.00–10.00	Runoff
Blai	Leaf area index for crop	0.00–1.00	Crop
Slope	Average slope steepness (m/m)	0.0001–0.6	Geomorphology
Canmx	Maximum canopy index	0.00–10.00	Runoff
Epco	Threshold depth of water in the shallow aquifer to percolation to the deep aquifer (mmH_2O)	0.00–1.00	Evaporation

2.5. Performance Assessment

We applied a coefficient of determination (R^2), Nash-Sutcliffe efficiency (NSE), and root mean square error (RMSE) to evaluate the model performance. R^2 evaluated how accurate the simulated values were compared to the observed values and is defined as the squared value of Bravais-Pearson's coefficient of correlation (r) [59,60]. It depicts the strength between the simulated and observed data and the direction of the linear relation. R^2 is expressed as the squared ratio between the covariance and the multiplied standard deviation of the observation and simulated values [61]. NSE measures the goodness of fit and describes the variance between the simulated and observed values [62]. NSE values can differ from negative values up to less than one [63]. Generally, the calibration and validation values of the SWAT model are considered to be acceptable or satisfactory performances when NSE is within the range of 0.5 and 0.65, 0.65–0.75 is considered satisfactory, while 0.75–1.00 indicate a very good performance [64–66]. Table 3 presents the performance ratings for NSE, as suggested by Moriasi *et al.* (2007). Lastly, RMSE was used to assess the validity of the model in this study. It measures the square root of the distance between the observed and predicted values and gives an estimate of the variability of the model compared to the observations [49]. The desired value for RMSE is 0, which depicts a perfect simulation, with lower values representing better performance [67].

Table 3. General performance rating for the recommended statistics for monthly time step [68].

Performance Rating	NSE
Very good	$0.75 < \text{NSE} \leq 1.00$
Good	$0.65 < \text{NSE} \leq 0.75$
Satisfactory	$0.50 < \text{NSE} \leq 0.65$
Unsatisfactory	$\text{NSE} \leq 0.50$

2.6. Climate Change Scenarios

We applied a Special Report on Emission Scenario (SRES) to assess the hydrological response using a future climate change scenario. We applied the Commonwealth Scientific and Industrial Research Organisation (CSIRO) Mark 3.5 (Mk3.5) (CSIRO Atmospheric Research, Melbourne, Victoria, Australia)from the CSIRO Marine and Atmospheric Research in Australia as the future climate change model since it provides daily meteorological data for a long period for each climate change scenario. The future climate scenarios from the Intergovernmental Panel on Climate Change (IPCC) (A2, A1B, and B1) were chosen as its outputs (greenhouse gas emission scenarios) [69,70]. Mk3.5 is based on CSIRO Mark 3.0 (Mk3.0), a prior model version that has a fully coupled ocean-atmosphere system [71]. It is a spectral model that was developed to use the horizontal spectral resolution T63 [$1.875°$ EW $\times$ $1.875°$ NS] with 18 vertical levels, and to remove the cold bias of Mk3.0 due to the rising of global air temperature [72].

Four different qualitative storylines were developed by IPCC that represented the different demographic, social, economic, technological, and environmental developments of the communities [73,74]. SRES scenarios A1B, A2, and B1 were chosen for this study [75,76]. A1B is under the A1 storyline and scenario family, which emphasizes globalization. A1B is described as a balance across all energy sources that do not rely heavily on a specific source [40]. The A2 scenario describes a very heterogeneous world with slower economic growth and technological advancement compared to the other storylines. Lastly, the B1 scenario stresses on rapid economic change towards service and information, social and environmental sustainability, improved equity, and global solutions without additional climate initiatives [77].

The potential future climate included a set of gridded map layers for the daily precipitation and air temperature for 2051–2059, based on the output from a set of SRES. GCM produced the climate data applied in this study; however, the coarse resolution of GCM will reduce the accuracy of the results. An interpolation method was implemented to convert the GCMs of the 12 meteorological stations to finer regional resolutions. The converted climate data was then downscaled and used as weather inputs for the SWAT model.

The change factor method, a downscaling technique, was used to adjust the observed daily temperature and daily precipitation using Equations (1) and (2), respectively. The daily temperature at the 2059 horizon was obtained by adding the difference of mean daily temperature in 2059 horizon predicted by the climate model and mean temperature of the reference period (2003–2011) to the observed daily temperature, obtained using Equation (1) [78]. The daily precipitation was calculated by multiplying the observed daily precipitation with the ratio of the mean projected daily precipitation at the 2059 horizon, and the mean precipitation of the reference period, as described in Equation (2) [79].

$$T_{adj,2059,d} = T_{obs,d} + \left(\overline{T}_{CM,2059,m} - \overline{T}_{CM,ref,m} \right) \tag{1}$$

$$P_{adj,2059,d} = P_{obs,d} \times \left(\overline{P}_{CM,2059,m} \div \overline{P}_{CM,ref,m} \right) \tag{2}$$

where $T_{obs,d}$ and $P_{obs,d}$ are the observed daily temperature and precipitation, $T_{CM,2059,m}$ and $P_{CM,2059,m}$ are the projected daily temperature and precipitation at the 2059 horizon obtained by the climate model, $T_{CM,ref,m}$ and $P_{CM,ref,m}$ are the temperature and precipitation during the reference period (2003–2011), and $T_{adj,2059,d}$ and $P_{adj,2059,d}$ are the daily temperature and precipitation at the 2059 horizon. The averages of the temperature and precipitation from the 12 meteorological stations were the data used for the $T_{CM,ref,m}$ and $P_{CM,ref,m}$, respectively. The strength of the change factor (CF) method is similar to the results of factors derived from GCM or Regional Climate Model (RCM).

Table 4 shows the conditions of the hypothetical climate sensitivity scenarios. There were zero changes in precipitation and temperature at the reference point, with CO_2 concentrations at 330 ppm. Scenarios 1 to 3 have twice the CO_2 concentration (660 ppm) and high variation in the precipitation and air temperature. Scenarios 4 to 7 show variations in the precipitation, while maintaining constant CO_2 concentrations and air temperatures. Scenarios 8 to 10 have variations air temperature, while CO_2 concentrations and precipitations remain constant. The change factors of the SRES scenarios, A1B, A2, and B1, were also included in the table.

Table 4. Conditions of climate sensitivity scenarios and intergovernmental panel on climate change (IPCC) special report on emissions scenarios relative to the baseline. Scenarios 1–3 are referred to as the CO_2 scenarios, 4–7 as precipitation change scenarios, and 8–10 as temperature increase scenarios.

Scenario	CO_2 Concentration (ppm)	Precipitation Change (%)	Temperature ($^\circ$C)	
Baseline	330	0	0	
1	$CO_2 \times 2 = 660$	0	0	
2	$CO_2 \times 2 = 660$	+20	0	
3	$CO_2 \times 2 = 660$	0	+6	
4	330	+10	0	
5	330	+20	0	
6	330	−10	0	
7	330	−20	0	
8	330	0	+1	
9	330	0	+3	
10	330	0	+6	
A1B	330	+1.0644	Max	+2.0621
			Min	+2.4954
A2	330	+1.0338	Max	+1.8729
			Min	+2.2905
B1	330	+1.0054	Max	+0.7926
			Min	+0.6106

3. Results and Discussion

3.1. Model Evaluation

Thirty-five hydrological model parameters of the SWAT model underwent sensitivity and uncertainty analyses (e.g., Parameter solution (Para Sol) and Sequential Uncertainty Fitting (SUFI-2) in SWAT-CUP) to determine the optimal model parameters [80]. The top 11 parameters having sensitivity indices greater than or equal to 0.05 were then selected, as shown in Table 5 [81]. The result of the sensitivity analysis shows that the initial SCS runoff curve number for moisture condition II (CN2) and baseflow alpha factor-baseflow recession (Alpha_Bf) were the most sensitive parameters. They are followed by the deep aquifer percolation fraction (Rchrg_Dp), soil evaporation compensation factor (Esco), threshold depth of water in the shallow aquifer for percolation to the deep aquifer (Revapmn), effective hydraulic conductivity in main channel alluvium (Ch_K2), available water capacity of the soil layer (Sol_AWC), threshold depth of water in the shallow aquifer required for return flow to occur (Gwqmn), depth from soil surface to bottom of layer (Zol_Z), groundwater "revap"coefficient (Gw_Revap), and surface runoff lag coefficient (Surlag). Most parameters are related to either groundwater or soil process. The sensitive flow discharge parameters were then used to calibrate the model.

One of the limitations of this research was having a large watershed model with only 12 meteorological stations and one gauge station to calibrate the streamflow. The available data allowed nine years for the simulation period of the SWAT model. The comparison of daily observed and simulated streamflow during the nine-year (2003–2011) simulation period included a one-year spin up time (2003), a five-year (2004–2008) calibration period, and a three-year (2009–2011) validation period. Results from the statistical evaluation with the two numeric criteria including NSE, R^2, and RMSE, are listed in Table 6. Figure 2 shows how well-matched the daily streamflow simulations with the observations were. The NSE values of the daily streamflow simulations for the calibration and validation were 0.74 and 0.81, respectively. On the other hand, the NSE values of the monthly time step were 0.54 and 0.66 for the five-year calibration and three-year validations, respectively. The model evaluation statistics for the streamflow prediction show that there was a fair agreement between the measured and simulated flows, which were confirmed by the R^2 and NSE being greater than 0.5 [62].

Table 5. Top 11 parameters yielded by the sensitivity analysis.

Rank	Name	Definition	Sensitivity	Process
1	Cn2	SCS runoff curve number for moisture condition 2	1.49	Runoff
2	Alpha_Bf	Baseflow alpha factor (days)	1.42	Groundwater
3	Rchrg_Dp	Deep aquifer percolation fraction	0.66	Groundwater
4	Esco	Soil evaporation compensation factor	0.48	Evaporation
5	Revapmn	Threshold depth of water in the shallow aquifer for percolation to the deep aquifer (mm H_2O)	0.22	Groundwater
6	Ch_K2	Effective hydraulic conductivity in main channel alluvium (mm/h)	0.20	Channel
7	Gwqmn	Threshold depth of water in the shallow aquifer required for return flow to occur (mm)	0.18	Soil
8	Sol_Awc	Available water capacity of the soil layer (mm/mm soil)	0.14	Soil
9	Sol_Z	Maximum canopy index Soil depth	0.078	Soil
10	Gw_Revap	Groundwater "revap" coefficient	0.06	Groundwater
11	Surlag	Surface runoff lag coefficient	0.05	Runoff

Table 6. Prediction accuracy for the monthly streamflow in terms of R^2, Nash-Sutcliffe efficiency (NSE) and root mean square error (RMSE).

Statistical Index	Calibration	Validation
R^2	0.81	0.89
NSE	0.54	0.66
RMSE (m^3/s)	2.5466×10^3	3.0224×10^3

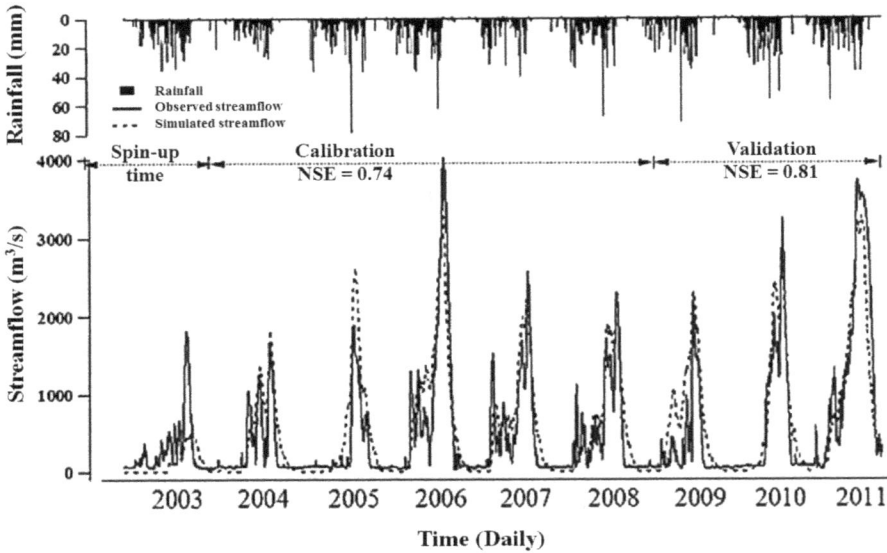

Figure 2. Observed and simulated streamflow and the corresponding daily rainfall during a nine-year period (2003–2011). Figure includes a one-year spin-up period (2003), five-year calibration period (2004–2008), and three-year validation period (2009–2011).

3.2. Climate Sensitivity Scenario

3.2.1. CO_2 Concentration

Scenarios 1 to 3 have atmospheric CO_2 concentrations of 660 ppm and a change in precipitation and air temperature (Table 7). These three scenarios were related to the stomatal conductance variable, which depends on the atmospheric CO_2 concentration [82]. For climate change simulations in the SWAT model, the reduction in stomatal conductance and increase in leaf area index were attributed to an increase in the CO_2 concentration [83]. According to the increasing atmospheric CO_2 concentration, variation in the streamflow was seen to be affected by the change in evapotranspiration, whereas the hydrological response depends on crop variables [84]. When the stomatal conductance of the vegetation decreases, the evapotranspiration also decreases, thus disturbing the water efficiency of the crops [84,85]. Table 7 shows the predicted relative change (percentage of baseline) in annual average streamflow with respect to the climate change scenario. In the calibration of the Chao Phraya subbasin, the annual average streamflow changed by 16.4% in calibration, another changed by 18.4%, with a maximum of 52.3% and a minimum of 1.6% in the whole basin. Scenario 2 reflects an increase of almost 48% in the annual average streamflow change of the subbasin; another subbasin increased by 52.6%, while the maximum was 128.9% and the minimum was 15.1% in the whole basin. Scenario 3, on the other hand, showed an annual average streamflow change of -5.3% at Chai Nat Station, whereas another site changed by -6.2%, with a maximum of 1.2%, and a minimum of -14.3% in the whole basin.

Table 7. Predicted relative percentage change of annual average streamflow from the baseline under different climate sensitivity scenarios and special report on emission scenario (SRES) (B1, A1B, and A2). It also includes the streamflow change at the Chai Nat Station.

Terms	Ref Stream-Flow	Climate Sensitivity Scenario										SRES		
		CO$_2$ (%)			Precipitation (%)				Air Temperature (%)			GCM (%)		
		1	2	3	4	5	6	7	8	9	10	B1	A1B	A2
Chai Nat Station	562.8	16.4	48	−5.3	15.6	30	−15.8	−32.8	−3.1	−9.3	−19.2	24.7	41.9	49.8
Max % change of the basin	671.8	52.3	128.9	1.2	35.9	70	−7	−14.5	8.2	8.2	−1.3	107.8	136.5	146.4
Min % change of the basin	1.3	1.6	15.1	−14.3	5.9	11.6	−37.3	−71.4	−8.2	−23.8	−53	−17.5	−1.1	4.1
Average % change of the basin	68.5	18.4	52.6	−6.2	16.6	32.2	−16.7	−34.5	−3.5	−10.6	−21.4	19.7	37.7	47

Figure 3a–c show streamflow variations of the basin, mostly ranging from -2% to 20%. The spatial variation for CO_2 concentration scenario displayed an increasing streamflow change in the southern area of Chao Phraya River basin that is within the range of 20% to 42% (Figure 3a). Scenario 2 (Figure 3b) showed a total streamflow increase and a more sensitive response in the southern area of the basin, while Scenario 3 (Figure 3c) indicated a decrease in streamflow in the whole basin area. Higher concentrations of CO_2 directly caused the stomata of plants to close, which then decreased their rate of transpiration and increased their water use, efficiency leading to a reduction in evapotranspiration [86]. The latter increased runoff and led to an increase in the streamflow. However, when increased CO_2 concentration is paired with an increase in temperature, as shown by Scenario 3, the streamflow will decrease—as also stated by the negative change in the annual average streamflow of Scenario 3 [87].

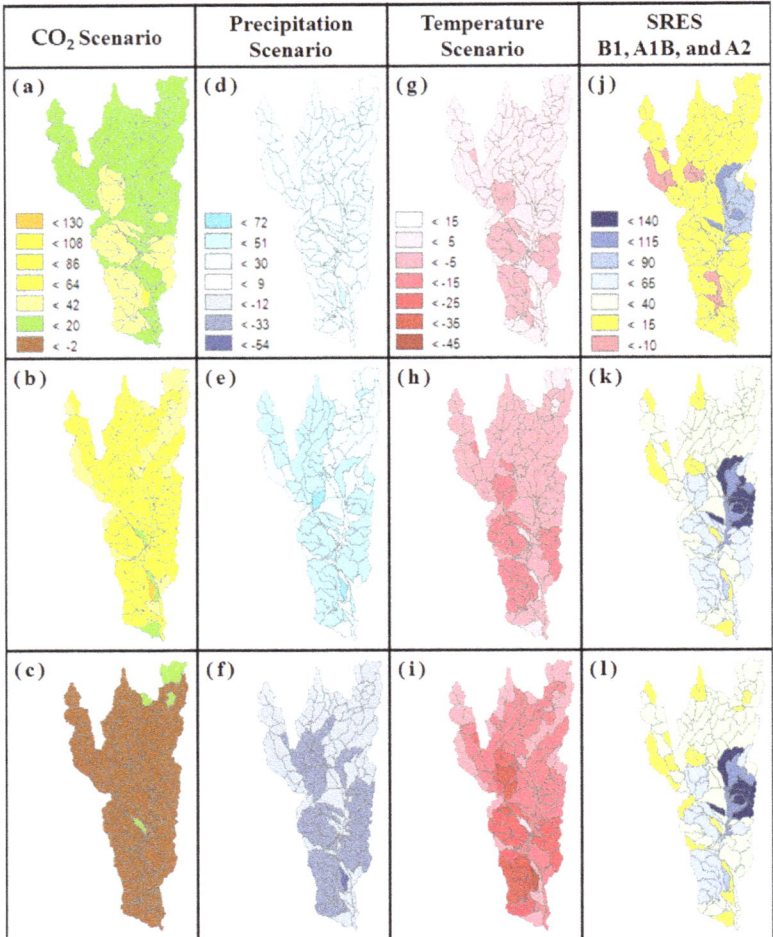

Figure 3. Spatial distributions of streamflow ratio under the climate sensitivity scenarios during the reference period (2003–2011) and SRES B1, A1B, and A2 (2051–2059). (**a–c**) CO_2 scenarios; (**d–f**) Precipitation scenarios; (**g–i**) Temperature scenarios; and (**j–l**) SRES.

3.2.2. Precipitation

Scenarios 4 through 7 represent precipitation changes of +10%, +20%, −10%, and −20%, while holding the baseline CO_2 concentration (330 ppm) and air temperature constant (Table 7). Annual average streamflow changes of 15.6%, 30%, −15.8, and −32.8% correspond to the changes implemented for the annual precipitation. On the other hand, changes in the annual average streamflow in the entire Chao Phraya River basin, 16.6%, 32.2%, −16.7%, and −34.5%, corresponded to the changes at Chai Nat

198

Station. Figure 3d–f shows the spatial distributions of streamflow ratio of Scenarios 4 to 6, which reflect the results of the average streamflow changes. Figure 3f, which shows Scenario 6, represents the spatial distribution of the decrease in precipitation. It can be concluded that Scenario 7 will have a darker shade, signifying an overall decrease of streamflow ratio in the basin. Based on these results, streamflow and precipitation have a positive linear relationship. One of the significant factors that affect streamflow is soil water content. Figure 4 shows the plots of soil water content of different scenarios. Compared to the others, the precipitation scenarios have notable differences from the baseline; as precipitation was increased, the soil water content also steadily increased. Increases or decreases in precipitation directly lead to corresponding directional changes in the streamflow [88,89].

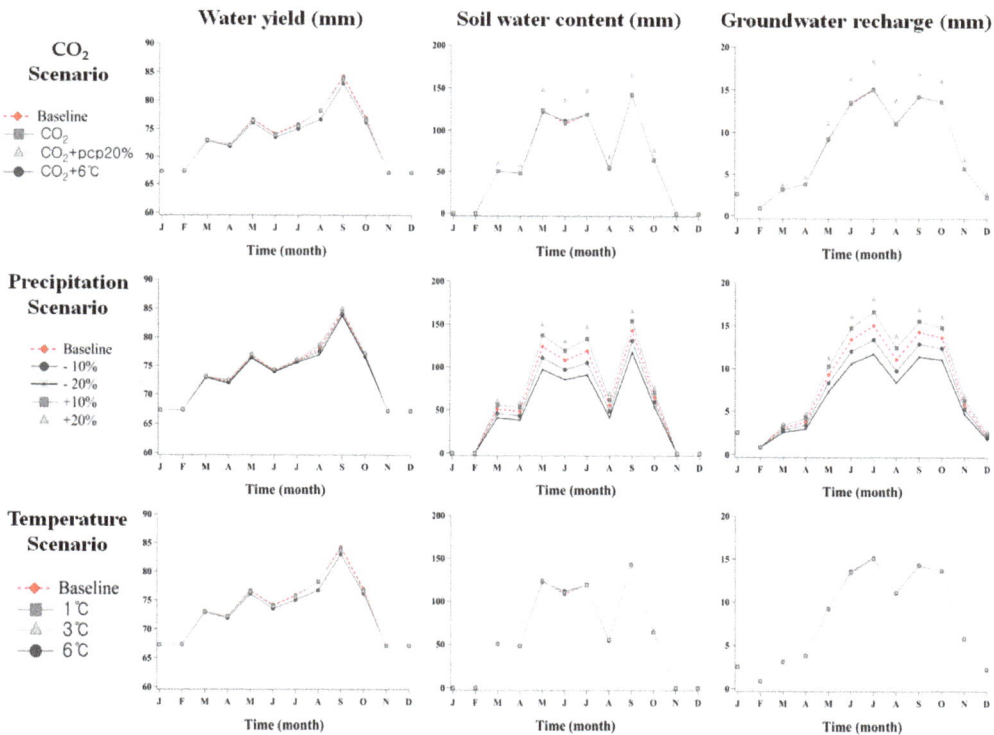

Figure 4. Water yield, soil water content, and groundwater recharge of Chao Phraya River basin under climate sensitivity scenarios.

In terms of precipitation scenarios, the annual streamflow of Chao Phraya River have maximum changes of 35.9%, 70%, −7%, and −14.5%, and minimum changes of 5.9%, 11.6%, −37.3%, −71.4%. Figure 3g–i shows the variation of upstream flow, middle streamflow, and down streamflow from the Chao Phraya River. It can be seen that an increase in precipitation led to an increase in the annual average streamflow in the Chao Phraya Watershed, as indicated by Scenarios 4 and 5. These results coincide with other literature that investigated the hydrological impact of climate change and hydrological scenarios, which stated that the mean annual river discharge increases due to an increase in precipitation [90,91]. The decreased streamflow in Scenarios 6 and 7 also confirmed that precipitation plays a major role in streamflow variations.

During the wet season, a precipitation change was responsible for the streamflow variation [92]. In the upstream, the annual average streamflow was 249.2 m^3/s in September, while the middle and down streams had around 562.8 m^3/s and 671.8 m^3/s, respectively. Figure 5 compares the seasonal variation of the peak streamflow for dry and wet seasons. The apparent increase in the streamflow in the increasing trend of the precipitation scenarios could lead to flooding events during the wet season. Monsoonal rains occur in the Chao Phraya River every year from May to October, making the watershed vulnerable to flood-related disasters during this season [93–95]. In this study, the streamflow increased in May and continued to rise until reaching their peak in September. The peak streamflow of the baseline and scenarios are highlighted in blue in Figure 5.

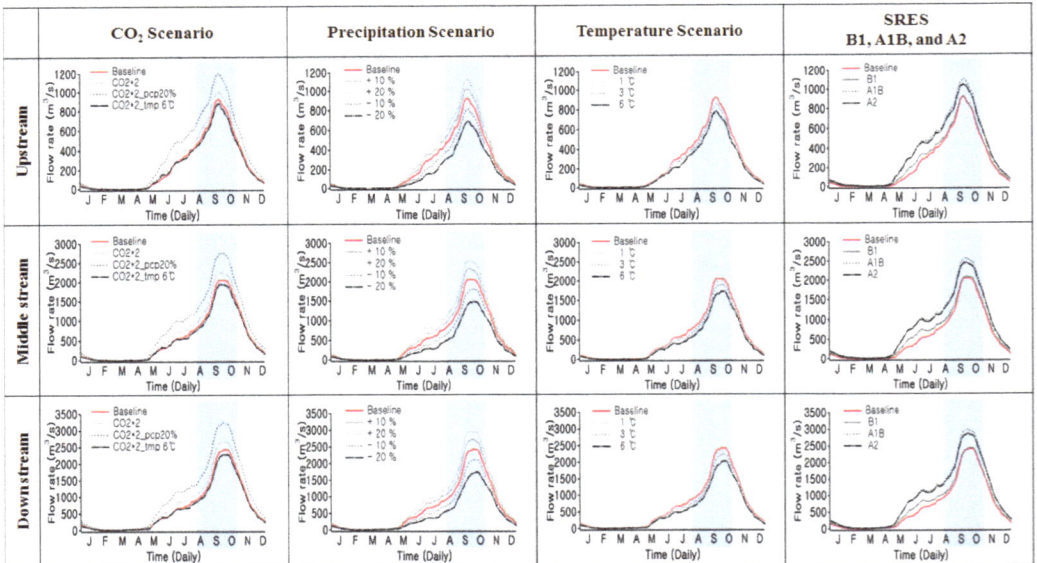

Figure 5. Seasonal variations of the peak stream flow under different climate scenarios in a year. The peak streamflow of each scenario is highlighted in blue to emphasize the difference from the baseline. It includes the upstream, middle stream, and downstream regions of the Chao Phraya River.

3.2.3. Air Temperature

Scenarios 8–10 represent air temperature increases of 1 °C, 3 °C, and 6 °C, respectively, from the baseline, while maintaining the CO_2 concentration at 330 ppm with no change in precipitation. Table 7 shows the variation of streamflow at the Chai Nat Station, which decreased by −3.1%, −9.3%, and −19.2% for Scenarios 8, 9, and 10. The table also includes the annual average streamflow and maximum and minimum changes in the annual average streamflow of the whole basin for each scenario. It was found that as the temperature increases, the average streamflow significantly decreases. These results indicate that a rise in temperature due to global warming may lead to a serious water shortage [96].

These results are similar to the results of spatial variations of previous scenarios, such as Scenarios 3, 6, and 7, which either increased the temperature or decreased the precipitation. In Scenario 6, Figure 3f shows that the southern area has a more sensitive response due to the significant decrease in its streamflow level. The southern part of the Chao Phraya Watershed also represents Central Thailand, which is described as a lush, fertile valley. This area is a perfect catchment basin of the mountainous Northern Thailand; thus, a drastic decrease in the streamflow will be first noticed in this area then followed by the rest of the watershed [97]. Figure 3g–i also shows a decrease in streamflow change in the entire river basin.

3.2.4. Climate Change Effects of SRES

We applied downscaled GCM outputs to modify the meteorological data in the SWAT model to predict the hydrological effect of potential future climate for the mid-21st century. Figure 3j–l illustrates baseline conditions of the monthly average precipitation and air temperature for the baseline period (2003–2011). Future climate change scenarios under three greenhouse gas emission scenarios from 2050–2059 are then projected.

Figure 3j–l show the variation of locational streamflow under different emission scenarios. The predicted streamflow displayed a higher frequency of flood events, which was expected due to a shift in the peak flow runoff in early May. Relative to the baseline conditions, the annual average percentage changes of the projected climate scenarios in streamflow under B1, A1B, and A2 were 12.3%, 45.7%, and 40.8% respectively. The maximum values of annual average percentage changes were 95.7%, 138.3%, and 132.2%, whereas the minimum values were −33.3%, 5.3%, −0.7% in the basin. The variations in streamflow for the upstream, middle stream, and downstream follow similar baseline patterns under the projected climate scenarios.

The previous spatial streamflow variations of Scenarios 8 to 10 yielded a decrease in streamflow at the southern part of the basin, whereas the middle-eastern area had a more significant increase in percentage change of the streamflow when the projected climate change scenario under SRES was applied. The middle-eastern area represents Northeast Thailand, an arid region having a rolling surface and undulating hills that often experiences harsh climatic conditions. These results are justified, as Chinvanno [98] previously stated that this part of Thailand will experience a significant shift in season with increased rainfall and a longer late season rain peak, consequently increasing its water quantity. The streamflow variation of three emission scenarios commonly displayed an increase of streamflow in this area of the river basin.

4. Conclusions

The SWAT model was used to create a hydrological model of the Chao Phraya Watershed to investigate the effect of climate sensitivity and greenhouse gas emission scenarios on its streamflow. The model yielded percentage increases of the streamflow that revealed a need to create safety measures during flood events: daily average streamflow (72.3%), during the wet season in early May (22.7%), and after May (70.1%). This study also achieved its objectives.

1. The SWAT model showed a satisfactory performance in terms of calibration and validation, with R^2 and NSE values greater than 0.5.
2. Precipitation scenarios yielded streamflow variations that corresponded to the change of rainfall intensity and amount of rainfall, while scenarios with increased air temperature yielded a decrease in water level leading to a water shortage. However, the three greenhouse gas emission scenarios from 2051–2059 had streamflow variations that increased from the baseline (2003–2011).
3. Scenarios 1 to 3 were related to an increase in CO_2 concentration scenarios, which reduced stomatal conductance and increased the leaf area index. The results showed an increase in streamflow levels; however, a negative change in streamflow was also observed when the air temperature increased.
4. Variations under three SRES indicate low streamflow values compared to those of the southern Chao Phraya Watershed. Hence, flood measures should be performed in the main streamline of Chao Phraya River and the southern area of the basin. As such, further water resource management will be needed in the northeastern area of the Chao Phraya river basin in the future.

Further increasing the uncertainty of climate change brings a corresponding uncertainty into the predictions of severe flood and drought. In addition, change in the landuse of the Chao Phraya River subbasins may result in a different distribution, which also depends on the changes of climate conditions such as climate sensitivity scenarios and the three emission scenarios.

Acknowledgments: This project was partially funded by the Asia-Pacific Network for Global Change Research (APN) (ARCP2013-23NMY-Sthiannopkao) and by the Basic Science Research Program through the National Research Foundation (NRF) of Korea funded by the Ministry of Education (NRF-2014R1A1A2059680).

Author Contributions: Mayzonee Ligaray arranged and wrote the paper while Hanna Kim developed the methodology and built the SWAT model. Suthipong Sthiannopkao, Seungwon Lee, and Kyung Hwa Cho prepared the climate data. Joon Ha Kim made the overall conclusion.

Conflicts of Interest: The authors declare no conflict of interest.

References

1. Graiprab, P.; Pongput, K.; Tangtham, N.; Gassman, P.W. Hydrologic evaluation and effect of climate change on the at Samat watershed, northeastern region, Thailand. *Int. Agric. Eng. J.* **2010**, *19*, 12–22.

2. Immerzeel, W.W.; van Beek, L.P.H.; Bierkens, M.F.P. Climate change will affect the Asian water towers. *Science* **2010**, *328*, 1382–1385.

3. Kundzewicz, Z.W.; Mata, L.J.; Arnell, N.W.; DÖLl, P.; Jimenez, B.; Miller, K.; Oki, T.; ŞEn, Z.; Shiklomanov, I. The implications of projected climate change for freshwater resources and their management. *Hydrol. Sci. J.* **2008**, *53*, 3–10.

4. Tebaldi, C.; Smith, R.L.; Nychka, D.; Mearns, L.O. Quantifying uncertainty in projections of regional climate change: A bayesian approach to the analysis of multimodel ensembles. *J. Clim.* **2005**, *18*, 1524–1540.

5. Allen, M.R.; Stott, P.A.; Mitchell, J.F.B.; Schnur, R.; Delworth, T.L. Quantifying the uncertainty in forecasts of anthropogenic climate change. *Nature* **2000**, *407*, 617–620.

6. Le, T.; Sharif, H. Modeling the projected changes of river flow in central Vietnam under different climate change scenarios. *Water* **2015**, *7*, 3579–3598.

7. Kay, A.L.; Jones, R.G.; Reynard, N.S. Rcm rainfall for UK flood frequency estimation. II. Climate change results. *J. Hydrol.* **2006**, *318*, 163–172.

8. Delpla, I.; Jung, A.V.; Baures, E.; Clement, M.; Thomas, O. Impacts of climate change on surface water quality in relation to drinking water production. *Environ. Int.* **2009**, *35*, 1225–1233.

9. Wang, H.; Gao, J.E.; Zhang, M.J.; Li, X.H.; Zhang, S.L.; Jia, L.Z. Effects of rainfall intensity on groundwater recharge based on simulated rainfall experiments and a groundwater flow model. *CATENA* **2015**, *127*, 80–91.

10. Routschek, A.; Schmidt, J.; Kreienkamp, F. Impact of climate change on soil erosion—A high-resolution projection on catchment scale until 2100 in Saxony/Germany. *CATENA* **2014**, *121*, 99–109.

11. Nearing, M.A.; Jetten, V.; Baffaut, C.; Cerdan, O.; Couturier, A.; Hernandez, M.; le Bissonnais, Y.; Nichols, M.H.; Nunes, J.P.; Renschler, C.S.; *et al.* Modeling response of soil erosion and runoff to changes in precipitation and cover. *CATENA* **2005**, *61*, 131–154.

12. Wang, H.; Chen, L.; Yu, X. Distinguishing human and climate influences on streamflow changes in luan river basin in China. *CATENA* **2015**. in press.

13. Wang, S.; Wang, Y.; Ran, L.; Su, T. Climatic and anthropogenic impacts on runoff changes in the songhua river basin over the last 56 years (1955–2010), Northeastern China. *CATENA* **2015**, *127*, 258–269.

14. Knox, J.C. Large increases in flood magnitude in response to modest changes in climate. *Nature* **1993**, *361*, 430–432.

15. Clarvis, M.H.; Fatichi, S.; Allan, A.; Fuhrer, J.; Stoffel, M.; Romerio, F.; Gaudard, L.; Burlando, P.; Beniston, M.; Xoplaki, E.; *et al.* Governing and managing water resources under changing hydro-climatic contexts: The case of the upper Rhone basin. *Environ. Sci. Policy* **2014**, *43*, 56–67.

16. Costanza, R.; Bohensky, E.; Butler, J.R.A.; Bohnet, I.; Delisle, A.; Fabricius, K.; Gooch, M.; Kubiszewski, I.; Lukacs, G.; Pert, P.; *et al.* A scenario analysis of climate change and ecosystem services for the Great Barrier Reef. In *Treatise on Estuarine and Coastal Science*; Wolanski, E., McLusky, D., Eds.; Academic Press: Waltham, MA, USA, 2011; pp. 305–326.

17. López-Moreno, J.I.; Zabalza, J.; Vicente-Serrano, S.M.; Revuelto, J.; Gilaberte, M.; Azorin-Molina, C.; Morán-Tejeda, E.; García-Ruiz, J.M.; Tague, C. Impact of climate and land use change on water availability and reservoir management: Scenarios in the Upper Aragón river, Spanish Pyrenees. *Sci. Total Environ.* **2014**, *493*, 1222–1231.

18. Pingale, S.M.; Jat, M.K.; Khare, D. Integrated urban water management modelling under climate change scenarios. *Resour. Conserv. Recycl.* **2014**, *83*, 176–189.

19. Nam, W.H.; Choi, J.Y.; Hong, E.M. Irrigation vulnerability assessment on agricultural water supply risk for adaptive management of climate change in South Korea. *Agric. Water Manag.* **2015**, *152*, 173–187.

20. United Nations/World Water Assessment Programme. *United Nations World Water Development Report: Water for People, Water for Life*, 1st ed.; United Nations Educational, Scientific and Cultural Organization (UNESCO): Paris, France; Berghahn Books: Brooklyn, NY, USA, 2003.

21. Asian Development Bank. *The Economics of Climate Change in Southeast Asia: A Regional Review*; Asian Development Bank: Metro Manila, Philippines, 2009.

22. Chaturongkasumrit, Y.; Techaruvichit, P.; Takahashi, H.; Kimura, B.; Keeratipibul, S. Microbiological evaluation of water during the 2011 flood crisis in Thailand. *Sci. Total Environ.* **2013**, *463–464*, 959–967.

23. Vachiramon, V.; Busaracome, P.; Chongtrakool, P.; Puavilai, S. Skin diseases during floods in Thailand. *J. Med. Assoc. Thail.* **2008**, *91*, 479–484.

24. Assanangkornchai, S.; Tangboonngam, S.; Edwards, S.G. The flooding of Hat Yai: Predictors of adverse emotional responses to a natural disaster. *Stress Health* **2004**, *20*, 81–89.

25. Nara, P.; Mao, G.G.; Yen, T.B. Climate change impacts on agricultural products in Thailand: A case study of thai rice at the Chao Phraya River basin. *APCBEE Procedia* **2014**, *8*, 136–140.

26. Ziegler, A.D.; Lim, H.; Tantasarin, C.; Jachowski, N.R.; Wasson, R. Floods, false hope, and the future. *Hydrol. Process.* **2012**, *26*, 1748–1750.

27. Komori, D.; Nakamura, S.; Kiguchi, M.; Nishijima, A.; Yamazaki, D.; Suzuki, S.; Kawasaki, A.; Oki, K.; Oki, T. Characteristics of the 2011 Chao Phraya River flood in central Thailand. *Hydrol. Res. Lett.* **2012**, *6*, 41–46.

28. Raungratanaamporn, I.S.; Pakdeeburee, P.; Kamiko, A.; Denpaiboon, C. Government-communities collaboration in disaster management activity: Investigation in the current flood disaster management policy in Thailand. *Procedia Environ. Sci.* **2014**, *20*, 658–667.

29. Thaipadungpanit, J.; Wuthiekanun, V.; Chantratita, N.; Yimsamran, S.; Amornchai, P.; Boonsilp, S.; Maneeboonyang, W.; Tharnpoophasiam, P.; Saiprom, N.; Mahakunkijcharoen, Y.; *et al.* Leptospira species in floodwater during the 2011 floods in the Bangkok Metropolitan Region, Thailand. *Am. J. Trop. Med. Hyg.* **2013**, *89*, 794–796.

30. Ngaosuwankul, N.; Thippornchai, N.; Yamashita, A.; Vargas, R.E.M.; Tunyong, W.; Mahakunkijchareon, Y.; Ikuta, K.; Singhasivanon, P.; Okabayashi, T.; Leaungwutiwong, P. Detection and characterization of enteric viruses in flood water from the 2011 Thai flood. *Jpn. J. Infect. Dis.* **2013**, *66*, 398–403.

31. Thepent, V. *Agricultural Engineering and Technology for Food Security and Sustainable Agriculture in Thailand*; Department of Agriculture Thailand: Bangkok Thailand, 2005.

32. Molle, F.; Sutthi, C.; Keawkulaya, J.; Korpraditskul, R. Water management in raised bed systems: A case study from the Chao Phraya Delta, Thailand. *Agric. Water Manag.* **1999**, *39*, 1–17.

33. Teamsuwan, V.; Satoh, M. Comparative analysis of management of three water users' organizations: Successful cases in the Chao Phraya Delta, Thailand. *Paddy Water Environ.* **2009**, *7*, 227–237.

34. Bossa, A.; Diekkrüger, B.; Agbossou, E. Scenario-based impacts of land use and climate change on land and water degradation from the meso to regional scale. *Water* **2014**, *6*, 3152–3181.

35. Malanson, G.P.; Verdery, A.M.; Walsh, S.J.; Sawangdee, Y.; Heumann, B.W.; McDaniel, P.M.; Frizzelle, B.G.; Williams, N.E.; Yao, X.; Entwisle, B.; *et al.* Changing crops in response to climate: Virtual nang rong, Thailand in an agent based simulation. *Appl. Geogr.* **2014**, *53*, 202–212.

36. Molle, F. Scales and power in river basin management: The Chao Phraya River in Thailand. *Geogr. J.* **2007**, *173*, 358–373.

37. World Water Assessment Programme. *Water for People, Water for Life: A Joint Report by the Twenty-Three un Agencies Concerned with Freshwater*; UNESCO Pub.: Paris, France, 2003.

38. Bachelet, D.; Brown, D.; Böhm, M.; Russell, P. Climate change in Thailand and its potential impact on rice yield. *Clim. Chang.* **1992**, *21*, 347–366.

39. Arias, R.; Rodríguez-Blanco, M.; Taboada-Castro, M.; Nunes, J.; Keizer, J.; Taboada-Castro, M. Water resources response to changes in temperature, rainfall and CO_2 concentration: A first approach in NW Spain. *Water* **2014**, *6*, 3049–3067.

40. Wu, Y.; Liu, S.; Gallant, A.L. Predicting impacts of increased CO_2 and climate change on the water cycle and water quality in the semiarid james river basin of the midwestern USA. *Sci. Total Environ.* **2012**, *430*, 150–160.

41. Varanou, E.; Gkouvatsou, E.; Baltas, E.; Mimikou, M. Quantity and quality integrated catchment modeling under climate change with use of soil and water assessment tool model. *J. Hydrol. Eng.* **2002**, *7*, 228–244.

42. Stone, M.C.; Hotchkiss, R.H.; Hubbard, C.M.; Fontaine, T.A.; Mearns, L.O.; Arnold, J.G. *Impacts of Climate Change on Missouri River Basin Water Yield*; Wiley Online Library: Hoboken, NJ, USA, 2001.

43. Zhang, X.; Srinivasan, R.; Hao, E. Predicting hydrologic response to climate change in the Luohe River basin using the SWAT model. *Trans. ASABE* **2007**, *50*, 901–910.

44. Ficklin, D.L.; Luo, Y.; Luedeling, E.; Zhang, M. Climate change sensitivity assessment of a highly agricultural watershed using SWAT. *J. Hydrol.* **2009**, *374*, 16–29.

45. Abbaspour, K.C.; Faramarzi, M.; Ghasemi, S.S.; Yang, H. Assessing the impact of climate change on water resources in Iran. *Water Resour. Res.* **2009**, *45*.

46. Hanratty, M.P.; Stefan, H.G. Simulating climate change effects in a minnesota agricultural watershed. *J. Environ. Qual.* **1998**, *27*, 1524–1532.

47. Gosain, A.; Rao, S.; Basuray, D. Climate change impact assessment on hydrology of Indian River basins. *Curr. Sci.* **2006**, *90*, 346–353.

48. Githui, F.; Gitau, W.; Mutua, F.; Bauwens, W. Climate change impact on SWAT simulated streamflow in Western Kenya. *Int. J. Climatol.* **2009**, *29*, 1823–1834.

49. Jha, M.; Pan, Z.; Takle, E.S.; Gu, R. Impacts of climate change on streamflow in the upper Mississippi River basin: A regional climate model perspective. *J. Geophys. Res. Atmos.* **2004**, *109*.

50. Wang, S.; Kang, S.; Zhang, L.; Li, F. Modelling hydrological response to different land-use and climate change scenarios in the Zamu River basin of Northwest China. *Hydrol. Process.* **2008**, *22*, 2502–2510.

51. Fontaine, T.A.; Klassen, J.F.; Cruickshank, T.S.; Hotchkiss, R.H. Hydrological response to climate change in the black hills of south dakota, USA. *Hydrol. Sci. J.* **2001**, *46*, 27–40.

52. SRTM 90m Digital Elevation Data. Available online: http://srtm.csi.cgiar.org/ (accessed on 2 September 2015).

53. Global Land Cover Characterization. Available online: http://edc2.usgs.gov/glcc/glcc.php (accessed on 2 September 2015).

54. Leon, L.F. Step by step geo-processing and set-up of the required watershed data for MWSWAT (mapwindow SWAT). Available online: http://www.waterbase.org/docs/Geo_Process.pdf (accessed on 1 December 2015).

55. Qiang, C.; Si, G.; Dayong, Q.; Zuhao, Z. Analysis of SWAT 2005 parameter sensitivity with LH-OAT method. *HKIE Trans.* **2010**, *17*, 1–7.

56. McKay, M.D.; Beckman, R.J.; Conover, W.J. Comparison of three methods for selecting values of input variables in the analysis of output from a computer code. *Technometrics* **1979**, *21*, 239–245.

57. Morris, M.D. Factorial sampling plans for preliminary computational experiments. *Technometrics* **1991**, *33*, 161–174.

58. Feng, G.; Sharratt, B. Sensitivity analysis of soil and PM_{10} loss in Weps using the LHS-OAT method. *Trans. ASAE* **2005**, *48*, 1409–1420.

59. Croitoru, A.E.; Minea, I. The impact of climate changes on rivers discharge in Eastern Romania. *Theor. Appl. Clim.* **2014**, 1–11.

60. Grillakis, M.G.; Tsanis, I.K.; Koutroulis, A.G. Application of the HBV hydrological model in a flash flood case in Slovenia. *Nat. Hazards Earth Syst. Sci.* **2010**, *10*, 2713–2725.

61. Krause, P.; Boyle, D.; Bäse, F. Comparison of different efficiency criteria for hydrological model assessment. *Adv. Geosci.* **2005**, *5*, 89–97.

62. Moriasi, D.N.; Arnold, J.G.; van Liew, M.W.; Bingner, R.L.; Harmel, R.D.; Veith, T.L. Model evaluation guidelines for systematic quantification of accuracy in watershed simulations. *Trans. Asabe* **2007**, *50*, 885–900.

63. Mishra, A.; Kar, S. Modeling hydrologic processes and NPS pollution in a small watershed in subhumid subtropics using SWAT. *J. Hydrol. Eng.* **2011**, *17*, 445–454.

64. Gitau, M.W.; Chaubey, I. Regionalization of SWAT model parameters for use in ungauged watersheds. *Water* **2010**, *2*, 849–871.

65. Singh, J.; Knapp, H.V.; Arnold, J.G.; Demissie, M. Hydrlogical modeling of the Iroquois River watershed using HSPF and SWAT. *J. Am. Water Resour. Assoc.* **2005**, *41*, 343–360.

66. Setegn, S.G.; Melesse, A.M.; Haiduk, A.; Webber, D.; Wang, X.; McClain, M.E. Modeling hydrological variability of fresh water resources in the Rio Cobre watershed, Jamaica. *CATENA* **2014**, *120*, 81–90.

67. Chu, T.W.; Shirmohammadi, A.; Montas, H.; Sadeghi, A. Evaluation of the SWAT model's sediment and nutrient components in the Piedmont physiographic region of Maryland. *Trans. Am. Soc. Agric. Eng.* **2004**, *47*, 1523–1538.

68. Moriasi, D.; Arnold, J.; van Liew, M.; Bingner, R.; Harmel, R.; Veith, T. Model evaluation guidelines for systematic quantification of accuracy in watershed simulations. *Trans. ASABE* **2007**, *50*, 885–900.

69. Watson, R.T.; Zinyowera, M.C.; Moss, R.H. *The Regional Impacts of Climate Change: An Assessment of Vulnerability*; Cambridge University Press: Cambridge, UK, 1997.

70. Dix, M.; Gordon, H. *Csiro mk3.5 Climate System Model Output: Tasmanian Partnership for Advanced Computing*; Centre for Australian Weather and Climate Research: Victoria, Australia, 2012.

71. Watterson, I.; O'Farrell, S. Climate change simulated by full and mixed-layer ocean versions of CSIRO Mk3.5 and Mk3.0: Large-scale sensitivity. *Asia Pac. J. Atmos. Sci.* **2013**, *49*, 375–387.

72. Gordon, H.; O'Farrell, S.; Collier, M.; Dix, M.; Rotstayn, L.; Kowalczyk, E.; Hirst, T.; Watterson, I. *The CSIRO Mk3.5 Climate Model*; CAWCR Technical Report No. 21; Centre for Australian Weather and Climate Research: Victoria, Australia, 2010.

73. Intergovernmental Panel on Climate Change (IPCC). *Summary for Policymakers: Emission Scenarios*; IPCC: Geneva, Switzerland, 2000.

74. Girod, B.; Wiek, A.; Mieg, H.; Hulme, M. The evolution of the IPCC's emissions scenarios. *Environ. Sci. Policy* **2009**, *12*, 103–118.

75. Chien, H.; Yeh, P.J.F.; Knouft, J.H. Modeling the potential impacts of climate change on streamflow in agricultural watersheds of the midwestern United States. *J. Hydrol.* **2013**, *491*, 73–88.

76. Ribalaygua, J.; Pino, M.R.; Pórtoles, J.; Roldán, E.; Gaitán, E.; Chinarro, D.; Torres, L. Climate change scenarios for temperature and precipitation in Aragón (Spain). *Sci. Total Environ.* **2013**, *463–464*, 1015–1030.

77. Gaffin, S.R.; Rosenzweig, C.; Xing, X.; Yetman, G. Downscaling and geo-spatial gridding of socio-economic projections from the ipcc special report on emissions scenarios (SRES). *Glob. Environ. Chang.* **2004**, *14*, 105–123.

78. Arnell, N.W.; Livermore, M.J.L.; Kovats, S.; Levy, P.E.; Nicholls, R.; Parry, M.L.; Gaffin, S.R. Climate and socio-economic scenarios for global-scale climate change impacts assessments: Characterising the SRES storylines. *Glob. Environ. Chang.* **2004**, *14*, 3–20.

79. Chen, J.; Brissette, F.P.; Leconte, R. Uncertainty of downscaling method in quantifying the impact of climate change on hydrology. *J. Hydrol.* **2011**, *401*, 190–202.

80. Schuol, J.; Abbaspour, K.C.; Srinivasan, R.; Yang, H. Estimation of freshwater availability in the West African sub-continent using the SWAT hydrologic model. *J. Hydrol.* **2008**, *352*, 30–49.

81. Lenhart, T.; Eckhardt, K.; Fohrer, N.; Frede, H.G. Comparison of two different approaches of sensitivity analysis. *Phys. Chem. Earth A B C* **2002**, *27*, 645–654.

82. Jarvis, P. The interpretation of the variations in leaf water potential and stomatal conductance found in canopies in the field. *Philos. Trans. R. Soc. Lond. B Biol. Sci.* **1976**, *273*, 593–610.

83. Morison, J.I.L. Intercellular CO_2 concentration and stomatal response to CO_2. In *Stomatal Function*; Zeiger, E., Farquhar, G.D., Cowan, O.R., Eds.; Stanford University Press: Redwood City, CA, USA, 1987; pp. 229–252.

84. Li, F.; Kang, S.; Zhang, J. Interactive effects of elevated CO_2, nitrogen and drought on leaf area, stomatal conductance, and evapotranspiration of wheat. *Agric. Water Manag.* **2004**, *67*, 221–233.

85. Bernacchi, C.J.; Kimball, B.A.; Quarles, D.R.; Long, S.P.; Ort, D.R. Decreases in stomatal conductance of soybean under open-air elevation of CO_2 are closely coupled with decreases in ecosystem evapotranspiration. *Plant Physiol.* **2007**, *143*, 134–144.

86. Wigley, T.; Jones, P. Influences of precipitation changes and direct CO_2 effects on streamflow. *Nature* **1985**, *314*, 149–152.

87. Eckhardt, K.; Ulbrich, U. Potential impacts of climate change on groundwater recharge and streamflow in a central European low mountain range. *J. Hydrol.* **2003**, *284*, 244–252.

88. Kiely, G. Climate change in ireland from precipitation and streamflow observations. *Adv. Water Resour.* **1999**, *23*, 141–151.

89. McCabe, G.J.; Wolock, D.M. A step increase in streamflow in the conterminous United States. *Geophys. Res. Lett.* **2002**, *29*.

90. Kure, S.; Tebakari, T. Hydrological impact of regional climate change in the Chao Phraya River basin, Thailand. *Hydrol. Res. Lett.* **2012**, *6*, 53–58.

91. Ogata, T.; Saavedra Valeriano, O.C.; Yoshimura, C.; Liengcharernsit, W.; Hirabayashi, Y. Past and future hydrological simulations of Chao Phraya River basin. *J. Jpn. Soc. Civil Eng. Ser. B1 Hydraul. Eng.* **2012**, *68*, 97–102.

92. Groisman, P.Y.; Knight, R.W.; Karl, T.R. Heavy precipitation and high streamflow in the contiguous United States: Trends in the twentieth century. *Bull. Am. Meteorol. Soc.* **2001**, *82*, 219–246.

93. Sayama, T.; Tatebe, Y.; Iwami, Y.; Tanaka, S. Hydrologic sensitivity of flood runoff and inundation: 2011 Thailand floods in the Chao Phraya River basin. *Nat. Hazards Earth Syst. Sci. Discuss.* **2014**, *2*, 7027–7059.

94. Lee, D.; Oh, B.; Kim, H.; Lee, S.; Chung, G. Comparison of the hydro-climatological characteristics for the extra-ordinary flood induced by tropical cyclone in the selected river basins. *Trop.Cyclone Res. Rev.* **2013**, *2*, 45–54.

95. Tamagno, B. Recent floods in Southeast Asia. *Geodate* **2012**, *25*, 2.

96. Christensen, N.; Wood, A.; Voisin, N.; Lettenmaier, D.; Palmer, R. The effects of climate change on the hydrology and water resources of the Colorado River basin. *Clim. Chang.* **2004**, *62*, 337–363.

97. Arnell, N.W. Climate change and global water resources. *Glob. Environ. Chang.* **1999**, *9*, 31–49.

98. Chinvanno, S. *Information for Sustainable Development in Light of Climate Change in Mekong River basin*; Southeast Asia START Regional Centre: Bangkok, Thailand, 2004.

Potential Impacts of Climate Change on Water Resources in the Kunhar River Basin, Pakistan

Rashid Mahmood, Shaofeng Jia and Mukand S. Babel

Abstract: Pakistan is one of the most highly water-stressed countries in the world and its water resources are greatly vulnerable to changing climatic conditions. The present study investigates the possible impacts of climate change on the water resources of the Kunhar River basin, Pakistan, under A2 and B2 scenarios of HadCM3, a global climate model. After successful development of the hydrological modeling system (HEC-HMS) for the basin, streamflow was simulated for three future periods (2011–2040, 2041–2070, and 2071–2099) and compared with the baseline period (1961–1990) to explore the changes in different flow indicators such as mean flow, low flow, median flow, high flow, flow duration curves, temporal shift in peaks, and temporal shifts in center-of-volume dates. From the results obtained, an overall increase in mean annual flow was projected in the basin under both A2 and B2 scenarios. However, while summer and autumn showed a noticeable increase in streamflow, spring and winter showed decreased streamflow. High and median flows were predicted to increase, but low flow was projected to decrease in the future under both scenarios. Flow duration curves showed that the probability of occurrence of flow is likely to be more in the future. It was also noted that peaks were predicted to shift from June to July in the future, and the center-of-volume date—the date at which half of the annual flow passes—will be delayed by about 9–17 days in the basin, under both A2 and B2 scenarios. On the whole, the Kunhar basin will face more floods and droughts in the future due to the projected increase in high flow and decrease in low flow and greater temporal and magnitudinal variations in peak flows. These results highlight how important it is to take cognizance of the impact of climate change on water resources in the basin and to formulate suitable policies for the proper utilization and management of these resources.

Reprinted from *Water*. Cite as: Mahmood, R.; Jia, S.; Babel, M.S. Potential Impacts of Climate Change on Water Resources in the Kunhar River Basin, Pakistan. *Water* **2016**, *8*, 23.

1. Introduction

The concentration of greenhouse gases (GHGs) has dramatically increased during the last few decades because of anthropogenic forces such as burning of fossil fuels and biomass, land use changes, rapid industrialization, and deforestation. This increased GHGs concentration has resulted in global warming and a global energy imbalance [1,2]. According to the Fifth Assessment Report of the Intergovernmental Panel on Climate Change (IPCC-AR5), the global average temperature has increased by 0.85 °C (0.65 °C–1.06 °C) over the period of 1800–2012, relative to 1961–1990 [3], and 0.74 °C $\pm$ 18 °C has been detected during the last hundred years (1906–2005) [4]. There is a very high likelihood of this trend in global warming to only exacerbate. The global average temperature is projected to increase by 1.7 °C–4.8 °C in 2081–2100 (relative to 1986–2005) under different Representative Concentration Pathways (RCPs).

The projected increase in global warming is likely to intensify the hydrological cycle of the world, and hence, disturb the existing hydrological system. As a consequence of global warming, hydrological systems are likely to experience changes in the average availability of water, as well as changes in extreme events such as increase in precipitation intensity, frequency, and/or amount of heavy precipitation [5–7]. For example, the average annual precipitation is likely to increase in the high latitudes and the equatorial Pacific Ocean by the end of the 21st century under RCP8.5. However, in several mid-latitude areas and subtropical dry regions, mean precipitation is likely to decrease. The areas of the globe with increasing precipitation are much more than those with decreasing precipitation [3]. This disturbance in a hydrological system can pose problems for public health, industrial and municipal water demand, water energy exploitation, and the ecosystem. However, as stated above, the impact of climate change on hydrological systems may vary from region to region [2,3,5,8].

Hydrological systems are of great importance as they greatly affect the environmental and economic development of a region and are highly complex because they comprise the atmosphere, cryosphere, hydrosphere, biosphere, and geosphere. The hydrological cycle of a basin (catchment) is mainly influenced by the physical characteristics of the basin, climatic conditions in the basin, and human activities. Most studies on climate change have focused on temperature, precipitation, and evaporation [9] since these are considered to be the key indicative factors of climate change and variability in a river basin [10]. There is an increasing consensus that changing trends in climatic variables, especially temperature and precipitation, can change the hydrological and ecological conditions of a river basin [11,12]. Given all these conditions, it is of great importance to assess the possible impacts of climate change on the water resources of a region, which can help in the proper utilization and management of these water resources.

The economy of Pakistan relies greatly on agriculture, which is heavily dependent on the Indus River Irrigation System. Water issues in Pakistan are crucial challenges for the policymakers and managers of water resources in the country [13]. Today, Pakistan is one of the most water-stressed countries in the world as water availability in the country has reduced from about 5000 m^3 per capita per year in 1952 to 1100 m^3 per capita per year in 2006, which is an alarming situation [14]. According to the United Nations report, a country with water availability of less than 1000 m^3 per capita per year is considered a water scarce country [15]. The reduction in water availability has created tensions among different provinces due to the increasingly unequal distribution of water. Climate change and variability are likely to affect water availability for, and magnitude of, irrigation and hydropower production in the country. These changing water conditions are likely to increase the tensions among the provinces, especially at the downstream (Sindh province) [13]. Therefore, a clear estimation of future water resources under changing climatic conditions is of great importance for the planning, operation, and management of hydrological installations in any watershed in Pakistan.

In the last few years, outputs from General Circulation Models (GCMs)—the most advanced and numerical based coupled models—are fed into hydrological models to explore the impacts of climate change on water resources in various parts of the world in the future. However, these models are coarse in spatial resolution (about 200–500 km) [16] and might not be suitable at the basin level, especially small basins which require very fine spatial resolution [17,18]. To bridge GCM resolution and basin scales, downscaling—dynamical and statistical—techniques have been developed. In dynamical downscaling, a high-resolution numerical based Regional Climate Model (RCM), with a resolution of about 5-50 km, uses the course outputs of a GCM and provides detailed information or high resolution outputs at the basin level [2]. On the other hand, Statistical downscaling (SD) methods (*i.e.*, stochastic weather generator, regression, and weather typing) create empirical/statistical relationships between the GCM scale and basin scale variables (e.g., temperature and precipitation). Compared to dynamical downscaling, SD methods are faster and computationally inexpensive, and thus offer approaches that have been rapidly adopted by a wider community of scientists [17]. For the present study, downscaled data (maximum temperature, minimum temperature, and precipitation) was obtained from [1]. They used a Statistical Downscaling Model (SDSM), a combination of multiple linear regression and a stochastic weather generator, to downscale temperature and precipitation over the period of 2011–2099 under A2 and B2 scenarios of HadCM3, a global climate model.

Different studies such as Akhtar *et al.* [13], Ahmad *et al.* [19], Shrestha *et al.* [20], and Bocchiola *et al.* [21] have assessed the impacts of climate change on the water resources of Pakistan [13,19–21]. These studies were mostly conducted in the Upper Indus basin using hydrological models such as Snowmelt Runoff model (SRM), Hydrologiska Byråns Vattenbalansavdelning (HBV), Soil and Water Assessment Tool (SWAT), and the WEB-DHM-S model. However, to the best of our knowledge, no studies have been conducted to explore the potential impacts of climate change on the water resources of the Kunhar River basin, which is one of the main tributaries of the transboundary Jhelum River and is located entirely in Pakistan. The Kunhar River originates from the Greater Himalayas and contributes to the Mangla Reservoir after joining the Jhelum River. The Mangla Reservoir's water is used for irrigation and hydropower production. It is the second largest reservoir of Pakistan. Although HEC-HMS, a well-known hydrological model, has been successfully used for small to large and plains to mountainous areas of the world [22–27], almost no studies have been reported in Pakistan that use HEC-HMS for the assessment of climate change impacts on water resources.

Thus, the main objectives of the present study were: (1) the development of HEC-HMS, a hydrological model, in the mountainous Kunhar River basin which is greatly influenced by winter snowfall; and (2) to assess the possible impacts of climate change on the water resources of the Kunhar River basin, located in Pakistan, under A2 and B2 scenarios of HadCM3. Description of the data and the study area are given in Sections 2 and 3 of this paper. A brief introduction of the hydrological model used in the study, and the methods used for the analysis of streamflow, is given in Section 4. Sections 5 and 6 include the results/discussion and conclusions respectively. This study will be very useful for policies that govern the utilization and management of the water resources of the Kunhar basin, which contributes significantly to the Mangla reservoir, under the effects of climate change.

2. Data Description

2.1. Hydro-Meteorological Data

Observed daily historical data of three climate stations (maximum temperature, minimum temperature, and precipitation) were collected from the Pakistan Meteorological Department (PMD) and the streamflow data of two hydrometric stations, spanning 1961 to 2000, from the Water and Power Development Authority (WAPDA). The geographical and basic information of the hydro-meteorological stations is presented in Table 1 and Figure 1.

2.2. HadCM3 Downscaled Data

Downscaled data of maximum temperature (TX), minimum temperature (TN), and precipitation (PP) for the period of 2011-2099 was taken from [1]. They downscaled TX, TN, and PP with a well-known downscaling model, Statistical Downscaling Model (SDSM), under A2 and B2 scenarios of HadCM3 in the Jhelum River basin. They selected HadCM3 for two reasons: (a) it showed better results during the evaluation of various GCMs (CGCM2, CSISRO, CCSR/NIES, and HadCM3); and (b) it has been used by a majority of studies for statistical downscaling of climate variables around the Jhelum basin [1]. They also applied the bias correction technique on the downscaled data to get closer to realistic results.

Table 1. Basic information about hydro-climatic stations in the Kunhar River basin.

SR	Station Meteorological	Lat (° N)	Long (° E)	Altitude (m ASL)	Period (year)	PP Annual (mm)	TX Mean (°C)	TN Mean (°C)
1	Balakot	34.55	73.35	995	1961–2000	1514	24.9	12.4
2	Muzaffarabad	34.37	73.48	702	1961–2000	1477	28.4	13.5
3	Naran	34.9	73.65	2421	1961–2000	1522	12.3	3.2
						1504	**21.8**	**9.7**

SR	Hydrometric	Lat	Long	Altitude	Period	Mean Discharge Annual (mm)	m³/s	Catchment Area km²
1	Naran	34.90	73.65	2362	1961–2000	1441	47	1011
2	Gari Habibullah	34.40	73.38	810	1961–2000	1349	101	2335

PP, precipitation; *TX*, maximum temperature; *TN*, minimum temperature; *ASL*, above sea level.

SDSM is a hybrid of multiple linear regression and a stochastic generator [28] and is widely used for downscaling climate variables [1]. HadCM3 is a Global Climate Model (GCM) developed by the Hadley Center of the UK Meteorological Office. It has a spatial resolution of 2.5° × 3.75° (latitude × longitude), with a surface spatial resolution of about 278 × 417 km, decreasing to 278 × 295 km at 45 degrees North and South [29]. A2 and B2 are the emission scenarios developed by the IPCC. The A2 scenario describes a very heterogeneous world: very slow fertility patterns across regions, continuous increase in global population, regionally oriented economic development, more fragmented and slower per capita economic growth, and more rapid technological changes as compared to other scenarios. The B2 scenario presents the world with an emphasis on social and environmental sustainability and local solutions to economic issues. Under this scenario, the population of the world increases at a rate lower than under A2 and economic development is intermediate. B2 also posits less rapid and more diverse technological changes relative to the B1 and A1 scenarios [30].

Mahmood and Babel [1] have shown that TX, TN, and PP are projected to increase by 0.91 °C–3.15 °C, 0.93 °C–2.63 °C, and 6%–12% respectively under A2, and 0.69 °C–1.92 °C, 0.56 °C–1.63 °C, and 8%–14% respectively under B2. According to them, in autumn, winter, summer, and spring, precipitation is projected to increase by 24%–32%, 11%–17%, 9%–12% and 4%–5% under A2 and 12%–25%, 12%–15%, 10%–14%, 0.2%–4% respectively under B2. The projected increases in temperature (TX and TN) are 2.7 °C–3.0 °C, 2.5 °C–3.2 °C, and 2.0 °C–2.6 °C, 3.2 °C–3.9 °C under A2 and 1.6 °C–1.73 °C, 2.0 °C–2.5 °C, 1.4 °C–1.7 °C, 1.4 °C–1.8 °C under B2 by the end of this century.

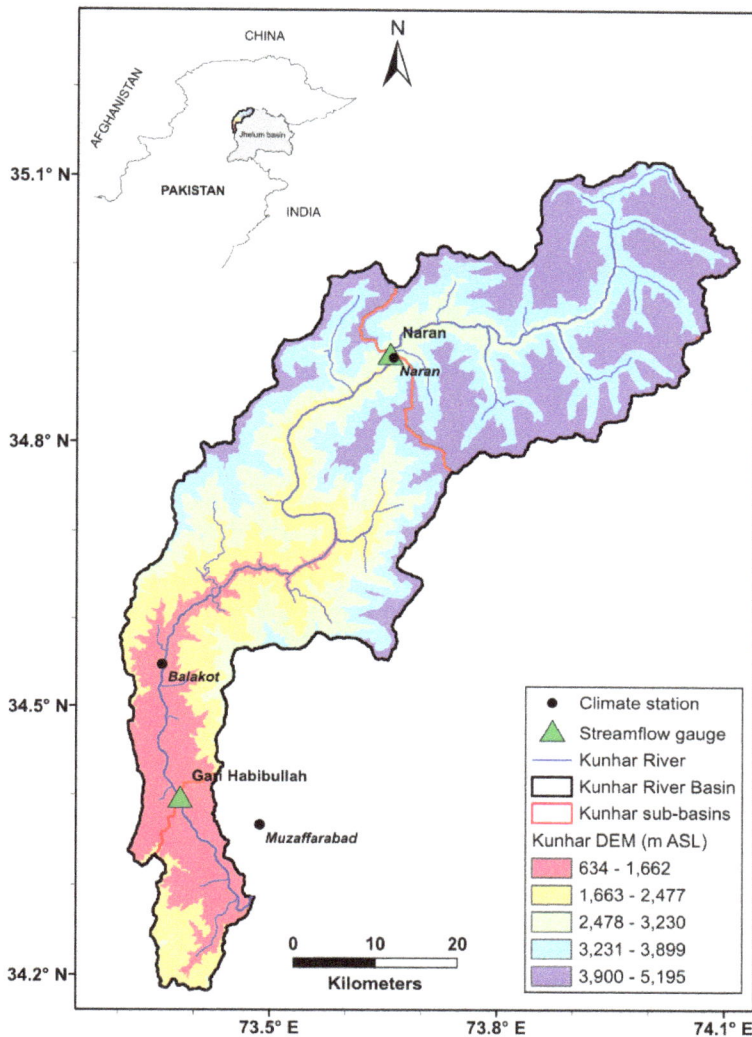

Figure 1. Location map of the Kunhar River basin showing hydro-climatic stations, streamlines, and altitude.

2.3. DEM Data

Elevation data is one of the primary sources used in hydrological studies. The most common form of elevation data is the digital elevation model (DEM) that is widely used to extract the topographical characteristics of a terrain [31,32]. In this study, a DEM of 30 m, created by the Advanced Spaceborne Thermal Emission and Reflection Radiometer (ASTER), was obtained from the U.S. Geological Survey (USGS) (http://gdex.cr.usgs.gov/gdex/). This DEM is shown in Figure 1.

ASTER-DEM was developed by the National Aeronautics and Space Administration (NASA) of USA, the Japan Space Systems (JSS), and the Ministry of Economy, Trade and Industry (METI) of Japan. The coverage area of ASTER-DEM ranges between 83° N and 83° S, covering 99% of the Earth's landmass.

In the present study, DEM was processed with HEC-GeoHMS, an extension of GIS, to extract basin parameters like slopes of rivers and sub-basins, areas of sub-basins, river lengths, longest flow paths, elevations in the basin, flow direction, streamlines, contour lines, and aspects during the process of watershed delineation. Among them, contours, aspects, slope, and the delineated sub-basin are shown in Figure 2.

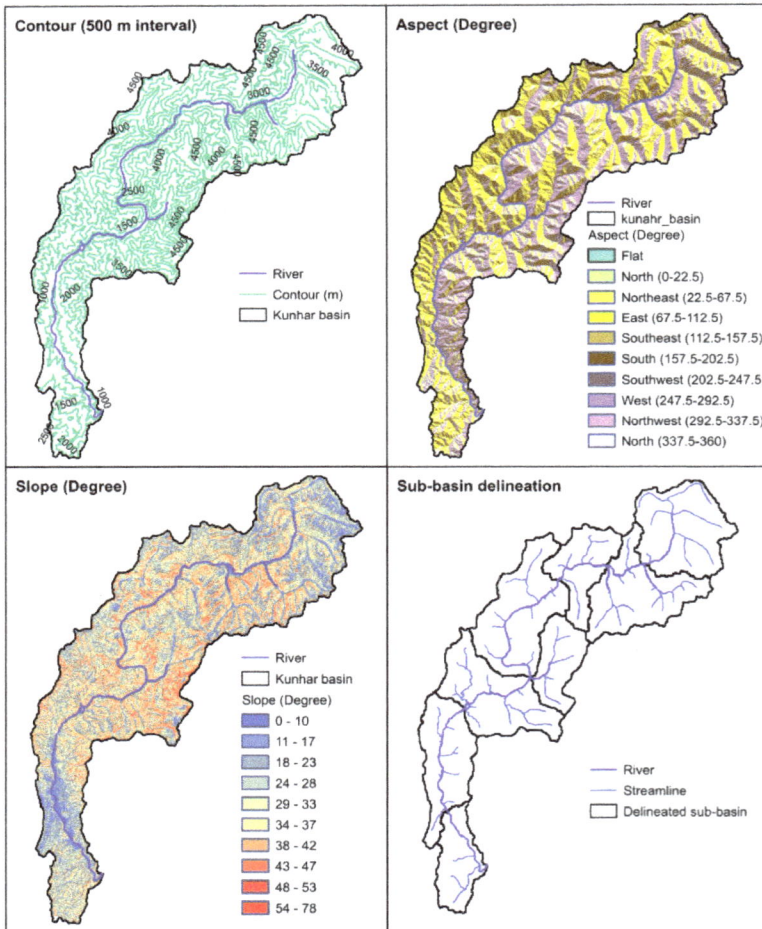

Figure 2. Topographical features and streamline delineation of the Kunhar River basin.

2.4. Land Cover and Soil Data

Land cover can strongly affect the hydrological processes in a region. These processes are mainly influenced by the density of plant cover and the morphology of plant species [33]. The land cover data for the Kunhar River basin was derived from the global land cover data (1 km resolution) developed by the Joint Research Center (JRC) of the European Commission (http://eusoils.jrc.ec.europa.eu/data.html). This data is shown in Figure 3.

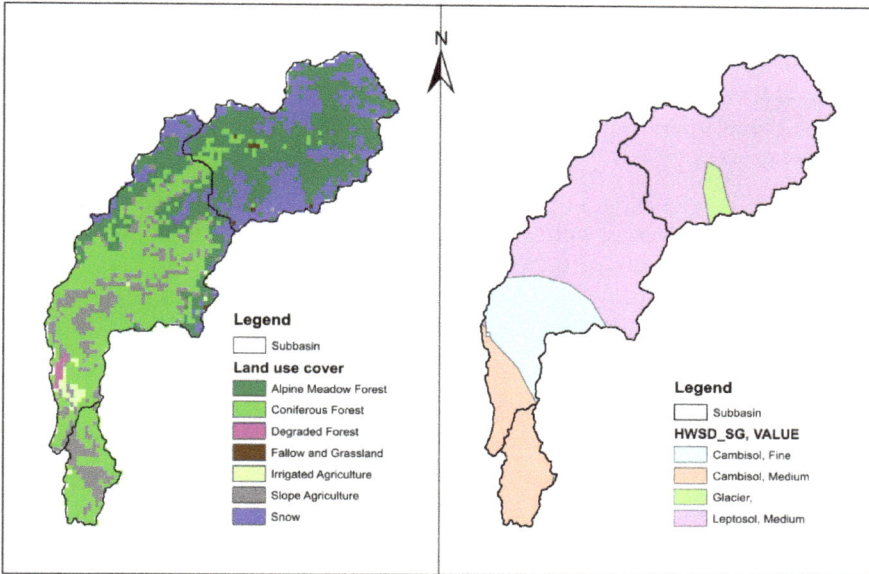

Figure 3. Land use and soil map of the Kunhar River basin.

The soil data for the Kunhar basin was obtained from the Harmonized World Soil Database, with a resolution of 1 km, as is shown in Figure 3. This database was developed by FAO in collaboration with the International Institute Of Applied Systems Analysis (IIASA), the International Soil Reference and Information Centre (ISRIC) of World Soil Information, the Institute of Soil Science of Chinese Academy of Sciences (ISSCAS), and the Joint Research Centre of the European Commission (JRC) (http://www.fao.org/soils-portal/soil-survey/soil-maps-and-databases/harmonized-world-soil-database-v12/en/).

The soil and land cover data were used to extract the initial estimations of the hydrological properties of the basin such as maximum moisture deficit, hydraulic conductivity, crop coefficient, percentage of imperviousness, and basin lag. However, the exact estimates of these parameters for each sub-basin were obtained only during the calibration process.

217

3. Study Area

The Kunhar River basin is located in the northern side of Pakistan and stretches between 34.2°–35.1° N and 73.3°–74.1° E, as is shown in Figure 1. The Kunhar drains the southern slope of the Greater Himalayas, located in the Khyber Pakhtunkhwa (KPK) province of Pakistan. It originates from the Lulusar Lake in the Kaghan valley of KPK. It passes through Jalkhand, Bata Kundi, Naran, Kaghan, Kawai, Balakot, and Gari Habibullah and finally joins the Jhelum River at Rara. The Kunhar's water is rich in algal flora, resulting in a great diversity of the aquatic life it harbors [34]. It has a drainage area of 2535 km^2, with elevation ranging from 600 to 5000 m (Figure 1).

The Kunhar is one of the biggest tributaries of the transboundary Jhelum River basin. This is the only main tributary which is entirely situated in Pakistan's territory and therefore, has great importance from the perspective of hydrological monitoring by the WAPDA of Pakistan. The Kunhar contributes about 11% of the total flow to the Mangla Reservoir constructed in the Jhelum River basin. The Mangla Reservoir is the second largest water storage site in the country. The water stored here is primarily used for irrigating 6 million hectares of land in the country and for generating hydropower. The installed capacity of the Mangla power plant is 1000 MW, and electricity is generated as a byproduct [1,35]. Snowmelt from the Kunhar basin contributes about 65% to the total discharge of the Kunhar River and 20%–40% to the Jhelum River at Manga [35]. The population in the Kunhar basin is almost entirely rural and their economy is generally agro-pastoral based. The principal occupation of the population is agriculture, although rearing livestock is also practiced in the adjacent mountainous areas. A small portion of the population is involved in trade, local labor, and employment in the bigger cities of the country [36]. In Pakistan, 93% of the total annual flow is used for agricultural purposes, 5% for industrial use, and 2% for domestic use [37]. The water of the Kunhar River is mostly used for irrigation, municipal use, power generation, and recreation [35]. Since no industries are located in the basin, most of the water of the Kunhar Basin (about 98%) is used for agriculture and the rest for domestic use.

The data of other major topographical characteristics such as slope, contours lines, aspects, and delineated sub-basins, which were extracted from DEM, are presented in Figure 2. This figure shows that the basin has undulating relief ranging from 0° to 78°. The plains along the course of the Kunhar River are located on a gentle slope (0°–10°). However, most parts of the basin have moderate (>10° and <30°) to steep slopes (>30°).

The basin has a great diversity of vegetation such as temperate coniferous forests, subtropical coniferous forests, alpine meadows, agricultural cover, and snow, as described in Table 2 and shown in Figure 3. These varied land covers are reclassified into seven main classes to explore the major land use covers in the basin. Forest, agriculture, and snow cover the maximum area of the basin with about 65%, 14%, and 20% respectively (Table 2). This information was derived from the land cover data of 1 km resolution in the basin.

Table 2. Basic characteristics of soil and land use data in the Kunhar River basin.

	HWSD-Soil Group	Area		Texture	Top Soil Fraction			TSGC
		%	km^2		Sand (%)	Silt (%)	Clay (%)	Gravel (%)
1	Cambisol	14	344	Medium	42	36	22	9
2	Cambisol	13	342	Fine	22	30	48	8
3	Leptosol	71	1801	Medium	43	34	23	26
4	Glacier	2	48	-	0	0	0	0
	Land use cover	**Area**		**Land Use Cover**				
	Re-classes	%	km^2	**Original**				
1	Coniferous Forest	33.1	838	Temperate Conifer/Subtropical Conifer/Tropical Moist Deciduous				
2	Degraded Forest	0.4	11	Degraded Forest				
3	Fallow and Grassland	0.3	7	Slope Grasslands/Sparse vegetation (cold)/Gobi/Desert (cold)				
4	Alpine Meadow Forest	31.8	805	Alpine Meadow				
5	Irrigated Agriculture	1.0	25	Irrigated Agriculture				
6	Slope Agriculture	13.0	329	Slope Agriculture				
7	Snow	20.5	521	Snow				

HWSD, Harmonized World Soil Database; *TSGC*, Top Soil Gravel Content.

There are three main groups of soils in the Kunhar River basin, as shown in Figure 3: 1) cambisol fine, 2) cambisol medium, and 3) leptosol. Cambisol characterizes weak to moderately developed soils and it (both fine and medium varieties) covers 27% of the basin, while leptosol is very shallow soil over hard rock and is unconsolidated and very gravelly material. Leptosol covers around 71% area of the basin. Some glacier patches also exist in the upper part of the basin. The basic information about soil types found in the basin is given in Table 2. The properties of these soils were derived from the soil data of 1 km resolution in the basin.

The Kunhar basin is located in a humid, subtropical zone. In the present study, hydro-climatic data was processed for the period of 1961–2000 to extract some basic information about the hydro-climatic conditions in the basin. This information is presented in Figure 4. The average annual temperature in the basin is about 13 °C (2 °C–23 °C). February is the coldest and July is the warmest month here. This was calculated from the data of the Naran and Balakot climate stations available for the period of 1961-2000. At Balakot, located in the lower part of basin (Figure 1), TN ranges from 2 °C (January) to 23 °C (July) and TX from 14 °C (January) to 35 °C (June). At Naran, located in the upper part of the basin, TN ranges from −9 °C (February) to 13 °C (July) and TX from −1 °C (February) to 23 °C (July) (Figure 4a). The Kunhar basin has an annual precipitation of about 1500 mm (Table 1) with two peaks (Figure 4b). The first peak occurs in the upper part of the basin in the month of March because of the Western Disturbances (WDs) system in winter. Most parts of Pakistan, and its northwestern parts primarily, obtain precipitation due to WDs. WDs are caused by depressions over Mediterranean regions, resulting in precipitation over central and southwest Asia in the months of December to March [38,39]. The second peak happens in the month of July and in the lower parts of the basin due to the summer monsoons, which are the result of the saturated south western winds from the Bay of Bengal and Arabian Sea. It can be concluded that the monsoons do not reach the upper part of the basin, although WDs affect the whole basin. On the other hand, there is only one big streamflow peak, both in the upper (Naran) and lower parts (Gari Habibullah) of the basin, that occurs in the month of July (Figure 4b). This means the precipitation from December to March (winter) accumulates as snow cover, especially in the upper parts of the basin, and then starts melting after March and lasts till July when it overlaps with monsoon precipitation and results in one big peak. An average flow of about 1350 mm ($103 \text{ m}^3/\text{s}$) has been measured at Gari Habibullah near the mouth of the basin for the period of 1961-2000. The same was reported by de-Scally as well [39].

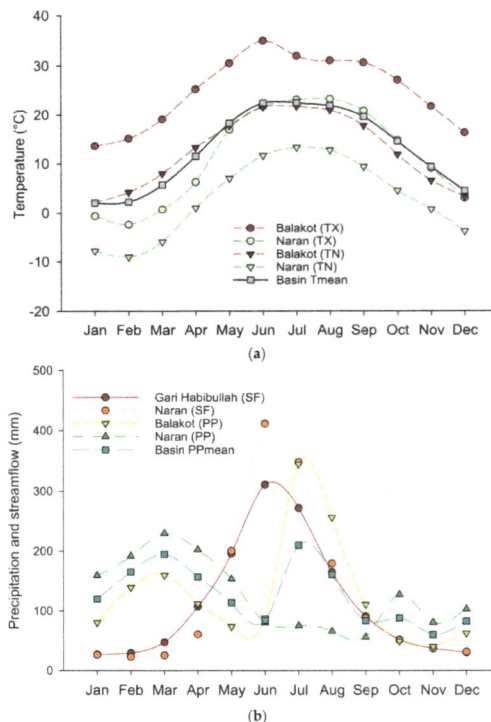

Figure 4. Monthly (**a**) max temperature (TX), min temperature (TN), (**b**) precipitation (PP), and streamflow (SF) in the Kunhar River basin for the period of 1961–2000.

4. Methodology

4.1. Description of HEC-HMS

The Hydrological Modeling system (HEC-HMS) is a rainfall-runoff simulation software used for a wide range of watersheds from large river basins to small urban areas. The model was formulated by the U.S. Army Corps of Engineers at the Hydrologic Engineering Center (HEC). HEC-HMS comprises different loss techniques such as SCS curve number, initial and constant, Green Ampt, one-layer deficit-constant, Smith Parlange, and five-layer soil moisture accounting. These techniques are used to estimate excess precipitation in fairly simple to very complex infiltration and evapotranspiration environments. This model can be used for both event and continuous modeling.

In order to calculate direct runoff from excess precipitation, seven methods including SCS, Clark, Snyder, and ModClark are available in the system. This model also consists of five base flow methods including recession method, constant monthly method, and linear reservoir method, and six channel routing methods including Muskingum, and modified pulse methods. It also has six kinds of meteorological models like Thiessen and inverse distance methods to analyze meteorological data such as precipitation, evapotranspiration, and snowmelt. The meteorological model extracts the precipitation for each sub-basin in the watershed. However, currently, only two methods (Temperature Index and Gridded Temperature Index) are available to compute runoff from snowfall in this modeling system [40,41].

A complete basin model setup for rainfall-runoff processes comprises a basin model, a meteorological model, control specification, and input time series. The basin model describes the physical characteristics of the study region such as the areas of sub-basins and river lengths of a watershed. Each basin model in HEC-HMS consists of a loss method, a transforming method, a base flow method, and a channel routing method [23]. Control specification is one of the main components of this model's setup and it controls the simulation period. For example, it controls when the model is to start and stop, and what the time interval for simulation should be. The input time series encompasses precipitation, temperature, evapotranspiration, and observed streamflow *etc.*, which have a direct link with the basin's model and the meteorological model. A detailed description of the model's formulation and its various processes is provided in the User's Manual and Technical Reference Manual of HEC-HMS [40] (p. 318 and p. 157).

In the present study, the basin model was developed using the deficit and constant loss (DCL) method for calculating excess precipitation, the SCS unit hydrograph for transforming direct runoff, Muskingum for channel routing, and the recession method for base flow. The meteorological model was established using the Thiessen polygon gauge weight method for precipitation calculation, the temperature index method for snowmelt modeling, and the monthly evapotranspiration method. The Thiessen polygons were created and their weights were calculated by HEC-GeoHMs in accordance with the precipitation gauges. The same combination has been used in different studies [15,17,18].

The DCL model computes excess precipitation in a watershed. It is a single layer continuous method used for calculating the changes in soil moisture content. It is similar to the initial and constant loss (ICL) method but this method recovers the initial losses after a long period of no precipitation. This method contains four main parameters: maximum deficit, initial deficit, constant rate, and impervious percentage. These parameters can be initially estimated using soil and land cover data as initial inputs for the model but are finalized only during the calibration process.

The excess precipitation calculated from DCL was transformed into direct surface runoff by the SCS unit hydrograph method. Basin lag is the only parameter of the SCS method, and it needs to be determined during calibration. It can also be estimated as an initial value for calibration by multiplying the time of concentration by 0.6. In this study, the recession method was used to calculate the base flow which contributes to the total flow from the watershed. Three parameters—initial discharge, recession constant, and threshold—were determined during calibration.

To transfer the total flow from one point to other, the Muskingum method was used. This method is a simple mass conservation scheme for routing flow through channels. There are two main parameters for this method: travel time (K), and the Muskingum coefficient (X). The Muskingum coefficient ranges between 0 and 0.5.

The Thiessen polygon method was used to assign weights to each gauge in the watershed during the development of the meteorological model. For snowmelt modeling, different elevation bands were used for each sub-basin in the temperature index method (TIM). TIM is an extension of the degree-day technique to calculate flow from snowpack. In the degree-day approach, a fixed amount of snowmelt is assigned for each degree above freezing point. This method is a conceptual representation of the cold energy stored in the snowpack. This also takes care of past conditions and some other climatic factors during the calculation of snowmelt. Different parameters such as base temperature, wet melt rate, rain rate limit, melt rate pattern, lapse rate, and antecedent temperature index are required for this method [32]. A lapse rate of -7.0 $^\circ$/km was calculated for the study area and kept constant for the entire Kunhar River basin. The FAO Penman-Monteith method—recommended as the standard method for computing potential evapotranspiration [24]—was carried out to calculate the potential evapotranspiration in the basin. A schematic diagram for the setup of HEC-HMS in the Kunhar River basin is shown in Figure 5.

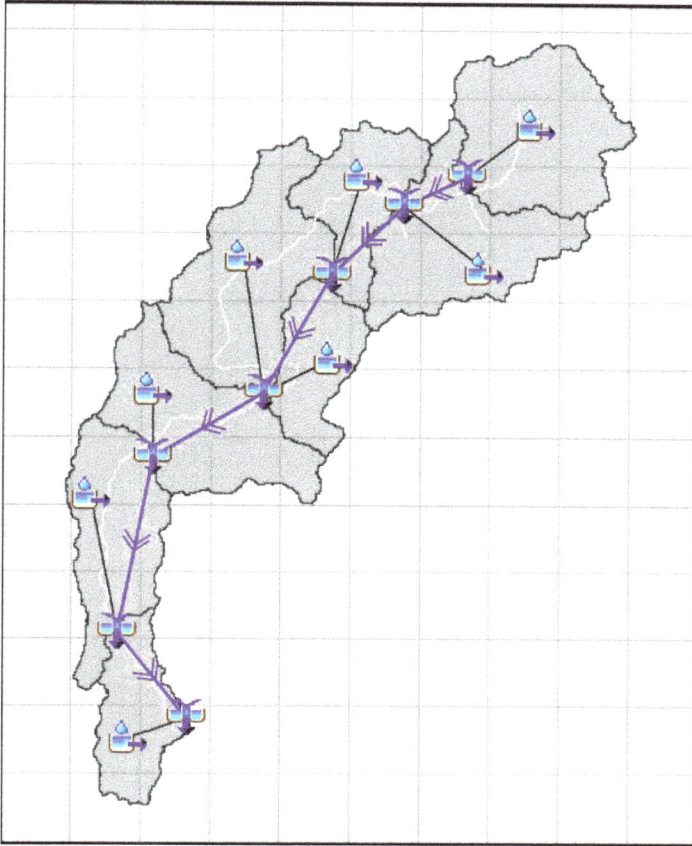

Figure 5. Schemetic diagram for the setup of HEC-HMS hydrological modeling system.

4.2. The Model's Calibration and Validation

The calibration of a model is a process in which the model's parameters are adjusted in such a way that the simulated flow captures the variations of the observed flow [25]. In this study, a split sample method was used for calibration and validation. In this method, the calibration period does not overlap with the validation period. A data period of eight years, from 1982 to 1989, was chosen as the calibration period and the period from 1978 to 1981 for validation because these periods had minimum missing values of both precipitation and streamflow. The physical properties—land use cover and soil properties—of the watershed were considered as constant during the simulation period.

Nelder Mead and Univariate Gradient are the two main algorithms to optimize the objective function. There are seven different kinds of objective functions in HEC-HMS. In this study, the sum of the squared residual was chosen and was minimized using the Nelder-Mead algorithm to explore the optimized model parameters in order to get the best results of the simulation. The simulated flows were compared with the observed flow using the coefficient of determination (R^2), percent deviation (D), and Nash-Sutcliffe efficiency (E). The R^2 values indicate how well the variations in the observed data are captured by the simulated data, D describes the mean percent deviation between observed and simulated flow, and E shows how well the observed plot fits with the simulated plot [22]. For more illustrative purposes, the simulated data was also compared with observed data graphically to explore how well the low and high observed flows were captured by simulated flow. In the present study, the model was calibrated and validated at both Naran and Gari Habibullah gauging stations.

The model's performance parameters—R^2, D, and E—were calculated using the following equations:

$$R^2 = \frac{\sum \left(Q_{obs} - \overline{Q_{obs}}\right) \times \left(Q_{sim} - \overline{Q_{sim}}\right)}{\sqrt{\sum \left(Q_{obs} - \overline{Q_{obs}}\right)^2 \times \left(Q_{sim} - \overline{Q_{sim}}\right)^2}} \tag{1}$$

Q_{obs} and Q_{sim} are observed and simulated values respectively. If the value of R^2 is close to 1, it indicates a good correlation between simulated and observed flows. The correlation is considered optimum if R^2 is exactly equal to 1 [22].

$$D\,(\%) = 100 \times \frac{\sum \left(Q_{sim} - Q_{obs}\right)}{\sum Q_{obs}} \tag{2}$$

The value of D should ideally be close to 0%. Positive and negative values are respectively indicators of over- and under-estimation by the model [22].

$$E = 1 - \frac{\sum \left(Q_{sim} - Q_{obs}\right)^2}{\sum \left(Q_{obs} - \overline{Q_{obs}}\right)^2} \tag{3}$$

The value of E lies between 0 and 1. A positive value close to 1 implies good calibration while a negative value close to 0 is not acceptable. If the value of E is greater than 0.75, then the results are considered to be good, and if it is between 0.36 and 0.75, the results are satisfactory [41].

4.3. Projected Changes in Streamflow

After successful calibration and validation, the downscaled daily time series (A2 and B2) of precipitation and temperature for the period of 2011 to 2099 were used as input for HEC-HMS to simulate daily flow data at both gauges (Naran and Gari Habibullah). The physical characteristics of the Jhelum basin were kept constant throughout the simulation period. However, it cannot be ignored that these characteristics do vary with time. The simulated data were divided into three periods: the 2020s (2011–2040), the 2050s (2041–2070), and the 2080s (2071–2099) and all three were compared with the baseline period (1961–1990) to assess changes in flow in the future. Different indicators such as mean flow, low flow, median flow, high flow, flow duration curves, temporal shift in peaks, and temporal shifts in center-of-volume dates were calculated for the three periods and the results were compared to the baseline period's data so as to explore the impact of climate change on the streamflow in the basin.

When analyzing streamflow to construct an installation such as a reservoir and headwork on a river, two questions are frequently asked: (1) how often will the streamflow occur in the future? and (2) what will be the magnitude of the streamflow? Flow duration curves are the main tools to deal with these two questions. These curves present the percentage of times that the flow in a stream is likely to exceed or be equal to a specified value of the flow. These curves can be applied in different kinds of studies such as hydropower management, reservoir sedimentation, water quality management, and low and high flow studies [42]. The following equation is used to construct the flow duration curves:

$$P\,(\%) = \frac{M}{(n+1)} \times 100 \qquad (4)$$

P or the probability of flow is equal to or exceeds a specified value (% of time), M is the rank of events, and N is the number of events in a specified period of time.

In the present study, the daily time series were used to construct the flow duration curves for the base period (1961–1990) and for the three future periods: the 2020s, 2050s, and 2080s.

5. Results and Discussion

5.1. Calibration and Validation

Table 3 shows the model's performance parameters (E, D, and R^2) that were calculated using the daily and monthly observed and simulated streamflows for calibration (1982–1989) and validation (1978–1981) at Naran and Gari Habibullah. In the case of daily data, the values of E and R^2 ranged from 0.55 to 0.74 during calibration and validation at both stations, and the D values lay between −9% and 12%. The D values show that the model underestimated values during the calibration and validation periods at Naran and also underestimated them at Gari Habibullah during calibration. However, the model overestimated the values during validation at the Gari Habibullah station. The results were greatly improved when these parameters were calculated from the monthly time series. The values of E and R^2 were improved to 0.63–0.88. These results are quite satisfactory and complement well some previous studies such as Meenu et al. [22] in India, Verma et al. [23] in India, Yimer et al. [24] in Ethiopia, and Garcia et al. [25] in Spain. All these studies also used HEC-HMS to simulate streamflow for climate change studies, with E ranging from 0.48–0.83 and R^2 from 0.63–0.84.

Table 3. Nash-Sutcliffe efficiency (E), coefficient of determination (R^2), and percent deviation (D) for calibration (1982–1989) and validation (1978–1981) for different stream gauges, calculated from daily and monthly data, in the Kunhar River basin.

Hydrometric Station	Calibration			Validation		
	E	R^2	D (%)	E	R^2	D (%)
From daily streamflow						
Naran	**0.71**	0.74	−7.78	0.71	0.73	−3.13
Gari Habibullah	0.66	0.67	−8.69	0.55	0.65	12.17
From monthly streamflow						
Naran	0.83	0.84	−11.02	0.83	0.84	−3.14
Gari Habibullah	0.88	0.85	−9.92	0.63	0.70	12.18

The graphical comparisons of observed against simulated flow for the calibration and validation periods are shown in Figures 6 and 7 respectively. At both Naran and Gari Habibullah gauging stations, patterns of observed flow were well matched by patterns of simulated flow during the calibration period. However, the peak and low flows were not captured well by the model. At Naran, in some years, peak and low flows were underestimated by the model but it was the reverse in the case of Gari Habibullah. This underestimation of flow might be due to the lack of enough precipitation gauges available in the basin or it might be due to the use of a simple temperature index model to calculate flow from snowmelt because about 65% of the total flow comes from snowmelt. Over- and under-estimation of flow must be considered during the discussion of this study's results. For example, at Gari Habibullah, the model overestimated the flow by 12%. This means that if the projected increase at Gari Habibullah is 50%, then the final projected increase will be 38%.

In the case of validation, the peaks were comparatively better captured by the simulated flow at both stations, especially at Naran. On the whole, the calibration and validation results were satisfactory and comparable with some previous studies like Meenu et al. [22], Verma et al. [23], Yimer et al. [24], and Garcia et al. [25].

The present study's results could be much more exact if the number of precipitation gauges were more than are currently available. In the present study, the vegetation cover (where 20% is snow cover) was considered constant for the entire calibration and validation periods. Vegetation cover, especially snow cover, could not have been the same for all the years of the calibration and validation periods.

It can be concluded that taking into consideration changes in vegetation cover will perhaps improve the results.

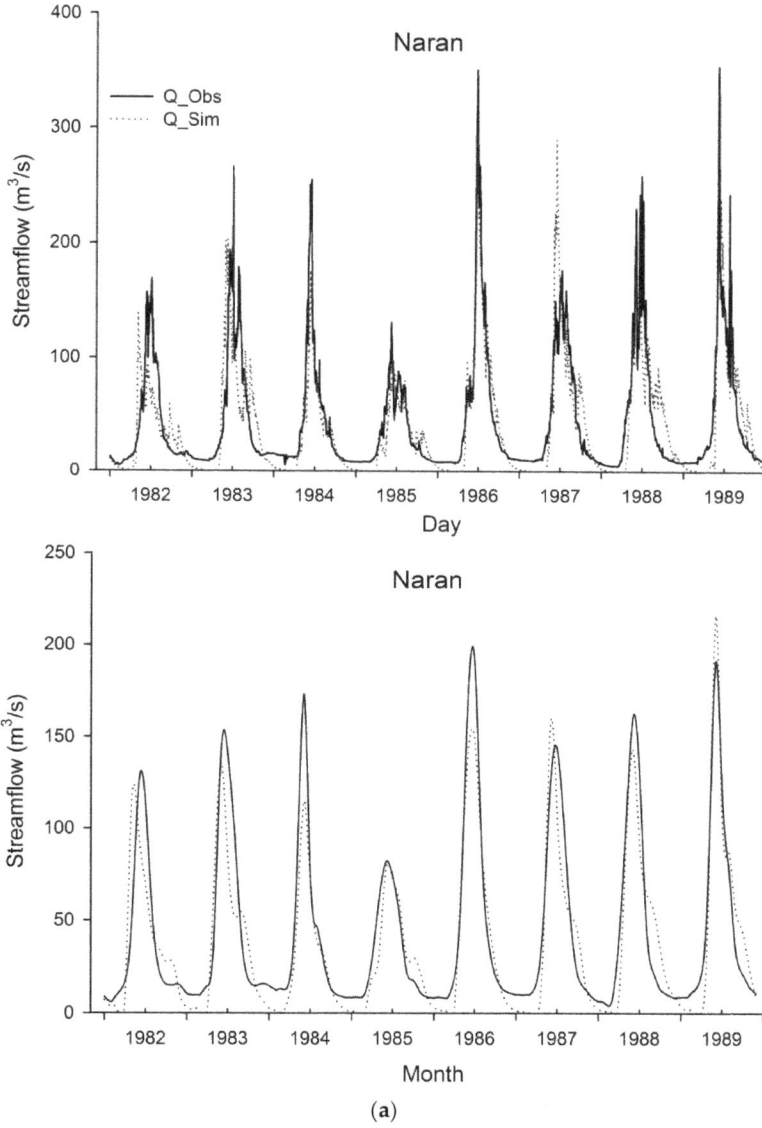

Figure 6. *Cont.*

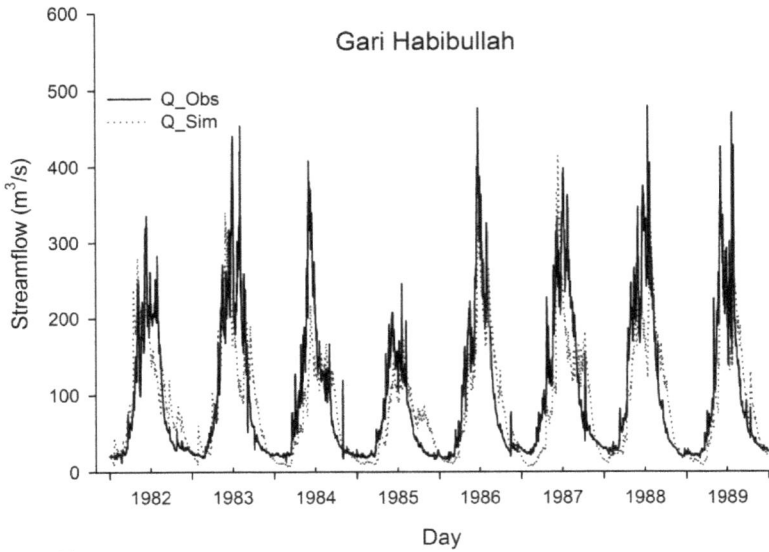

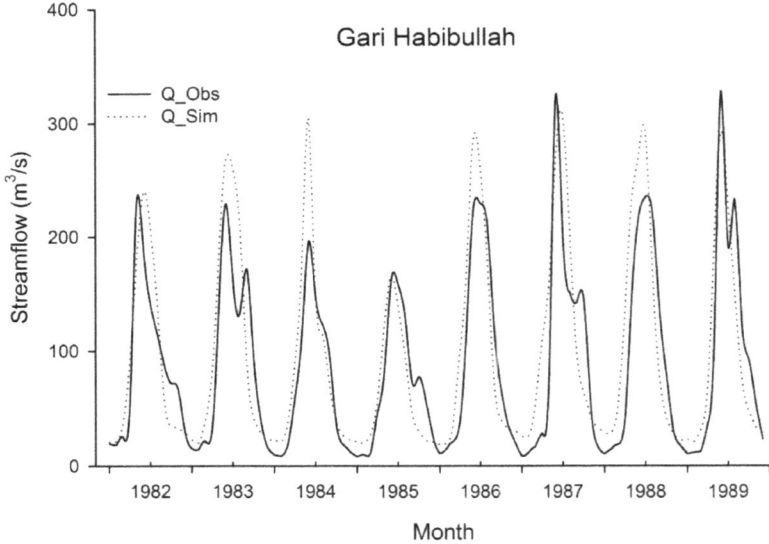

Figure 6. 3-dimensional (**a**) and 2-dimensional (**b**) space of the pressure measurements.

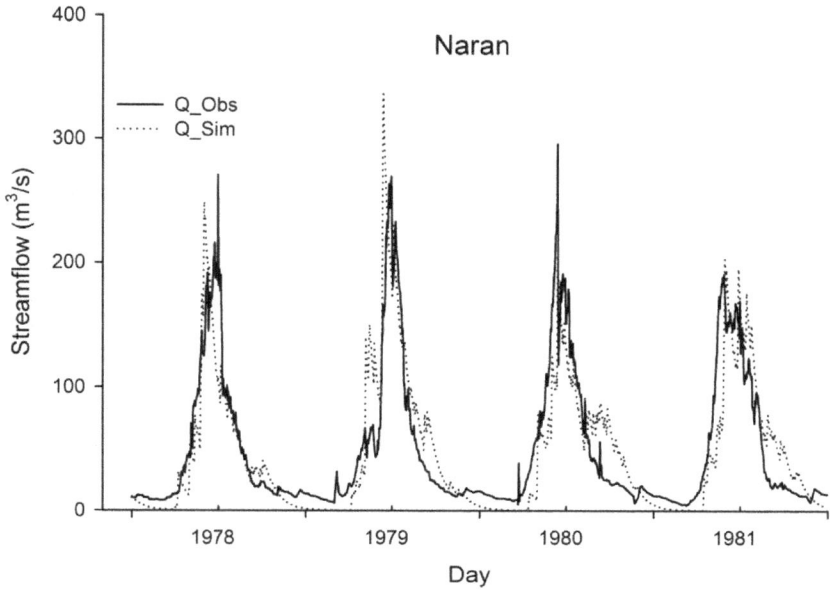

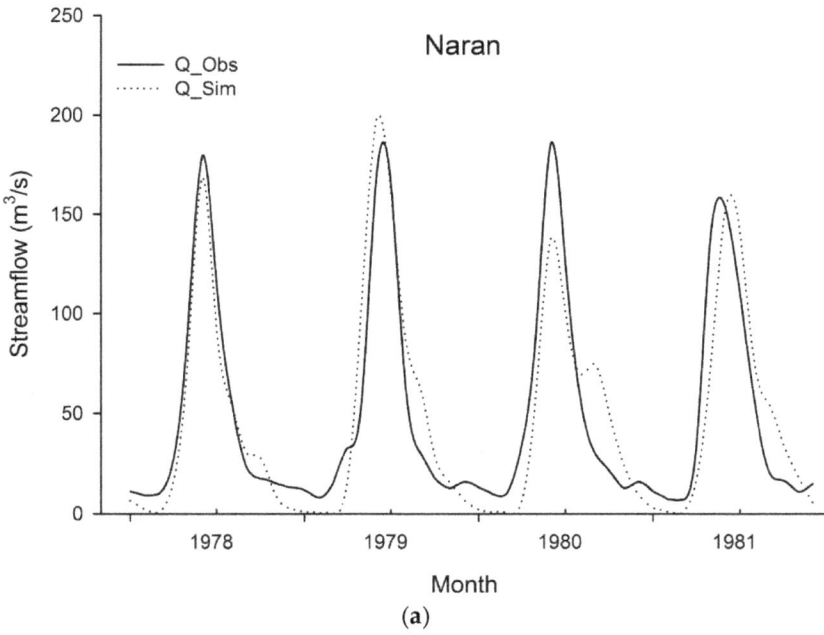

Figure 7. *Cont.*

229

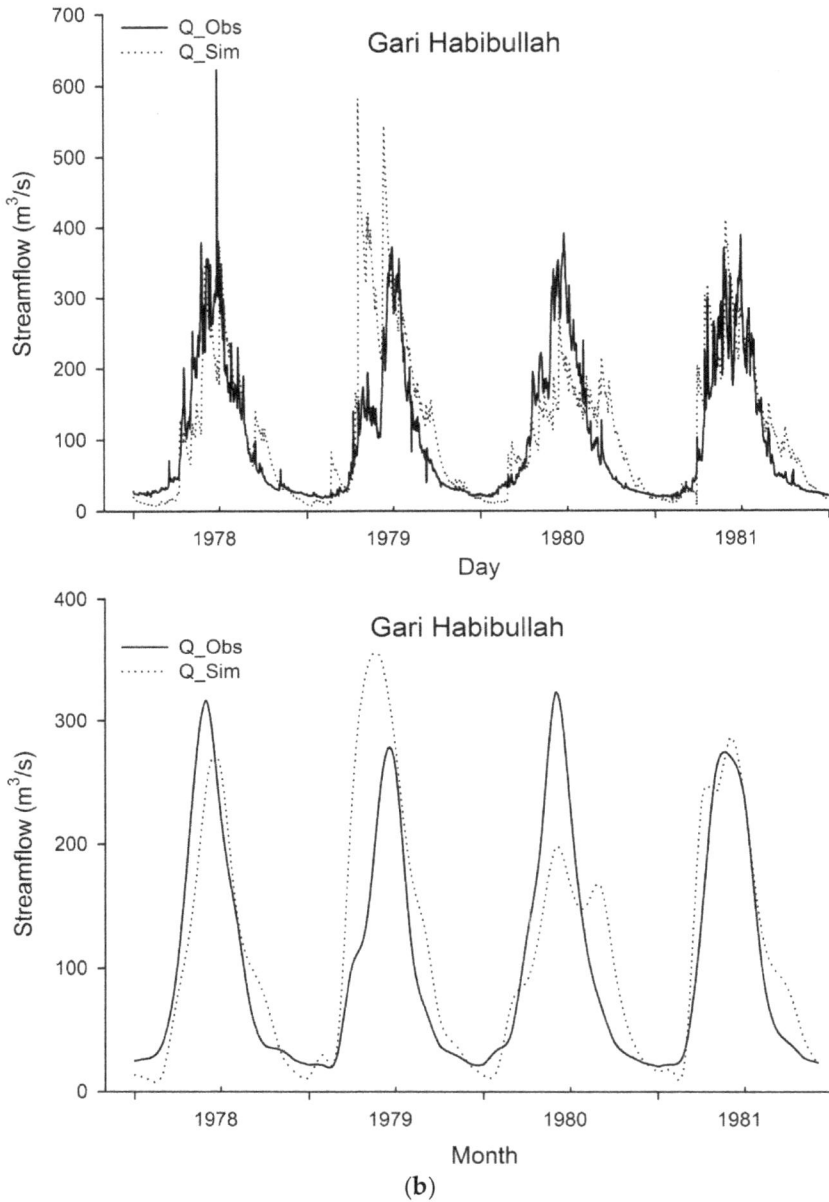

Figure 7. Comparison of daily and monthly observed and simulated streamflow at (**a**) Naran and (**b**) Gari Habibullah for validation (1978-1981) in the Kunhar River basin.

5.2. Projected Changes in Mean Streamflow

Table 4 outlines, in percentage, the projected changes in seasonal and annual flows in the 2020s, 2050s, and 2080s with respect to the baseline period (1961–1990) under A2 and B2 scenarios. In all three periods and under both scenarios, the flow in winter (DJF) and spring (MAM) seasons was projected to decrease at both gauges except in winter at Gari Habibullah in the 2080s. In both seasons, the negative changes at Naran (ranging from 40%–73%) were greatly higher than Gari Habibullah, (ranging from 6% to 28%) under both A2 and B2. Mahmood and Babel [43] conducted a study about extreme temperature events in the Jhelum River basin under A2 and B2, and they found that the intensity of cold days and cold nights will increase in the future. The reduction in flow (as indicated by the results of this study) is likely to be the case due to more accumulation of snowfall that would result from the increased intensity of cold days and cold nights in winter (based on the projections of Mahmood and Babel that indicate that precipitation will increase in winter and decrease in spring) [1].

Conversely, the summer (JJA) and autumn (SON) seasons were projected to have increased flow in all three future periods and under both scenarios but with different magnitudes. Autumn showed highest projected increase at both stations and under both scenarios, ranging from 91% to 131%. This high increase in autumn flow is most likely due to the most projected increase of 25%-32% in precipitation indicated by Mahmood and Babel [1]. The projected changes in this study appear much more than the projected changes in precipitation explored by Mahmood and Babel [1]. This could be because the vegetation cover was considered to be constant throughout the simulation period in this study. In the present state, snow cover (20%) dominates the basin and we used the same percentage of snow cover and constant proportions of other vegetation types for the simulation of projected flow. It is quite possible that snow cover will reduce in the future because of increasing temperature in the basin. So, along with validation results wherein the model overestimated by about 12%, if we also reduce snow cover, then the changes would be less than the projected changes.

On the whole, the mean annual flow was projected to increase by 42%–43% at the end of this century in the basin (at both sites) under both scenarios. The high increase in flow in summer and autumn and the decrease in winter and spring shows that it is hard to rely on only one GCM. So, a group of GCMs is required to give a clearer picture about these results.

The projected decrease in winter and spring can cause water shortage during these seasons not only for the agriculture sector but also for the domestic sector. On the other hand, the projected increase in summer (monsoon season) and autumn can result in flooding in the basin, and that can cause economic losses in the basin. These losses can be reduced by building reservoirs in the basin which can be used to store

231

water during the flooding season and the stored water can be used during the period of shortage.

Table 4. Future changes (%) in streamflow at different gauges relative to the baseline period (1961-1990) under A2 and B2 scenarios in the Kunhar River basin.

A2 Scenario	1961–1990 (m³/s)		2020s		2050s		2080s	
	Naran	G. Habib	Naran	G. Habib	Naran	G. Habib	Naran	G. Habib
Winter	10	25	−70	0	−71	−6	−70	8
Spring	37	110	−56	−14	−52	−26	−40	−12
Summer	121	226	53	38	51	46	61	49
Autumn	24	58	104	109	98	107	111	126
Annual	48	105	8	33	6	30	16	43
B2 Scenario								
Winter	10	25	−73	−2	−73	−11	−70	2
Spring	37	110	−62	−16	−58	−28	−49	−10
Summer	121	226	58	38	56	43	54	46
Autumn	24	58	88	97	92	91	97	131
Annual	48	105	3	29	5	24	8	42

5.3. Changes in Flow Duration Curve as Well as Low, Medium, and High Flows

Figure 8 displays the comparison of flow duration curves during the baseline period (1961–1990) and the three future periods (2020s, 2050s and 2080s) under A2 and B2 scenarios at the Gari Habibullah hydrometric station. This comparison shows that the probability of occurrence of flow and magnitudes of flow could be higher in the future in the Kunhar basin under both scenarios. Table 5 shows the projected changes in high flow (Q_5), median flow (Q_{50}), and low flow (Q_{95}) in the 2020s, 2050s, and 2080s with respect to the baseline under both scenarios. Under both scenarios, Q_5 and Q_{50} were projected to increase in all three future periods in the Kunhar basin, with 17%–52% (Q_5) and 43%–84% (Q_{50}). This increase in mean flow is most likely due to the increase in mean annual precipitation and increase in high flow is most likely due to the increase in summer precipitation (monsoon season), as also projected by Mahmood and Babel [1]. However, Q_{95} was predicted to decrease by 18%–99% in the future in the basin. This might be due to the increase in the intensity of cold days and cold nights (mentioned by Mahmood and Babel [43]) which may cause precipitation as snow accumulation in winter (low flow season).

These changes in high flow and flow duration curve show that the frequency of floods and their magnitudes will increase in the future which will create a lot of management problems in the basin. Flooding can not only cause economic losses but also loss of life. Nonetheless, with proper utilization and management of the increased flow, Pakistan can actually increase food and hydropower production in the basin.

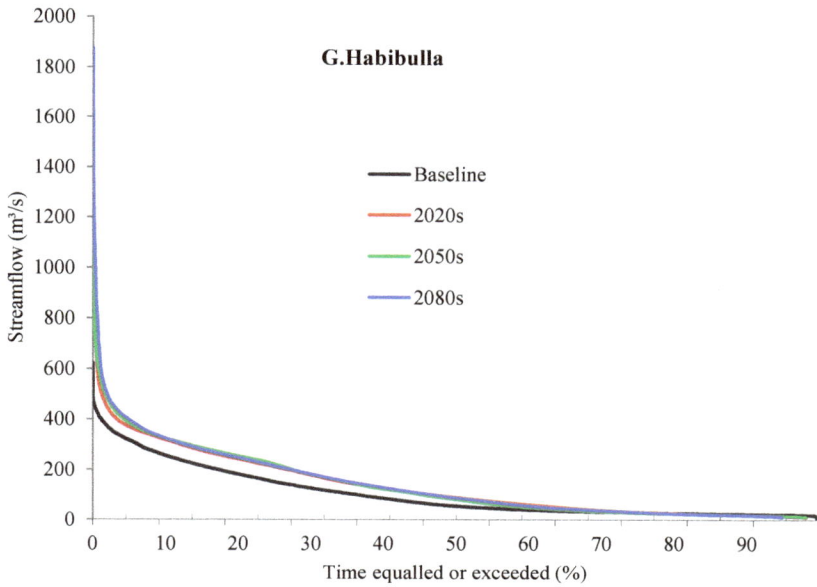

(a)

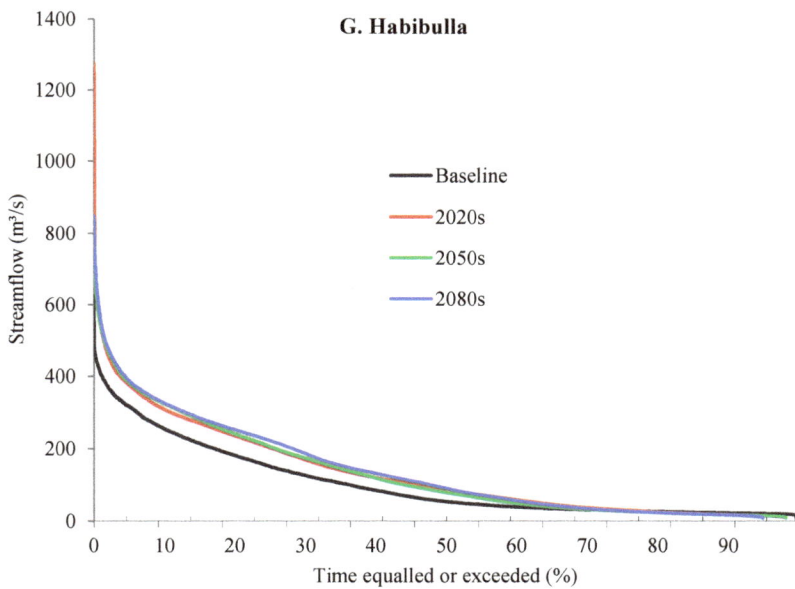

(b)

Figure 8. Flow duration curves under (**a**) A2 and (**b**) B2 scenarios at Gari Habibullah in the Kunhar River basin.

233

Table 5. Percent future changes in low, median, and high flows with respect to the baseline (1961–1990) under A2 and B2 scenarios in the Kunhar River basin.

A2 Scenario	1961–1990 (m³/s)		2020s		2050s		2080s	
	Naran	G. Habib	Naran	G. Habib	Naran	G. Habib	Naran	G. Habib
Q_5	170	316	41	17	39	22	52	28
Q_{50}	19	54	43	67	44	55	61	76
Q_{95}	8	21	−99	−25	−99	−22	−99	−18
B2 Scenario								
Q_5	170	316	42	19	46	21	43	24
Q_{50}	19	54	31	65	43	51	45	86
Q_{95}	8	21	−99	−23	−99	−25	−99	−21

Q_5, high flow; Q_{50}, median flow; Q_{95}, low flow.

5.4. Temporal Shifts in Peak Flows

In Figure 9, the mean monthly discharge of the baseline period (1961–1990) is plotted against the mean monthly discharge in the future periods (2020s, 2050s, and 2080s) to explore temporal shifts and magnitudes of peak flows at Gari Habibullah. At this site, a definite delay and increase in peak flows were projected in all three future periods under both scenarios. The peak flows were projected to shift from June to July, with about a 10%–25% increase under both scenarios. This shows that the basin will not only face an increase in frequency and magnitude of floods (mentioned in previous sections) but will also face the shift of these floods from June to July.

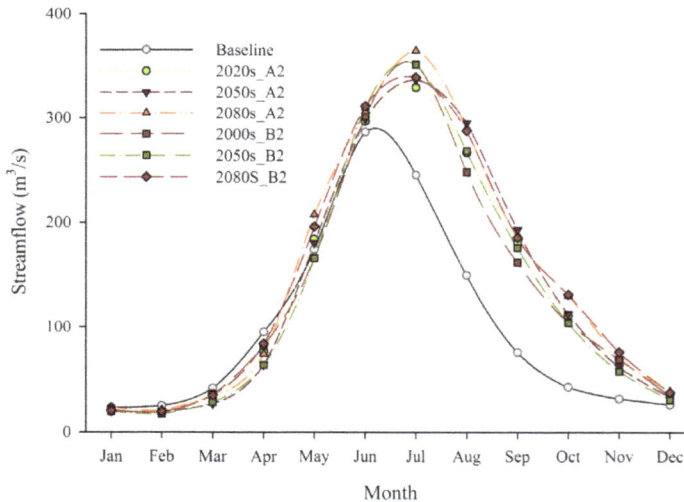

Figure 9. Future shift of peak flows in timing and magnitude under A2 and B2 in the Kunhar River basin.

5.5. Temporal Shifts in Center-of-Volume Date (CVD)

To determine the impact of climate change on the timing of streamflows, an indicator such as center-of-volume date (CVD)—a date at which half of the total volume of streamflow passes through at a gauging station for a specific time period—was used in the present study. CVD was calculated according to the equation described in [44]. Table 6 shows the changes in CVD under A2 and B2, with respect to the baseline period, in the three future periods at Naran and Gari Habibullah. The positive values show delay in flow, and the negative values show earlier in flows. The delay flows were projected at both stations in the Kunhar River basin under A2 and B2 in all three future periods, with 9–17 days delay. Table 6 shows that about half of the flow, on an average, of each year, in the Kunhar basin, passed through by 2–6 July for the baseline period (1961–1990) at both sites. However, this is predicted to shift to mid-July in the future.

Table 6. Future changes in center-of-volume dates (CVD) with respect to the baseline period (1961–1990) at different streamflow stations under both scenarios, A2 and B2, in the Kunhar River basin.

River	Station	CVD of Baseline		2020s	2050s	2080s
		Day of Year		A2 (B2) Day		
Kunhar	G. Habib	183	2 July	15(14)	17(16)	9(16)
Kunhar	Naran	187	6 July	13(14)	13(12)	11(10)

6. Conclusions

Pakistan is one of the most water-stressed countries in the world and its water resources are greatly vulnerable to changing climate. In the present study, the possible impacts of climate change on the water resources of the Kunhar River basin, Pakistan, were assessed under A2 and B2 scenarios of HadCM3. The Kunhar River originates in the Greater Himalayas and is one of the main tributaries of the transboundary Jhelum River. It is located entirely in Pakistan and contributes to the Mangla Reservoir after joining the Jhelum River.

The HEC-HMS hydrological model was used to simulate streamflow in the basin for the future. The model was calibrated and validated for the periods of 1982–1989 and 1978–1981 respectively, at two hydrometric stations (Naran and Gari Habibullah). Three indicators (*i.e.*, Nash Efficiency, coefficient of determination, and percentage deviation), and graphical representations of differences between observed and simulated data were used to check the performance of the model. Downscaled temperature and precipitation data for the period of 2011–2099 under A2 and B2 scenarios of HadCM3 were obtained from Mahmood and Babel [1] and fed into HEC-HMS to simulate the streamflow for the future. Mahmood and Babel projected an overall increase of 1.92 °C–3.15 °C in temperature and 5%–11% in precipitation in the basin. In this study, the simulated streamflow data was divided into three future periods (2011–2040, 2041–2070, and 2071–2099) and was compared with the baseline period (1961–1990). Different indicators like changes in mean flow, low flow, median flow, high flow, flow duration curves, temporal shift in peaks, and temporal shifts in center-of-volume dates were used to investigate the changes in streamflow under A2 and B2. The main conclusions of the study are the following:

1. Mean annual flow was projected to increase in the basin under both A2 and B2 scenarios. Noticeable increase in streamflow was predicted for summer and autumn. However, spring and winter showed decrease in flow.

2. High and median flows were predicted to increase but low flows were projected to decrease in the future under both scenarios. Flow duration curves showed that the probability of occurrence of flow will be more in the future, relative to the baseline.

3. Peaks were predicted to shift from June to July in the future. Similarly, center-of-volume date, a date at which half of the annual water passes, might get delayed by about 9–17 days in the basin under both A2 and B2.

The overall conclusion of the study is that the Kunhar basin is likely to face more floods and droughts in the future due to the projected increase in high flows and decrease in low flows. Many temporal and magnitudinal variations in peak flows would be faced in the basin. This can create many problems for the policy makers and managers of water resources if they do not consider the impacts of changing climate in the basin. For further studies of the basin, we recommend the use of different GCMs so as to cover the range of uncertainties related to GCMs and to explore a wider range of possible impacts of climate change on water resources in the basin.

Limitations of the Study

In the present study, the impact of climate change on the water resources of the Kunhar River basin were assessed by using only single GCM (HadCM3), although it is suggested that more than one GCM should be used to cover the uncertainties related to GCM outputs. Only two meteorological stations are available in the basin, an indication of the data scarcity of the basin. The scarcity of data can cause a lower level of performance by a hydrological model during the calibration and validation processes. Land cover and soil properties were considered constant throughout the simulation period; such an assumption can affect the projections of streamflow in the basin.

Acknowledgments: This study is part of the first author's doctoral research work, conducted at the Asian Institute of Technology (AIT), Thailand. The authors wish to acknowledge and offer gratitude to the Pakistan Meteorological Department (PMD) and the Water and Power Development Authority (WAPDA) of Pakistan for providing important and valuable data for the research. Heartfelt gratitude is also extended to the Higher Education Commission (HEC) of Pakistan for providing financial support to the first author for his doctoral studies at AIT. The authors are also grateful to the Natural Sciences Fund of China (41471463) and CAS (Chines Academy of Sciences) for providing financial support to complete this research.

Author Contributions: The first author conducted this research during his doctoral program under the supervision of the third author. The second author supervised the writing of this article and also reviewed it.

Conflicts of Interest: The authors declare no conflict of interest.

References

1. Mahmood, R.; Babel, M.S. Evaluation of sdsm developed by annual and monthly sub-models for downscaling temperature and precipitation in the jhelum basin, pakistan and india. *Theor. Appl. Climatol.* **2013**, *113*, 27–44.

2. Chu, J.; Xia, J.; Xu, C.Y.; Singh, V. Statistical downscaling of daily mean temperature, pan evaporation and precipitation for climate change scenarios in haihe river, China. *Theor. Appl. Climatol.* **2010**, *99*, 149–161.

3. Intergovernmental Panel on Climate Change (IPCC). *Climate Change 2013: The Physical Science Basis*; Contribution of Working Group I to the Fifth Assessment Report of the Intergovernmental Panel on Climate Change; Cambridge University Press: Cambridge, UK; New York, NY, USA, 2013; p. 1535.

4. Solomon, S., Qin, D., Manning, M., Chen, Z., Marquis, M., Averyt, K.B., Tignor, M., Miller, H.L., Eds.; *Climate Change 2007: The Physical Science Basis*; Contribution Of Working Group I to the Fourth Assessment Report of the Intergovernmental Panel on Climate Change; Cambridge University Press: Cambridge, UK, 2007.

5. Khattak, M.S.; Babel, M.S.; Sharif, M. Hydro-meteorological trends in the upper indus river basin in Pakistan. *Clim. Res.* **2011**, *46*, 103–119.

6. Ma, Z.Z.; Wang, Z.J.; Xia, T.; Gippel, C.J.; Speed, R. Hydrograph-based hydrologic alteration assessment and its application to the Yellow River. *J. Environ. Inform.* **2014**, *23*, 1–13.

7. Yang, W.; Yang, Z.F. Evaluation of sustainable environmental flows based on the valuation of ecosystem services: A case study for the Baiyangdian Wetland, China. *J. Environ. Inform.* **2014**, *24*, 90–100.

8. Zhang, X.C.; Liu, W.Z.; Li, Z.; Chen, J. Trend and uncertainty analysis of simulated climate change impacts with multiple GCMs and emission scenarios. *Agric. Forest Meteor.* **2011**, *151*, 1297–1304.

9. Wang, H.; Zhang, M.; Zhu, H.; Dang, X.; Yang, Z.; Yin, L. Hydro-climatic trends in the last 50 years in the lower reach of the Shiyang River Basin, NW China. *Catena* **2012**, *95*, 33–41.

10. Dhital, Y.; Tang, Q.; Shi, J. Hydroclimatological changes in the Bagmati River Basin, Nepal. *J. Geogr. Sci.* **2013**, *23*, 612–626.

11. Gleick, P. Regional hydrologic consequences of increases in atmospheric CO_2 and other trace gases. *Climatic Change* **1987**, *10*, 137–160.

12. Jordan, Y.C.; Ghulam, A.; Chu, M.L. Assessing the impacts of future urban development patterns and climate changes on total suspended sediment loading in surface waters using geoinformatics. *J. Environ. Inform.* **2014**, *24*, 65–79.

13. Akhtar, M.; Ahmad, N.; Booij, M.J. The impact of climate change on the water resources of Hindukush–Karakorum–Himalaya region under different glacier coverage scenarios. *J. Hydrol.* **2008**, *355*, 148–163.

14. Jetly, R. *Pakistan in Regional and Global Politics*; Taylor & Francis: New Dehli, India, 2012.

15. Connor, R. *The United Nations World Water Development Report 2015: Water for a Sustainable World*; UNESCO Publishing: Paris, France, 2015.

16. Gebremeskel, S.; Liu, Y.B.; de Smedt, F.; Hoffmann, L.; Pfister, L. Analysing the effect of climate changes on streamflow using statistically downscaled gcm scenarios. *Int. J. River Basin Manag.* **2005**, *2*, 271–280.

17. Wilby, R.L.; Hay, L.E.; Gutowski, W.J.; Arritt, R.W.; Takle, E.S.; Pan, Z.; Leavesley, G.H.; Martyn, P.C. Hydrological responses to dynamically and statistically downscaled climate model output. *Geophys. Res. Lett.* **2000**, *27*, 1199–1202.

18. Hay, L.E.; Wilby, R.L.; Leavesley, G.H. A comparison of delta change and downscaled GCM scenarios for three mountainous basins in The United States. *J. Am. Water Resour. Assoc.* **2000**, *36*, 387–397.

19. Ahmad, Z.; Hafeez, M.; Ahmad, I. Hydrology of mountainous areas in the upper Indus Basin, Northern Pakistan with the perspective of climate change. *Environ. Monit. Assess.* **2012**, *184*, 5255–5274.

20. Shrestha, M.; Koike, T.; Hirabayashi, Y.; Xue, Y.; Wang, L.; Rasul, G.; Ahmad, B. Integrated simulation of snow and glacier melt in water and energy balance-based, distributed hydrological modeling framework at Hunza River Basin of Pakistan Karakoram region. *J. Geophys. Res. Atmos.* **2015**.

21. Bocchiola, D.; Diolaiuti, G.; Soncini, A.; Mihalcea, C.; D'Agata, C.; Mayer, C.; Lambrecht, A.; Rosso, R.; Smiraglia, C. Prediction of future hydrological regimes in poorly gauged high altitude basins: The case study of the upper Indus, Pakistan. *Hydrol. Earth Syst. Sci.* **2011**, *15*, 2059–2075.

22. Meenu, R.; Rehana, S.; Mujumdar, P.P. Assessment of hydrologic impacts of climate change in Tunga-Bhadra river basin, India with HEC-HMS and SDSM. *Hydrol. Process.* **2012**, *27*, 1572–1589.

23. Verma, A.; Jha, M.; Mahana, R. Evaluation of HEC-HMS and WEPP for simulating watershed runoff using remote sensing and geographical information system. *Paddy Water Environ.* **2010**, *8*, 131–144.

24. Yimer, G.; Jonoski, A.; Griensven, A.V. Hydrological response of a catchment to climate change in the Upper Beles River Basin, Upper Blue Nile, Ethiopia. *Nile Basin Water Eng. Sci. Mag.* **2009**, *2*, 49–59.

25. García, A.; Sainz, A.; Revilla, J.A.; Álvarez, C.; Juanes, J.A.; Puente, A. Surface water resources assessment in scarcely gauged basins in the north of Spain. *J. Hydrol.* **2008**, *356*, 312–326.

26. Munyaneza, O.; Mukubwa, A.; Maskey, S.; Uhlenbrook, S.; Wenninger, J. Assessment of surface water resources availability using catchment modelling and the results of tracer studies in the mesoscale Migina Catchment, Rwanda. *Hydrol. Earth Syst. Sci.* **2014**, *18*, 5289–5301.

27. Babel, M.; Bhusal, S.; Wahid, S.; Agarwal, A. Climate change and water resources in the Bagmati River Basin, Nepal. *Theor. Appl. Climatol.* **2014**, *115*, 639–654.

28. Wilby, R.L.; Dawson, C.W.; Barrow, E.M. SDSM—A decision support tool for the assessment of regional climate change impacts. *Environ. Model. Softw.* **2002**, *17*, 145–157.

29. Gordon, C.; Cooper, C.; Senior, C.A.; Banks, H.; Gregory, J.M.; Johns, T.C.; Mitchell, J.F.B.; Wood, R.A. The simulation of sst, sea ice extents and ocean heat transports in a version of the hadley centre coupled model without flux adjustments. *Clim. Dyn.* **2000**, *16*, 147–168.

30. IPCC. *Emissions Scenarios: A Special Report of Working Group III*; Cambridge University Press: Cambridge, UK, 2000; p. 570.

31. Agarwal, C.S.; Garg, P.K. *Textbook on Remote Sensing: In Natural Resources Monitoring and Management*; A H Wheeler Publishing Co Ltd: New Delhi, India, 2002.

32. Singh, D.; Gupta, R.D.; Jain, S. Assessment of impact of climate change on water resources in a hilly river basin. *Arab. J. Geosci.* **2015**, *8*, 10625–10646.

33. Ghoraba, S.M. Hydrological modeling of the simly dam watershed (Pakistan) using GIS and SWAT model. *Alex. Eng. J.* **2015**, *54*, 583–594.

34. Leghari, M.K.; Waheed, S.B.; Leghari, M.Y. Ecological study of algal flora of Kunhar River, Pakistan. *Pak. J. Bot.* **2001**, *33*, 629–636.

35. De Scally, F.A. Relative importance of snow accumulation and monsoon rainfall data for estimating annual runoff, Jhelum Basin, Pakistan. *Hydrol. Sci. J.* **1994**, *39*, 199–216.

36. Hussain, M.; Shah, G.M.; Khan, M.A. Traditional medicinal and economic uses of Gymnosperms of Kaghan valley, Pakistan. *Ethnobot. Leafl.* **2006**, *10*, 72–81.

37. Pakistan Ministry of Water and Power (PMWP). *Executive Summary, Pakistan Water Sector Strategy*; PMWP: Islamabad, Pakistan, 2002; p. 24.

38. Ahmad, I.; Ambreen, R.; Sun, Z.; Deng, W. Winter-spring precipitation variability in Pakistan. *Am. J. Clim. Chang.* **2015**, *4*, 115–139.

39. Mahmood, R.; Babel, M.S.; Jia, S. Assessment of temporal and spatial changes of future climate in the Jhelum river basin, Pakistan and India. *Weather Clim. Extrem.* **2015**, *10*, 40–55.

40. *Hydrologic Modeling System HEC-HMS*; Institute for Water Resources: Davis, CA, USA, 2010.

41. Van Liew, M.W.; Garbrecht, J. Hydrologic simulation of the Little Washita river experimental watershed using SWAT1. *J. Am. Water Resour. Assoc.* **2003**, *39*, 413–426.

42. *HEC-ResSim Reservoir System Simulation User's Manual*; Hydrologic Engineering Center: Davis, CA, USA, 2007; p. 512.

43. Mahmood, R.; Babel, M.S. Future changes in extreme temperature events using the statistical downscaling model (SDSM) in the trans-boundary region of the Jhelum river basin. *Weather Clim. Extrem.* **2014**, *5–6*, 56–66.

44. Stewart, I.T.; Cayan, D.R.; Dettinger, M.D. Changes toward earlier streamflow timing across western North America. *J. Clim.* **2005**, *18*, 1136–1155.

Modeling Climate and Management Change Impacts on Water Quality and In-Stream Processes in the Elbe River Basin

Cornelia Hesse and Valentina Krysanova

Abstract: Eco-hydrological water quality modeling for integrated water resources management of river basins should include all necessary landscape and in-stream nutrient processes as well as possible changes in boundary conditions and driving forces for nutrient behavior in watersheds. The study aims to assess possible impacts of the changing climate (ENSEMBLES climate scenarios) and/or land use conditions on resulting river water quantity and quality in the large-scale Elbe river basin by applying a semi-distributed watershed model of intermediate complexity (SWIM) with implemented in-stream nutrient (N+P) turnover and algal growth processes. The calibration and validation results revealed the ability of SWIM to satisfactorily simulate nutrient behavior at the watershed scale. Analysis of 19 climate scenarios for the whole Elbe river basin showed a projected increase in temperature (+3 °C) and precipitation (+57 mm) on average until the end of the century, causing diverse changes in river discharge (+20%), nutrient loads (NO_3-N: -5%; NH_4-N: -24%; PO_4-P: +5%), phytoplankton biomass (-4%) and dissolved oxygen concentration (-5%) in the watershed. In addition, some changes in land use and nutrient management were tested in order to reduce nutrient emissions to the river network.

Reprinted from *Water*. Cite as: Hesse, C.; Krysanova, V. Modeling Climate and Management Change Impacts on Water Quality and In-Stream Processes in the Elbe River Basin. *Water* **2016**, *8*, 40.

1. Introduction

Changes of the world's and Europe's climate and increased anthropogenic pressure on natural resources have already been detected in the past, and this development is likely to continue in the future [1–6].

Looking at the climate aspect, a global rise in mean temperature, change in precipitation pattern as well as an increase in intensity and frequency of extreme events can be recognized [1,2,7], impacting the water cycle [8–10], vegetation and biodiversity [11–14] as well as human health [15,16] and economy [17,18]. The potential warming in Europe could reach values from +1 to +6 °C by the end of this century [19], depending on the location. The annual mean precipitation is expected to increase in Northern Europe and decrease in most parts of Southern Europe and Mediterranean regions up to ±20% [20]. The catchment of the Elbe river, mainly located in Germany and the Czech Republic in Central Europe, is already experiencing changes in climate conditions as well as changes in extreme temperature and precipitation values, and this trend is expected to continue. During the last century (1882–2013) the average temperature in Germany increased by 1.2 °C, whereas the precipitation amounts rose by 10.6% on average with a high spatial and temporal variability [21]. Application of ensembles of climate scenarios shows increasing trends in floods for the Elbe basin in Germany [22] as well as in the Czech Republic [23], especially in wintertime.

Climate change will have direct and indirect impacts on river water quantity and quality [24–28]. With the rising temperature, water availability will decrease due to increased evapotranspiration, and a reduced discharge will facilitate algal growth and reduce dilution of point source pollutants. Higher temperatures and lower flow velocities would additionally stimulate turnover processes and increase system respiration rates, causing oxygen deficits in river reaches. All these processes lead to the degradation of water quality and ecological status of a waterbody connected with a higher probability of algal blooms [24,29,30] and increasing problems for water supply and treatment [25].

As phytoplankton growth is a key factor for water quality in lowland river ecosystems [31], the algal processes should be included in evaluating impacts of global change on water quality. Light and nutrients are generally deemed to be the most important external drivers of phytoplankton dynamics in rivers, along with temperature which also has a positive relationship with phytoplankton, and discharge which has a negative relationship [29–31].

Additionally, climate change would influence nutrient turnover and transport processes (denitrification, nitrification, volatilization and leaching) in the catchments, due to altered temperature and precipitation [32–34], and the terrestrial processes will also influence the final river water quality at the outlet of the basin. River systems are also affected by anthropogenic impacts (land use composition, population and industry) causing nutrient pollution from point (factories, sewage treatment plants) and diffuse sources (agricultural fields), which lead to eutrophication processes and a decrease in river water quality [35–38].

Due to the high pressure of rising population, growing industry and increasing transport demand on landscapes and natural vegetation, many changes in land use could be recognized in Europe in the past. The current tendencies include a decreasing trend in crop and pasture acreages in Spain, Czech Republic and Northern Germany, slowly growing forested areas in Northern Europe and Portugal, and notably growing urban areas in France and Western Germany [5,39]. It is expected that these trends will continue in the coming 10–20 years.

Population density and human activities are important underlying factors for point and diffuse nutrient pollution entering rivers [40]. During the last decades, many efforts to reduce nutrient inputs to the rivers by construction and improvement of sewage treatment plants and optimization of fertilizer application on cropland were undertaken in Europe. They resulted in a remarkable reduction of phosphorus emissions (mainly from point sources), but only a small decrease of nitrogen inputs (mainly from diffuse sources) due to the lag time of diffuse nutrients in soils [41–43]. It is widely recognized that the control of diffuse source emissions is much more difficult. So it is expected that the inputs of nutrients from households and industry will be further reduced in the future, and diffuse inputs from fertilizers and manure will become the main sources of nutrient pollution in Europe [42].

Climate as well as land use change impacts on river water quality superimpose each other and create a very complex system of interactions and feedbacks [27,44,45]. The nitrate loads in the rivers, for example, are climate-dependent, and were likely influenced by former climate variations, so it is difficult to define and interpret the pure effects of management changes in the past [41]. Furthermore, adaptation measures and policy responses to projected climate change, e.g., subvention for bio-fuels or control of greenhouse gas emissions, affect freshwater quality as well [24]. A combined land use and climate change impact assessment would be an important step facilitating an integrated river basin management. The system characteristics and variable boundary conditions should be taken into account by default in modern management strategies [31] to support the implementation of adaptation measures in river basins.

The process-based eco-hydrological watershed models driven by climate and land use parameters can be useful for assessing potential future developments in a changing world. Watershed models including both landscape and in-stream nutrient processes, which are able to simulate nutrient turnover processes in a catchment and river network, may represent effective tools for the evaluation of river water quality at the basin scale [46,47]. However, it should be kept in mind that current water quality modeling at the watershed scale still has more weaknesses and uncertainties compared to pure hydrological modeling due to the higher complexity of modeled processes and the requirement to include more input data and parameters.

In former applications of the semi-distributed eco-hydrological Soil and Water Integrated Model (SWIM) [48] for water quality modeling in river basins in Germany, it was observed that nutrient retention and decomposition solely in the soils of the watershed is not sufficient for tackling nutrients coming from different sources, especially in larger basins [49]. Therefore, SWIM was extended by a new module representing nutrient and oxygen turnover processes and algal growth in rivers, which was already tested for the mesoscale Saale river, a sub-catchment of the Elbe river with an area of about 25,000 km^2 [50]. The aim of this study is to apply the new SWIM version for a combined climate and land use change impact assessment on the entire Elbe basin including the upstream part in the Czech Republic and the lower part in Germany (total drainage area about 150,000 km^2). This can support the development of management strategies and adaptation measures to potential changes in the future in this large-scale river basin.

In particular, the following objectives were pursued in this study:

- Testing the in-stream SWIM module for the large scale by modeling water quality parameters (nitrate nitrogen (NO$_3$-N), ammonium nitrogen (NH$_4$-N), phosphate phosphorus (PO$_4$-P), dissolved oxygen (DOX), and chlorophyll a (Chla)) at the basin outlet and at confluences of the large Elbe tributaries in the historical period;
- Analysis of climate scenarios for the region provided by the European ENSEMBLES project [51], and climate change impact assessment on water quantity and quality for two future periods (2021–2050 and 2071–2100) with unchanged management;
- Simulation of selected land use change and management scenarios aiming at the reduction of point and diffuse nutrient loads in the basin; and
- Analysis of the combined climate and land use change impacts on water quantity and quality to derive ideas for suitable measures for adaptation to climate change.

The model-based assessments of climate and land use change impacts on water quality are rare in literature so far compared to impact assessments on the hydrological cycle, especially at the large scale. The recently implemented in-stream module enables a more realistic representation of all interrelated processes for the impact study. Therefore, this study is an important step forward to large-scale application of water quality models with distributed calibration for impact studies in general, as well as towards a fully integrated water resources assessment in the Elbe catchment in particular.

2. Study Area: The Elbe Catchment

The Elbe river (1094 km) originates at 1386 m a.s.l. in the Giant Mountains located between Poland and the Czech Republic, drains an area of 148,268 km^2 and flows into the North Sea [52]. The Elbe has the fourth largest catchment area among the European rivers [31]. 65.5% of its catchment belongs to Germany, 33.7% to the Czech Republic, 0.6% to Austria and 0.2% to Poland (see Figure 1). The discharge regime (861 m$^3 \cdot$ s^{-1} on average) usually shows high water levels in winter and spring and low water levels in summer and autumn. In total, 24.5 million people live in the Elbe basin, which also includes the large cities Berlin, Hamburg and Prague [52].

Table 1 gives an overview of the main characteristics of the Elbe basin until the gauge Neu Darchau, and its six main tributaries, covering catchment areas above 5000 km^2. In this table some topography-specific differences can be detected between the tributary subbasins, namely in regard to climate parameters, soil conditions and, as a consequence, land use distribution, which also affect nutrient concentrations in the rivers. So, in the sub-catchments with dominating agricultural land use due to fertile loess soils (e.g., Saale and Mulde) the nitrate and nitrogen concentrations are higher (see Table 1), resulting from fertilizer application and leaching. In contrast, the catchment of the Havel river has the highest share of low fertile soils consisting of almost two-thirds of sandy grained particles (about 70% of the total area) and shows the lowest nitrogen pollution but the highest phosphorus level. The high phosphate concentrations of the Havel river can be mainly explained by desorption from historically polluted sediments [53].

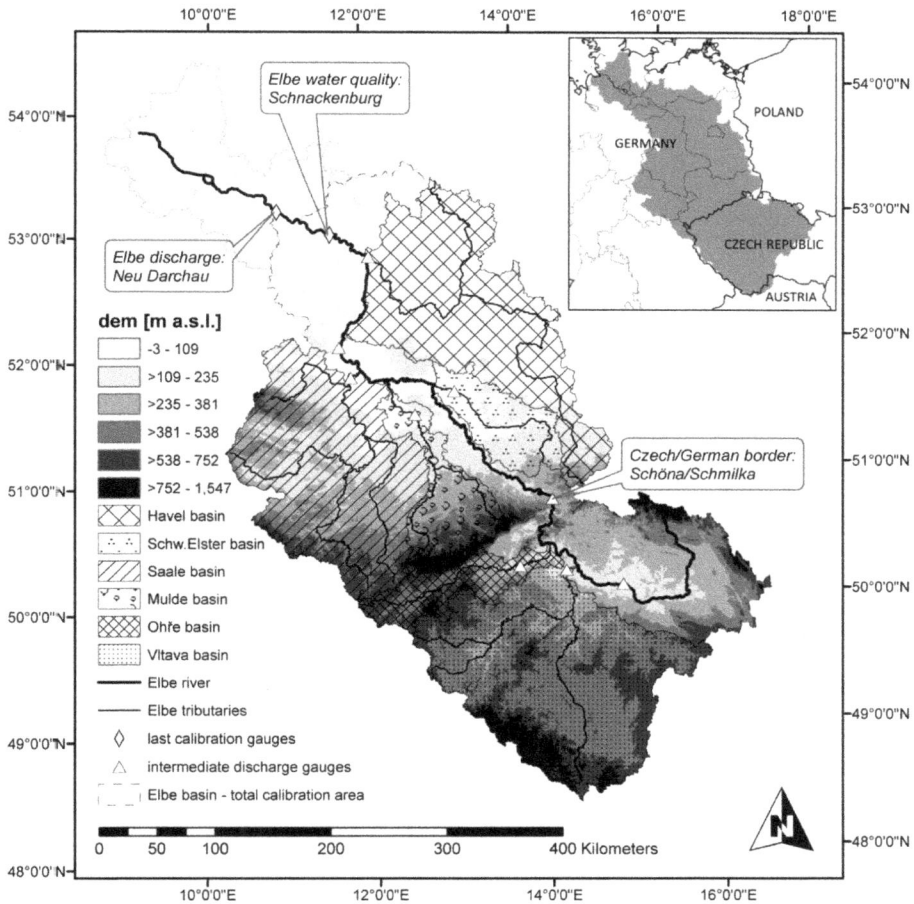

Figure 1. Location and digital elevation model of the Elbe river basin and six catchments of its main tributaries, as well as location of the observation gauges used for calibration.

245

Table 1. Characteristics of the Elbe river catchment and its main tributaries of second (classical) order for the time period 2001–2010.

River gauges (discharge/water quality)		Elbe Neu Darchau/Schmackenburg	Vltava Vraňany/Zelčín	Ohře Louny/Terezín	Schwarze Elster Löben/Gorsdorf	Mulde Bad Düben/Dessau	Saale Calbe-Grizehne/Groß Rosenburg	Havel Havelberg/Toppel	Unit
Length [1]		907	430	305	179*	314	434	334	km
Mean discharge [1]		711	145	38	21	67	117	114	$m^3 \cdot s^{-1}$
Catchment area [1]		131,950	28,090	5614	5705	7400	24,079	23,858	km^2
Average altitude		281	523	507	131	394	287	74	m a.s.l.
Average temperature [2]		8.9	7.8	7.6	9.7	8.9	9.2	9.6	°C
Average sum of precipitation [2]		698	713	771	652	822	680	616	$mm \cdot y^{-1}$
Land use [3]	Agriculture	51.3	49.7	42.2	48.1	53.3	63.0	38.6	%
	Forest	31.7	36.8	37.7	35.0	28.8	23.3	38.2	
	Grassland	8.4	7.8	13.6	7.2	6.9	4.6	11.1	
	Settlements	6.3	4.3	3.9	5.9	9.4	7.6	7.9	
Soil texture [4]	clay	16.2	20.3	22.1	8.5	17.1	20.0	9.0	%
	silt	38.2	37.4	39.6	29.8	47.9	54.9	26.3	
	sand	45.6	42.3	38.3	61.7	35.0	25.1	64.7	
Point sources [5]	TN	22318	4704	570	183	1673	3557	2768	$t \cdot y^{-1}$
	TP	1870	564	73	29	155	357	167	
Nutrients [6]	NO_3-N av.	3.17	3.73	2.38	2.31	4.35	4.68	0.82	$mg \cdot L^{-1}$
	NO_3-N 90-perc.	4.60	4.91	3.20	4.30	6.10	6.15	1.56	
	NH_4-N av.	0.16	0.31	0.08	0.20	0.16	0.21	0.10	
	NH_4-N 90-perc.	0.33	0.94	0.11	0.53	0.36	0.46	0.23	
	PO_4-P av.	0.07	0.12	0.03	0.02	0.06	0.09	0.13	
	PO_4-P 90-perc.	0.11	0.22	0.05	0.03	0.09	0.13	0.24	
	DOX av.	11.7	11.7	10.6	9.7	10.6	10.3	10.6	
	DOX 10-perc.	9.7	9.4	8.0	7.8	8.5	7.8	6.7	
Chlorophyll [6]	CHLA av.	77.1	36.7	8.0	9.3	10.7	21.8	37.6	$\mu g \cdot L^{-1}$
	CHLA 90-perc.	184.0	96.1	14.2	18.0	28.5	61.7	73.0	

Notes: * Wikipedia. Data sources: [1] [55], [2] DWD/PIK, [3] Corine2000, [4] Germany: BÜK1000, Czech Republic: [56], [5] Germany: [57], Czech Republic: [58], [6] German gauges: FIS, Czech gauges: IKSE.

According to the German classification of water quality [54], which uses the 90th percentile for nutrients and the 10th percentile for dissolved oxygen to compare with certain water quality thresholds, the highest nitrate level results in water quality class III (Mulde and Saale), the highest ammonium value belongs also to class III (Vltava), the maximum phosphate phosphorus level represents water quality class II–III (Vltava and Havel), and the lowest dissolved oxygen concentration results in water quality class II (Havel). There is quite high diversity between the rivers in this respect, and no river exists which has the worst or best status for all components.

The Elbe river is the most important river draining the eastern part of Germany. The natural flow regime is influenced by reservoirs and regulation of small rivers, drainage of wetlands and brown coal mining [59]. Due to former political and socio-economic conditions, the Elbe was one of the most polluted rivers in Europe with a low ecological potential. The water quality improved after the German reunification in 1990 due to closure or upgrading of sewage treatment plants and industrial enterprises in the basin, as well as to a substantial decrease in fertilization rates on agricultural land [58,60]. However, contamination problems still exist, especially looking at sediments, which are characterized by a high adsorption of heavy metals and other polluting substances [61]. There are also no significant improvements regarding chlorophyll *a* concentrations in the Elbe river [60].

In general, the Elbe river is characterized by a strong phytoplankton growth in the free-flowing section due to inputs from the reservoirs of the upper Elbe and Vltava and high nutrient loads from tributaries [62]. The high primary productivity leads to substantial differences in nutrient concentrations along the river with remarkable intra- and interannual variations [62], and the season of main biological activity is between March and October [31]. Low flow velocities in the lowland tributaries with many natural lakes in the river course (e.g., Havel) and in rivers influenced by weirs and barrages (e.g., Vltava, Saale) facilitate good conditions for algal growth and cause high concentrations of chlorophyll *a*.

The middle course of the Elbe river in Germany contains several protected natural areas with a high diversity of flora, fauna and landscape types. Large parts of the river in Germany are free-flowing and not influenced by barrages. However, the original floodplain areas have often been cut off by flood protection measures for settlements, agriculture and industry during the last centuries. Approximately 84% of the floodplain along the Elbe river course in Germany is protected by dikes [63]. The reduced flooding area not only causes problems in times of very high water levels (e.g., during the last decades when immense flood events and damage occurred), but also hinders the natural nutrient retention capacity of the river ecosystems. This induces an intensification of nutrient pollution problems in the river waters.

3. Materials and Methods

3.1. Soil and Water Integrated Model (SWIM)

The Soil and Water Integrated Model (SWIM) is an eco-hydrological model of intermediate complexity simulating the hydrological cycle and vegetation growth integrated with nutrient turnover processes within a river basin driven by climate parameters and taking soil conditions and land use management into account. SWIM was developed on the base of the Soil and Water Assessment Tool (SWAT) [64] and the MATSALU model [65] specifically as a tool for the analysis of climate and land use change impacts on hydrological processes, agricultural production and water quality at the regional scale. More details can be found in [48].

Being a spatially semi-distributed dynamic model working with a daily time step, SWIM calculates all hydrological, vegetation and nutrient processes on a hydrotope level (set of elementary units in a subbasin with the same land use class and soil type). Lateral fluxes (surface, subsurface and groundwater flow with associated nutrients) are summarized at the subbasin level and routed through the river network to the outlet of the catchment.

Hydrological processes on the hydrotope level are based on the water balance equation, taking precipitation, evapotranspiration, percolation, surface and subsurface runoff, capillary rise and ground water recharge into account.

The available water content in soil is influenced by crop and vegetation types, which are parameterized in a database connected to SWIM [48]. The crop database is the same as in SWAT [66], and only some parameters were adapted during calibration. The vegetation affects nutrient turnover as well, as plants are important nutrient consumers as well as sources (from plant residue).

The nitrogen module of the applied SWIM version (compare Hesse *et al.* [50]) calculates nutrient processes in the soil profile and includes several pools: nitrate and ammonium nitrogen, active and stable organic nitrogen, and organic nitrogen in plant residues. They are influenced by fertilization, mineralization, volatilization and (de-)nitrification processes, plant uptake, wet deposition, wash-off, leaching and erosion. Leaching is calculated differently for nitrate and ammonium nitrogen, as the latter has much higher bonding capacity to soil particles.

The soil phosphorus module includes labile phosphate phosphorus, active and stable mineral phosphorus, organic phosphorus and phosphorus in the plant residue. The phosphorus pools are influenced by fertilization, (de-)sorption, mineralization, plant uptake, erosion, and leaching. The equation applied to calculate leaching of phosphate phosphorus through the soil profile can be found in Hesse *et al.* [49].

Processes related to diffuse source nitrogen and phosphorus flows to the river network are surface runoff, subsurface runoff, groundwater flow, wash-off, leaching, erosion and retention of nutrients in the landscape. After simulating all nutrient-specific processes in the soil profile, nitrogen and phosphorus are transported with surface, subsurface and groundwater flows to the rivers. During their passage through the basin, nutrients are subject to retention and transformation processes in soils, wetlands and in the river system. These processes and model equations, as well as the testing of different retention methods, were described in detail in previous publications [50,67,68].

Additional information about the general SWIM model concept, necessary input and output data, calibration parameters, process equations as well as the GIS interface for model setup can be found in the User Manual [48].

3.2. Data Preparation and Model Setup for Calibration

SWIM model setup requires spatial and temporal data sets as well as major water and land use management information. The spatial maps include a digital elevation model (DEM), a soil map with soil parameterization, a land use map and a subbasin map. The temporal data sets include the daily historical observed or projected future climate parameters (temperature, precipitation, solar radiation and relative air humidity) as external drivers of the model. The observed river discharge and nutrient concentrations, at least close to the catchment's outlet, are necessary for the model calibration and validation. Additional monitoring data at intermediate gauges and tributaries allow a multi-site calibration, which is more reliable, especially for large-scale catchments. Necessary management data include water abstraction, storage or transfer, major crops with their planting and harvesting dates, as well as fertilization rates and dates and effluents from industrial sites or waste water treatment plants.

The model setup for the Elbe river was based on spatial maps with a 250 m resolution. The DEM map was resampled from the data provided by the NASA Shuttle Radar Topographic Mission (SRTM). The general German soil map "BÜK1000" delivered by the Federal Institute for Geosciences and Natural Resources (BGR) was combined with the soil map and soil parameterization of the Czech Republic [59] and the European Soil Database (ESDB) provided by the Joint Research Centre of the European Commission to cover the entire Elbe river catchment. The land use map was obtained from the CORINE land cover (CLC2000) data set of the European Environment Agency (EEA) and reclassified to the 15 SWIM land use classes required by the model. The subbasin map was combined from the standard maps of the Federal Environment Agency (UBA) for Germany and the T.G.M. Water Research Institute for the Czech Republic, and included 2268 subbasins.

The historical climate data of 348 climate observation stations located within and 20 km around the Elbe catchment were used to interpolate the climate parameters to the centroids of all subbasins by an inverse distance method for calibration and validation of the SWIM model, taking climate information of at least one to maximum four neighboring stations into account. The station density with available climate data was higher in the German than in the Czech part of Elbe river catchment.

The observed discharge and water quality data for selected gauges located at the Elbe river and its main tributaries in Germany originated from the Data Information System (FIS) of the River Basin Community Elbe (FGG-Elbe) and were downloaded in December 2012. The Czech monitoring data with a monthly time step were taken from the publications of the International Commission for the Protection of the Elbe river (IKSE). In case the observed nutrient concentrations were indicated to be below the detection limit, the minimum detectable concentration was halved and assumed for this day. Data on nutrient inputs from point sources at the German part of the basin were taken from FGG-Elbe [60]. For the Czech part, assumed data on nutrient emissions from point sources were derived from a report of the IKSE for the year 2000 [58]. As there were only temporally aggregated data available, the point source emissions were implemented in the model as daily constant values.

The standard SWIM does not consider crop rotation management on agricultural fields so that only one main crop type could be assumed on the entire arable land. According to data in the statistical yearbooks of the German federal states in Germany considerably overlaying with the Elbe basin (Thuringia, Saxony-Anhalt, Brandenburg, Saxony and Mecklenburg-West Pomerania), winter wheat was selected to be the main crop. Assuming some nutrient storage in the soils, 100 kg N/ha and 12 kg P/ha were assumed as an average fertilization level in accordance with recommendations of the federal agriculture agencies. However, fertilization is recommended to be increased with increasing yield expectations [69]. To implement this option, arable land was classified according to the expected yield as simulated by SWIM (as a function of soil quality, water availability and climate conditions under constant fertilization). Then the medium yield class received the average fertilization, and fertilization of the low/high yield classes was reduced/increased by 20%.

In order to better represent specific behavior of vegetation in lowland areas with its connection to groundwater and the increased evapotranspiration potential, the simpler of two approaches for wetland simulation as described in Hattermann *et al.* [70] was used in SWIM. In total, 22.6% of the entire Elbe river basin belongs to wetlands, with especially high shares in the Schwarze Elster catchment (41%), the lower Elbe reaches (40%), and the Havel river catchment (33%). In the catchments of the other large tributaries (Saale, Mulde, Ohře and Vltava), wetlands make up 10%–16% of their total areas.

The model calibration and validation for the whole basin was performed for five years, each within the period 2001–2010, considering observed data at the last gauges at Neu Darchau (discharge, Elbe, km 536.4) and Schnackenburg (water quality, Elbe, km 474.5), which are undisturbed by tidal influences. The nutrient loads at the gauge at Schnackenburg were calculated as products of concentration and discharge using the discharge of the gauge at Wittenberge (km 453.9) with the correction factor 1.001 [71].

The calibration of water discharge (Q) and nutrient loads was done by adjusting the main model calibration parameters described in the SWIM manual [48], and listed in former SWIM model applications [49,50,68,72,73]. During the model calibration it was realized that a global calibration parameter set was not sufficient to represent the basin- and river-specific water and nutrient processes for the several catchments of the Elbe tributaries, which can be highly variable due to different combinations of elevation, soil, land use and river characteristics. Therefore, it was decided to use the most important calibration parameters spatially distributed for the seven large river catchments, which were calibrated individually. Table 2 lists and describes those parameters for water quantity and quality calibration used in a distributed mode within the Elbe river basin.

Table 2. SWIM calibration parameters applied spatially distributed in the Elbe river basin.

Module	Parameter	Description	Unit
Hydrology	bff	baseflow factor used to calculate return flow travel time	-
	delay	time needed for water leaving root zone to reach shallow aquifer	day
	roc2/roc4	coefficients to correct the storage time constants for surface and subsurface flows	-
Soil nutrients	ret	retention times of nitrate nitrogen (NO_3-N), ammonium nitrogen (NH_4-N) and phosphate phosphorus (PO_4-P) in the lateral subsurface and groundwater flows (6 parameters)	day
	deg	degradation rates of NO_3-N, NH_4-N and PO_4-P in the lateral subsurface and groundwater flows (6 parameters)	day^{-1}
	deth	soil water content threshold for denitrification of NO_3-N	%
	dad/dkd	ratios of adsorbed NH_4-N and PO_4-P to that in soil water	-
In-stream processes	mumax	maximum specific algal growth rate	day^{-1}
	tempo	optimal temperature for algal growth	°C
	lio	optimal radiation for algal growth	ly
	pr20	predation rate in the reach at 20 °C	day^{-1}
	ai1/ai2	fractions of algal biomass that is nitrogen and phosphorus	$mg \cdot mg^{-1}$
	rs1	local algal settling rate in the reach at 20 °C	$m \cdot day^{-1}$
	rs2/rs3	benthic source rates for PO_4-P and NH_4-N in the reach at 20 °C	$mg \cdot (m^2 \cdot day)^{-1}$
	rs5	organic phosphorus settling rate in the reach at 20 °C	day^{-1}
	rk2	oxygen reaeration rate in the reach at 20 °C	day^{-1}
	bc3	rate constant for hydrolysis of organic nitrogen to NH_4-N at 20 °C	day^{-1}
	bc4	rate constant for mineralization of organic phosphorus to PO_4-P at 20 °C	day^{-1}

3.3. Evaluation of Model Results

The ability of SWIM to simulate water and nutrient processes in the Elbe catchment and to reproduce the observed monitoring values was evaluated in different ways for water quantity and quality.

The simulated daily and/or monthly discharges were assessed using the Nash-and-Sutcliffe efficiency (NSE, [74]) as well as the deviation in water balance (DB) (compare [49]). The non-dimensional NSE is a measure to analyze the squared differences between the observed and simulated values, and DB describes the long-term differences of the observed values against the simulated ones for the whole simulation period in percent.

The model's efficiency to represent the water quality parameters was evaluated at the long-term average monthly basis using three criteria, $\Delta\mu$, $\Delta\sigma$ and r, according to Gudmundsson *et al.* [75]. Here $\Delta\mu$ is a balance measure defined as the relative bias of the mean annual observed and simulated values. Criterion $\Delta\sigma$ evaluates the amplitude or the spread from the lowest to the highest monthly values of the mean annual cycle by comparing the relative difference in standard deviations of the observed and the simulated values. Also, the usual Pearson's correlation coefficient r, which is sensitive to differences in the shape as well as in the timing of the mean annual cycle, was applied.

Table 3 lists the possible ranges, optima and aspired results of the different performance criteria used in this study.

Table 3. Description of performance criteria used in this study to evaluate model results.

Criterium	Range	Optimum	Aim in This Study
NSE	$-\infty$ to 1	1	>0.65
DB	$-\infty$ to $+\infty$	0	>−20% to <20%
$\Delta\mu$	$-\infty$ to $+\infty$	0	>−0.2% to <0.2%
$\Delta\sigma$	$-\infty$ to $+\infty$	0	>−0.2% to <0.2%
r	−1 to 1	1	>0.5

3.4. Description, Evaluation and Processing of Climate Scenario Data

The ENSEMBLES project [51] delivered projections for a possible climatic future of Europe by running an ensemble of different Regional Climate Models (RCMs) using the boundary conditions produced by several General Circulation Models (GCMs). All models assumed the A1B emission scenario with a balanced use of fossil and non-fossil fuels in a world with a rapidly growing economy, population growth until 2050 and a decline afterwards, and fast development of new and effective technologies [76]. According to this scenario an average global temperature rise of 2.8 °C (with a range between 1.8 and 4.4 °C) is estimated [1] until the end of the 21st century.

The resulting ENSEMBLES climate scenarios differ in resolution (25 or 50 km grids) and simulation period (1951/1961–2050/2100). For our study, 19 scenarios covering the period until 2100 were chosen (Table 4). As climate data necessary for SWIM modeling were not available for all scenarios until 2100, only data until 2098 were considered in all cases. A scenario-specific number of grid cells with data were treated as virtual climate stations for climate interpolation to the centroids of the 2268 subbasins within the Elbe basin using an inverse distance method.

Table 4. Numbering of the chosen climate scenarios as combinations of General Circulation Models (GCMs) and Regional Climate Models (RCMs), the responsible institute, resolution [km], starting year, and number of grid cells used for interpolation of the projected climate in the Elbe catchment.

ID	Institute	GCM	RCM	Resolution	Start Year	Number of Grid Cells
1	SMHI	HadCM3Q3	RCA	25	1951	316
2	HC	HadCM3Q0	HadRM3Q0	25	1951	316
3	HC	HadCM3Q3	HadRM3Q3	25	1951	316
4	HC	HadCM3Q16	HadRM3Q16	25	1951	316
5	C4I	HadCM3Q16	RCA3	25	1951	316
6	ETHZ	HadCM3Q0	CLM	25	1951	316
7	KNMI	ECHAM5-r3	RACMO	25	1951	316
8	SMHI	BCM	RCA	25	1961	316
9	SMHI	ECHAM5-r3	RCA	25	1951	316
10	MPI	ECHAM5-r3	REMO	25	1951	316
11	CNRM	ARPEGE_RM5.1	Aladin	25	1951	300
12	DMI	ARPEGE	HIRHAM	25	1951	316
13	DMI	ECHAM5-r3	DMI-HIRHAM5	25	1951	316
14	DMI	BCM	DMI-HIRHAM5	25	1961	316
15	ICTP	ECHAM5-r3	RegCM	25	1951	282
16	KNMI	ECHAM5-r1	RACMO	50	1951	79
17	KNMI	ECHAM5-r2	RACMO	50	1951	79
18	KNMI	ECHAM5-r3	RACMO	50	1951	79
19	KNMI	MIROC	RACMO	50	1951	79

To analyze the projected trends of single climate scenarios, climate change signals were calculated for two future periods for temperature, precipitation and solar radiation. Climate change signals describe the differences between the mean climate parameter values in a future period and in the reference period of the same scenario. The signals were derived taking all scenario grid cells into account and were evaluated for the annual mean climate parameter values as well as for their seasonal dynamics.

The 19 climate scenarios were used to drive the calibrated SWIM model, each for the reference period 1971–2000 (p0) and the two future periods 2021–2050 (p1) and 2071–2098 (p2).

It is very important which downscaling approach is used to generate climate scenarios, and whether it is statistical or dynamical. The choice of a hydrological model is less important in terms of its contribution to uncertainty, especially when only the long-term mean annual changes are compared [77]. Often it was detected that results achieved with one hydrological model under two or more climate scenarios differ more than the results of different hydrological models achieved by using only one climate scenario [78,79]. Hence, many authors suggest using an ensemble of climate change scenarios to get the full range of uncertainty between the different scenario projections [22,80,81]. The last two authors also mentioned that there is no direct link between the climate model performance in the historical period and the robustness of trends in the future, and thus the application of a smaller number of best fitting scenarios could not be recommended. Therefore, in our study we did not try to find the most probable future climate scenarios by their comparison with the historical measurements.

The observed climate data are also often used for bias correction of climate scenarios before applying them for impact assessment in order to avoid unrealistic simulations of runoff or nutrient loads. However, there is no consistent opinion on the usefulness of bias correction for impact assessments. While Teutschbein and Seibert [82] recommend an application of bias correction, other authors complain about the lack of physical justifications of corrections damaging the physical consistency between the variables [77,83]. The latter do not appreciate this method as a "valid procedure", and complain that an additional uncertainty is added to the model chain. In our study it was decided not to use bias correction and to simply compare the simulations driven by 19 RCMs between periods to detect trends and the relative changes caused by climate change.

3.5. Processing of Socio-Economic Change Experiments

In addition to climate change simulations, five land use change experiments were applied for testing the effects of specific socio-economic measures aimed at reducing point or diffuse nutrient emissions.

The applied land use change scenarios are listed in Table 5 together with the description of the changes implemented in model input data. Two scenarios are dealing with the direct reduction of nutrient emissions ("Point sources" and "Fertilization") by 10% or 20%. The decrease of point source emissions was assumed with different percentages for the two nutrients, as it was supposed that phosphorus reduction potential in sewage treatment is higher. The third scenario ("Retention") indirectly tested the effects of a possible increase of the retention potential and decomposition rate in the soils of the Elbe catchment by 10%. This could be achieved by different measures, for example by extension of wetland areas around the watercourses or intensified cultivation of plant communities with a high nutrient intake rate (mycorrhiza, legumes). In addition, two such measures were tested directly in the model ("Buffer" and "Slope") by changing the land use composition in the catchment. Due to the spatial resolution of the SWIM project with 250×250 m raster maps, water courses in agricultural areas were converted to 250 m raster cells in the "Buffer" experiment, containing extensive meadows without fertilization.

254

In the "Slope" scenario, all agricultural areas with a slope >4% were converted to extensively used meadows to study the effects on water quantity and quality in the catchment (see e.g., [84] where hillsides with slopes >4% are considered as being a risk of facilitating erosion).

Table 5. Description of applied land use change experiments in the Elbe river catchment.

Scenario Name	Description
Point sources	Reduction of emissions from point sources (nitrogen −10%, phosphorus −20%)
Fertilization	Reduction of N and P fertilizers on agricultural land by 20%
Retention	Increase of nutrient retention time and decomposition rate in soils by 10% each
Buffer	Conversion of all agricultural lands around water courses to extensive meadow
Slope	Conversion of agricultural lands to extensive meadows on hillsides with slopes >4%

The socio-economic experiments were run under the 19 ENSEMBLES climate scenarios to allow evaluation of the combined climate and land use change impacts on water quantity and quality in addition to the land use change impacts only. As 19 climate scenarios were applied with specific climate conditions, the results were different, not only for the combined impacts, but also for the land use change impacts. To show the possible effects of scenarios, the 19 single percental changes of the model outcomes were analyzed regarding their medians and 25th/75th percentile values, representing the most probable 50% range of all scenario results.

4. Results

4.1. SWIM Model Calibration and Validation

A successful calibration of a model for water quality requires a well-calibrated hydrological model. During the hydrological and water quality calibration, the observed and simulated values at the most downstream Elbe gauges, at the gauges located close to the German-Czech border, as well as at the main tributaries were compared and statistically evaluated for the period of 2001–2010.

Figure 2 presents the observed and simulated daily discharges for the 10-year period (left), and the long-term daily averages (right) at the two main Elbe gauges Schöna and Neu Darchau. The discharge dynamic is well reproduced, reaching the good to very good performance ratings. The performance criteria for the daily model results are better at the downstream gauge Neu Darchau. The long-term seasonal dynamics are reproduced well at both gauges.

However, not all simulation results at the tributaries reach the same model performance (Table 6). The most difficult river to simulate was the Schwarze Elster, which is highly influenced by human activities and regulation (opencast lignite mining, discharge regulation and stream straightening), so that the hydrological processes are no longer natural. As these site-specific management measures were not implemented in the model, the model does not perform well enough at the Löben gauge. Similar problems apply to the lowland catchment of the Havel river, which is characterized by a high number of wetland areas and stream lakes, and is also highly affected by lignite mining in its upper course, all this leading to lower NSE values at gauge Havelberg.

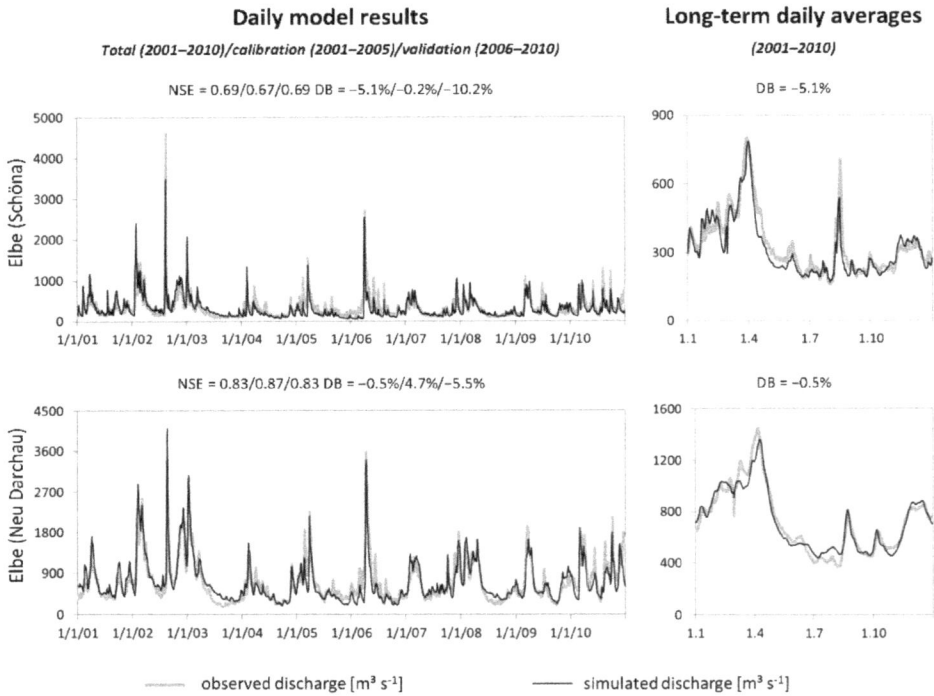

Figure 2. Calibration results for the Elbe river discharge at the most downstream gauge Neu Darchau and the intermediate Elbe gauge Schöna (Czech/German border) for the time period 2001–2010, separated into calibration and validation sub-periods.

Only monthly measurements for a shorter time period were available for the three gauges located in the Czech part of the Elbe basin. The best results here could be achieved for the smaller and mountainous river Ohře. The upper part of the Elbe river (gauge Nymburk), as well as the largest Elbe tributary, Vltava, show a slight underestimation of discharge. This could be explained by water regulation measures in the water course of these rivers, with a high number of barrages and dams to ensure water availability for shipping and for flood protection, which were not implemented in the model. However, the hydrological model performance in terms of NSE and DB for the Elbe and its tributaries mostly meet the aim (compare Table 3), so that it was used for the subsequent water quality calibration.

Table 6. Model performances for four discharge gauges of the Elbe river and six gauges of its main tributaries from the upstream to downstream location of tributaries.

River	Gauge	Time Period	NSE (−)		DB (%)
			Daily	Monthly	
Elbe	Nymburk	11/2002–10/2010		0.75	−13.5
Vltava	Vranany	11/2002–10/2010		0.64	−10.5
Ohře	Louny	11/2002–10/2010		0.86	−0.3
Elbe	Schöna	2001–2010	0.69	0.77	−5.1
Schwarze Elster	Löben	2001–2008	0.25	0.60	13.4
Mulde	Bad Düben	2001–2010	0.74	0.88	1.7
Saale	Calbe-Griezehne	2001–2010	0.61	0.84	1.5
Elbe	Magdeburg	2001–2010	0.82	0.86	1.1
Havel	Havelberg	2001–2010	0.54	0.68	−1.5
Elbe	Neu Darchau	2001–2010	0.83	0.86	−0.5

Figure 3 presents the results of water quality calibration for two main gauges in the Elbe river: Schmilka at the Czech-German border and the most downstream Elbe gauge Schnackenburg. The long-term average daily observed loads were calculated based on interpolated values between biweekly measurements and have some degree of uncertainty. The calibration was aimed at visually and statistically matching the inner-annual dynamics and minimizing the deviation in balance of the mean annual nutrient loads for the 10-year period of 2001–2010.

In Figure 3, a specific annual cycle of the three nutrients can be observed, which is reproduced quite well by the SWIM model. The nitrate nitrogen loads (mainly coming from diffuse sources) generally follow the discharge curve with a spring peak and low values in summer. Ammonium nitrogen and phosphate phosphorus are more algae-influenced. The periods with high concentrations of chlorophyll *a* especially result in ammonium depletion in the river due to the high ammonium preference factor of the algae defined in the model. Algal influences on the phosphate loads are less significant, but also obvious, especially during the spring algal bloom. The dissolved oxygen loads are highly connected to the water amounts and are simulated very well. The balance measure $\Delta\mu$ is low in all cases and is located within the aimed range, also reflecting sufficiently good calibration results.

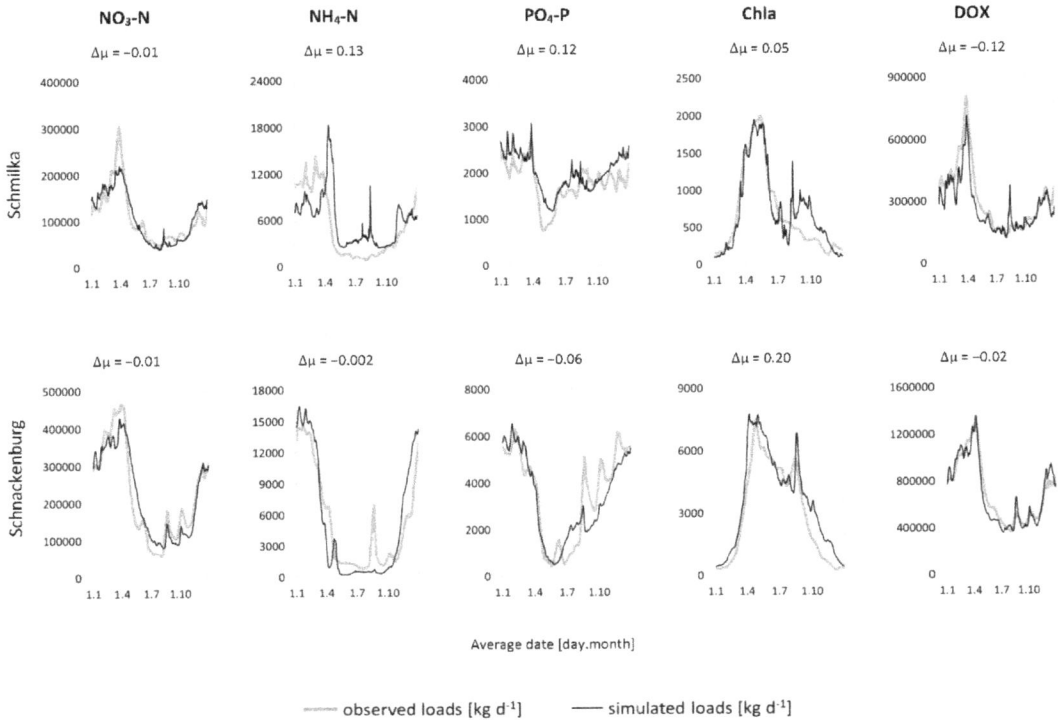

Figure 3. Long-term average daily observed and simulated loads of nitrate nitrogen (NO_3-N), ammonium nitrogen (NH_4-N), phosphate phosphorus (PO_4-P), chlorophyll *a* (Chla) and dissolved oxygen (DOX) at the two Elbe gauges Schmilka (corresponds to the total Czech loads) and Schnackenburg (most downstream gauge) for the time period 2001–2010.

Figure 4 and Table 7 show results on water quality for the main tributaries of the Elbe river and for selected Elbe gauges. Figure 4 plots the simulated *versus* observed long-term average monthly values and illustrates the variation of the long-term seasonal cycle ratios around a diagonal of perfect fit, and Table 7 analyzes the model's performance statistically.

Table 7. Model ability to simulate the long-term monthly average loads of water quality variables for six main tributaries as well as three selected Elbe water quality observation gauges in the time period 2001–2010 after spatially distributed model calibration. The model performance variables were calculated according to [75] and are described in Section 3.3.

Time period	Vltava Zelčín 2001–2004	Ohře Tereźin 2007–2010	Elbe Schmilka 2001–2010	Schwarze Elster Gorsdorf 2004–2010	Mulde Dessau 2001–2010	Saale Groß Rosenburg 2001–2010	Elbe Magdeburg 2001–2010	Havel Toppel 2001–2010	Elbe Schnackenburg 2001–2010
					NO$_3$-N				
Δμ	−0.01	−0.09	−0.01	−0.13	0.04	0.10	0.09	−0.01	−0.01
Δσ	−0.11	−0.23	−0.04	−0.27	−0.20	0.18	0.03	−0.42	−0.07
r	0.91	0.93	0.93	0.96	0.93	0.94	0.93	0.97	0.95
					NH$_4$-N				
Δμ	0.02	0.12	0.13	0.17	−0.08	−0.01	0.17	0.05	0.01
Δσ	0.17	−0.25	−0.23	0.02	−0.31	−0.19	0.08	0.13	0.26
r	0.64	0.74	0.68	0.53	0.85	0.97	0.92	0.94	0.95
					PO$_4$-P				
Δμ	−0.01	0.11	0.12	0.12 *	0.03	−0.11	0.10	−0.10	−0.06
Δσ	−0.40	0.01	−0.02	0.39 *	−0.37	−0.03	0.00	−0.19	−0.03
r	0.73	0.70	0.87	0.13 *	0.87	0.93	0.92	0.24	0.94
					Chla				
Δμ	0.08	−0.02	0.05	0.03	−0.08	0.04	0.06	0.12	0.20
Δσ	0.06	0.12	−0.07	0.05	0.17	0.04	−0.07	0.18	0.00
r	0.95	0.82	0.94	0.82	0.80	0.82	0.91	0.78	0.97
					DOX				
Δμ	−0.06	−0.12	−0.12	0.02	−0.01	0.06	0.04	0.03	−0.02
Δσ	−0.28	−0.39	−0.30	−0.50	−0.37	0.08	−0.02	−0.25	0.02
r	0.85	0.95	0.98	0.91	0.95	0.97	0.96	0.98	0.98

Note: * 2008–2010.

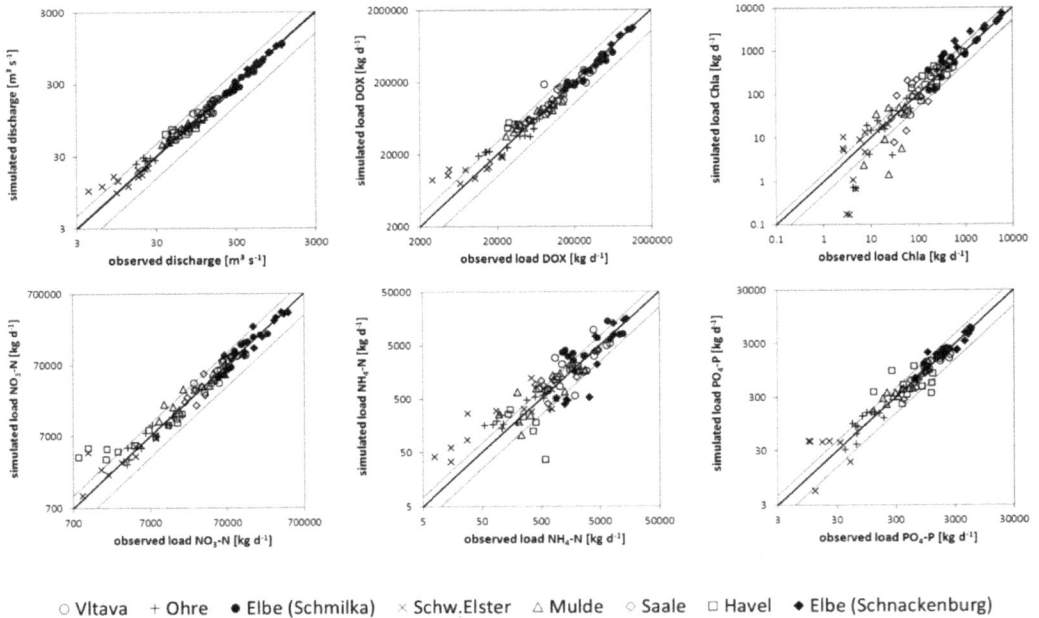

○ Vltava   + Ohre   ● Elbe (Schmilka)   × Schw.Elster   △ Mulde   ○ Saale   □ Havel   ◆ Elbe (Schnackenburg)

Figure 4. The long-term average monthly observed and simulated discharge and loads per tributary and at two selected Elbe gauges in the period 2001–2010 (diagonals: black—perfect fit, grey—± 50% intervals).

As already detected in the hydrological calibration, the largest discrepancies between the observed and simulated values can be seen for the Schwarze Elster river. The intensive human activities within this catchment (e.g., surface water management due to lignite mining) influence nutrient processes but are not fully implemented in the model, resulting in model outputs different from observations. Some problems can also be seen for the Havel and (partly) the Mulde tributaries. The largest dispersion around the diagonal of perfect fit is obvious for ammonium nitrogen, which is difficult to model as it is highly influenced by point source emissions involving input data uncertainty as well as by algal consumption processes (parameter uncertainty). The latter, due to their complex behavior influenced by many physical, chemical and biological interactions, are difficult to model, especially in large basins. The results in terms of statistical criteria (Table 7) with mostly high r and low $\Delta\mu$ and $\Delta\sigma$ confirm the visual impression.

Generally, the calibrated SWIM model for the large-scale Elbe river basin matches observations well, and can be used for climate and land use change impact assessment.

4.2. Climate Change Signals of the ENSEMBLES Scenarios

Before applying the 19 ENSEMBLES climate scenarios to the Elbe basin, they were analyzed for their trends in temperature, precipitation and solar radiation averaged over the whole basin by comparing two future scenario periods, p1 and p2, with the reference period p0. The comparison was done for the long-term average annual values as well as for the long-term average monthly values of all scenarios and periods.

Table 8. Climate change signals for temperature, precipitation and radiation of 19 analyzed ENSEMBLES climate scenarios and on average for the two future periods 2021–2050 (p1) and 2071–2098 (p2) compared to the reference period 1971–2000 (p0) for the Elbe basin.

Scenario	Temperature (°C)		Precipitation (mm)		Radiation (J·cm^{-2})	
	p1-p0	p2-p0	p1-p0	p2-p0	p1-p0	p2-p0
S1	1.5	2.9	67	95	−26	−76
S2	2.1	4.0	−2	16	27	28
S3	1.7	3.4	34	17	8	7
S4	2.2	5.0	48	−49	1	43
S5	1.8	4.1	104	94	−57	−67
S6	1.7	3.5	24	13	−22	−12
S7	0.9	2.6	35	110	−12	−20
S8	0.7	1.9	63	86	−30	−48
S9	0.8	2.4	47	112	−29	−65
S10	0.9	2.6	14	47	−18	−42
S11	1.1	2.8	−4	−68	5	36
S12	0.9	2.0	14	−31	−16	−111
S13	0.6	2.0	57	157	−21	−73
S14	0.9	2.5	37	99	−29	−47
S15	0.9	2.6	29	87	1	7
S16	1.0	3.1	52	63	−12	−5
S17	1.4	3.3	65	54	−22	−11
S18	0.9	2.6	36	99	−16	−20
S19	1.8	3.9	50	76	−26	−40
mean$_{all}$	1.3	3.0	41	57	−15	−27
stdev$_{all}$	0.5	0.8	26	60	18	42

The climate change signals per scenario can be found in Table 8. The results show an increase in temperature by 1.3 °C for the first period and by 3 °C for the second period on average, as well as an increase in precipitation by +41/+57 mm on average for all 19 climate scenarios. The increase in precipitation is accompanied by a decrease in solar radiation of −15/−27 J cm^{-2} on average, probably due to increased cloudiness with higher precipitation amounts. There is a wide spread in signals between the scenarios, which is increasing in the second period. Regarding temperature, all scenarios agree on increasing trend, but the increase in period p2 ranges between 2 and 5 °C depending on the scenario. The agreement of the single scenarios with the overall trends is lower for precipitation (15 of 19 scenarios agree with the trend) and solar radiation (14 scenarios agree). However, a majority of scenarios correspond to the average trends.

The seasonal climate change signals are visualized in Figure 5. Looking at the changes per month, it is obvious that the value as well as the spread of the climate change signals is higher in the second period. The increase in temperature is confirmed for the entire course of the year, and it is lowest in May and highest in winter months (December–February) and in August. The changes in precipitation and solar radiation vary around the zero-line and show an opposite behavior (probably due to connection of precipitation and cloudiness). In the first period precipitation is slightly decreasing in July and August, and in the second period negative changes in precipitation are projected from June to September. The changes in solar radiation show almost the opposite trends. In general, the 19 ENSEMBLES climate scenarios project a warmer and wetter climate with less sunshine hours from autumn to spring, but a warmer, dryer and sunnier climate in the summer months for the region.

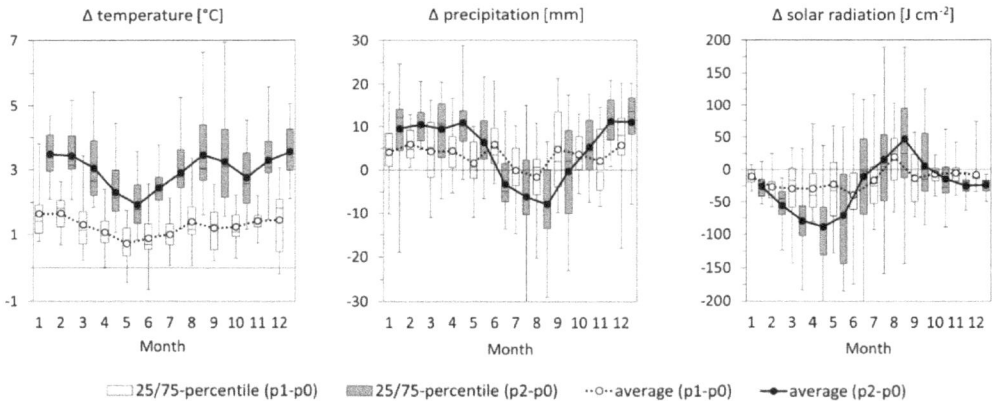

Figure 5. Ranges of seasonal climate change signals for temperature, precipitation and solar radiation of 19 ENSEMBLES climate scenarios for the two future periods compared to the reference period of the same scenario for the Elbe basin. The plots represent median (**line**), 25th/75th percentiles (**box**), min/max values (**whiskers**) and the average (**dots**) change of all 19 scenarios.

4.3. Climate Change Impacts

The projected changes in climate lead to changes in simulated water quantity and quality variables in the Elbe basin in future periods. The results are shown Figure 6 for the two Elbe river gauges Schöna and Neu Darchau. They present changes in the long-term average seasonal dynamics comparing the average and the 25th/75th percentile ranges of six variables from simulations driven by 19 climate scenarios in the future and the average of the reference period 1971–2000.

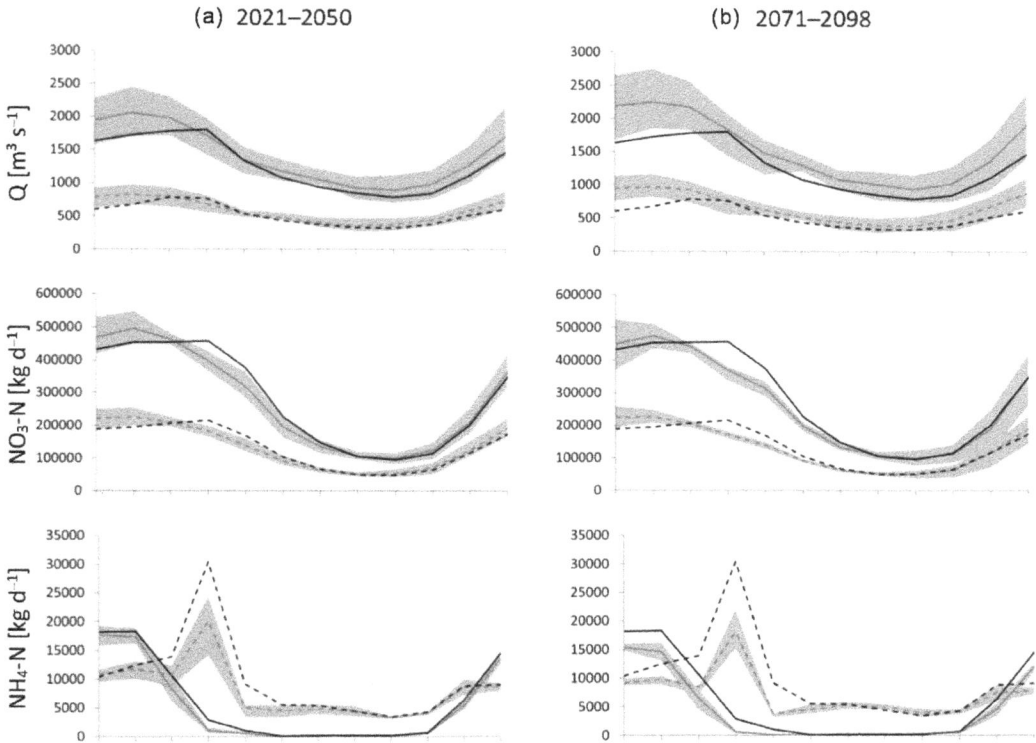

Figure 6. *Cont.*

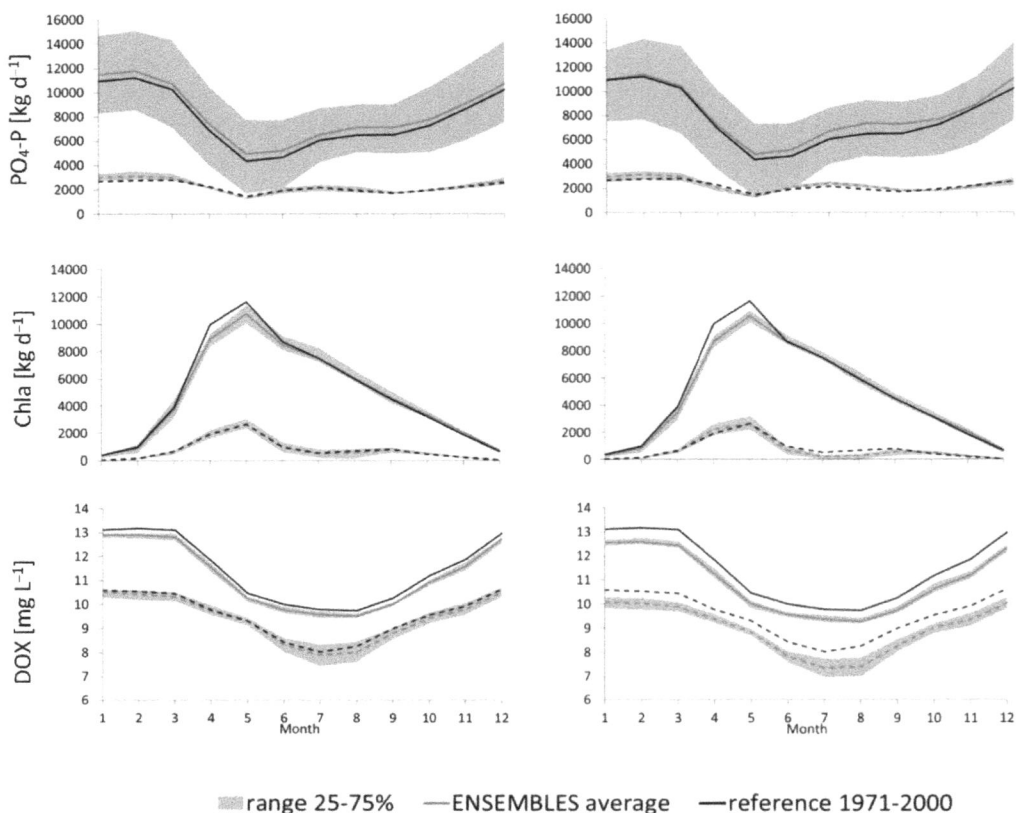

Figure 6. The long-term average monthly values of simulated discharge (Q), nutrient and chlorophyll *a* loads (NO$_3$-N, NH$_4$-N, PO$_4$-P, Chla) and dissolved oxygen concentrations (DOX) with uncertainty ranges (25th/75th percentiles corresponding to 19 simulations) at the two Elbe gauges Neu Darchau (**full lines**) and Schöna (**dashed lines**) for the future periods 2021–2050 (**p1, a**) and 2071–2078 (**p2, b**) in comparison to the corresponding average values of the reference period 1971–2000 (p0).

Following the increasing trend for precipitation in the Elbe basin, the discharge is projected to increase as well, both at the last Elbe gauge and at the gauge of the Czech-German border. The increase can be observed during almost the whole year, with the highest values in winter months (due to higher rainfall) and the lowest values, or even negative changes in the p1 period, in April (due to lower or missing snow melt peaks). Though a decrease in precipitation is projected in the summertime (compare Figure 5), the projected discharge in summer months is higher than in the reference period, probably due to the capability of soils to retain additional winter and spring water causing delayed subsurface and groundwater flows. However, the uncertainty ranges for the projected discharge are quite high, especially at the most downstream gauge.

The nitrate nitrogen load performs similarly to the discharge, as nitrate nitrogen comes to the river mainly dissolved in water from diffuse sources. A moderate increase can be observed in the first winter months, followed by some decrease in spring, whereas the second half of the season shows only minor changes on average compared to the reference period (due to higher retention time of nitrate nitrogen compared to water as well as impacts of vegetation).

The ammonium nitrogen loads are higher on average in the upstream part of the Elbe (gauge Schöna) than downstream (gauge Neu Darchau) due to higher loads in the Czech part of the catchment as well as to progressively increasing phytoplankton concentration downstream of the Elbe. The decrease in ammonium load caused by changes in climate conditions is obvious in the first half of the season (especially during spring flood). The decrease in NH_4-N loads is probably connected to the rising temperatures, as mineralization processes and the emergence of leachable ammonium in soils are temperature-related and occur mainly within a certain temperature range. The uncertainty ranges around the ENSEMBLES average, representing the most probable 50% of the 19 scenario results, are quite narrow.

The average phosphate phosphorus load shows a slight and almost constant increasing trend throughout the season, but the uncertainty ranges are the largest for this nutrient, caused by the high uncertainty and climate-dependence of phosphorus-related processes in the Havel catchment (compare with Figure 7). The increase in loads is probably connected to increasing erosion and leaching processes with higher precipitation in the future, washing more phosphorus from sandy and highly permeable soils. It could also be a result of less ingestion by a decreasing algae population in the future.

The chlorophyll *a* load is projected to decrease in the spring blossom time, when warmer temperatures (temperature stress) and lower solar radiation (below the optimum value) may hamper phytoplankton growth and less ammonium is available for algae consumption.

The dissolved oxygen concentration in the Elbe river is projected to decrease, and the changes remain almost constant throughout the season. This is probably connected to the increasing water temperature, resulting in lower values of oxygen saturation in the water. The uncertainty ranges for future dissolved oxygen concentrations are higher upstream, probably due to the generally higher ammonium loads modeled in the upper river reaches, where oxygen is used for nitrification in the water column.

In addition to the temporal analysis of climate impacts, Figure 7 illustrates some spatially distributed results for the Elbe and its tributaries. For that, average percental changes were calculated for six main tributaries of the Elbe and two Elbe gauges (the same as in Figure 6).

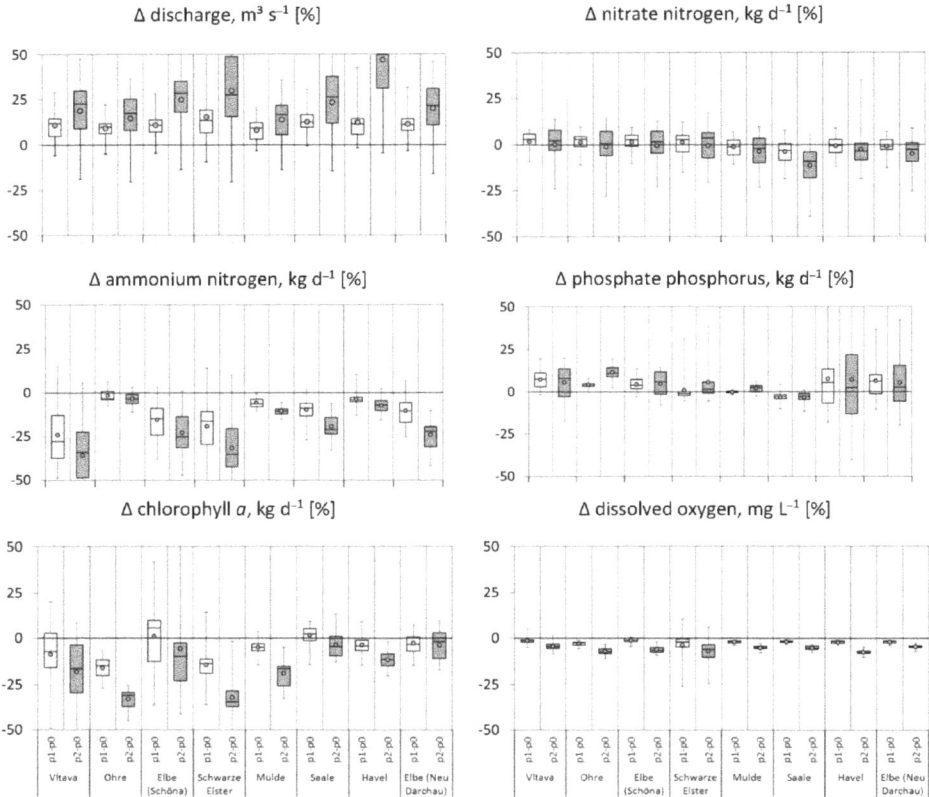

Figure 7. Ranges of the percental changes of 30-year-average river discharges, nutrients and chlorophyll *a* loads, as well as dissolved oxygen concentrations in the Elbe river and its main tributaries simulated with SWIM driven by 19 ENSEMBLES climate scenarios (in future periods p1 (**light**) and p2 (**dark**) compared to the reference period p0 of the same scenario). The plots visualize the following ranges: min/max (**whiskers**), 25th/75th percentiles (**boxes**), median (**line**) and average (**dots**) changes of all 19 scenarios.

The overall trend for the entire basin can be generally detected regarding different variables in Figure 7, though there are some outlying sub-catchments. For all gauges an increasing discharge is projected, which becomes higher in the second period. Also, the uncertainty ranges increase in p2. The differences between gauges are small.

266

The nitrate nitrogen load decreases on average for the entire Elbe river basin (Neu Darchau). The decrease is largest for the Saale catchment, which is characterized by the highest share of agricultural areas due to very fertile soils with a high nutrient retention capability. There are also some sub-catchments where a small increase (or no change) in nitrate load on average is simulated. This is probably connected to an increased diffuse pollution with increased precipitation in these sub-areas.

The impacts on ammonium nitrogen loads are almost all negative, and show a high diversity between the sub-catchments. The uncertainty ranges are highest in the Vltava and Schwarze Elster sub-catchments, where ammonium pollution is generally at its highest level, and have more space for variability due to climate change impacts.

Except for the Saale sub-catchment with its fertile soils and high nutrient retention potential, the climate change impact on phosphate phosphorus shows increasing loads due to increased leaching and erosion processes. The uncertainty ranges are extremely high in the Havel sub-catchment, where phosphorus contamination is the highest in the Elbe drainage area, and a high share of permeable and sandy soils causes a high phosphorus leaching potential with higher precipitation amounts.

Chlorophyll *a* demonstrates a decreasing trend on average almost everywhere. The uncertainty ranges, especially in the upper tributaries, are quite high, due to the high complexity of algae processes simulated in the model, which are influenced by many system-internal and external drivers.

Changes in the dissolved oxygen concentrations have a very small uncertainty range and show a decreasing trend on average for all gauges due to increased temperatures and lower oxygen saturation capacity. The highest range in average changes can be observed for the Schwarze Elster sub-catchment, which is quite heavily polluted with ammonium nitrogen. The latter is highly sensitive to climate change impacts and is connected to the oxygen processes in the river water.

4.4. Socio-Economic Change Impacts under Climate Change

In addition to the climate change impact assessment, five land use change experiments were run to test the model's reaction on certain management measures aimed at reducing nutrient inputs to the river network. The aim was to check whether such measures are able to be reversed, intensify or revoke climate change impacts. The land use change experiments were run 19 times, driven by the 19 ENSEMBLES climate scenarios for the near future period 2021–2050 (p1), and the results were compared with the results achieved under the reference management conditions for the period 1971–2000 (p0) of the same scenarios (combined impacts) as well as with the climate scenario-driven results with the reference management for period p1 (land use change impacts only).

The single and combined impacts were analyzed for the two Elbe gauges Schöna (Czech/German border) and Neu Darchau (Elbe outlet) as well as for the outlets of the two selected tributaries Saale and Havel (Figure 8). The results are shown as median values with a 25th/75th percentile range. In some cases, even the single land use change impact shows some range of relative changes caused by different behavior of temperature- and water-dependent nutrient processes under different climate conditions used as an external driver.

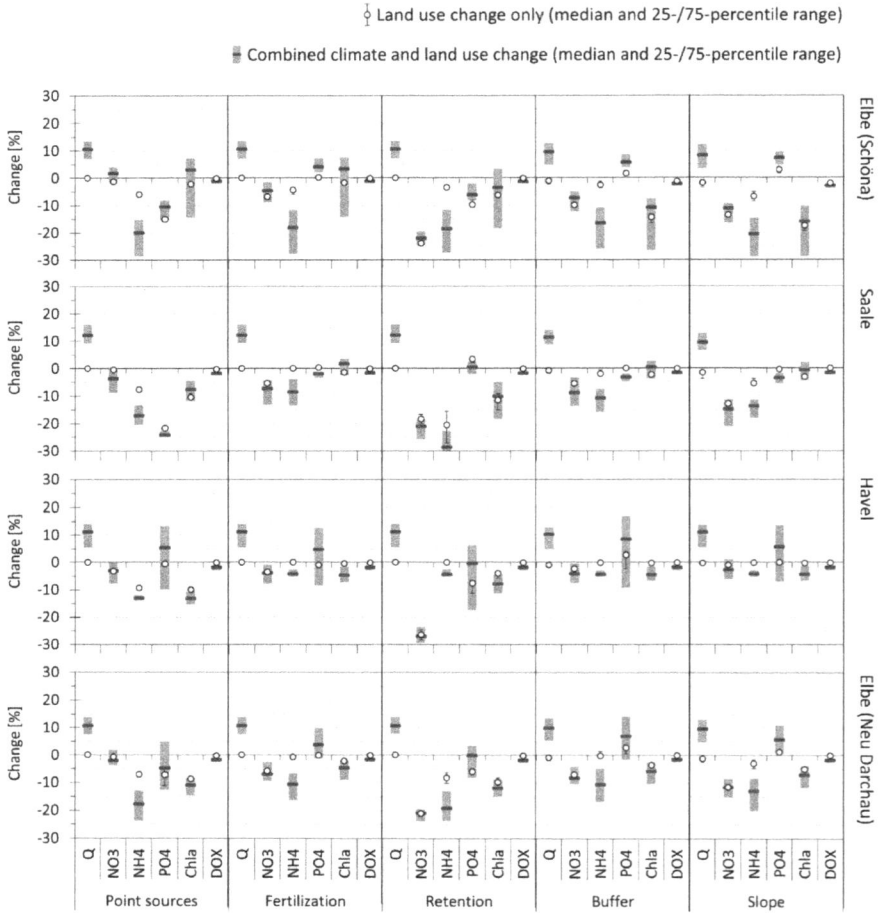

Figure 8. Impacts of socio-economic changes and combined climate and socio-economic changes on the average water discharge (Q), nutrient (NO_3-N, NH_4-N, PO_4-P) and chlorophyll a (Chla) loads and dissolved oxygen concentrations (DOX) of the Elbe river at two stations and at two main German tributaries. The dark grey bars and white dots show the median of 19 percental changes together with their 25th/75th percentile ranges (light grey ranges and black whiskers).

The socio-economic changes related to nutrient inputs to the river network (experiments "Point sources" and "Fertilization") and an increased nutrient retention potential in soils (experiment "Retention") have no influence on water discharge. Only the combined impacts show an increase in discharge of about 10% due to climate change. The solely socio-economic impacts of a changed land use composition ("Buffer" and "Slope") on river discharge show a decrease (due to increased evapotranspiration of the enlarged grassland areas), but it is quite low, and cannot compensate the increase in Q caused by the projected climate change, so that all combined impacts for these experiments have a positive direction.

The reduction of point source emissions has the highest influence on phosphate and ammonium loads, as these nutrients mainly originate from anthropogenic inputs of water treatment plants or industrial units. The projected climate change even intensifies the reduction of ammonium nitrogen loads in the rivers, whereas the decrease of phosphate phosphorus is reduced by climate change impacts (except for the Saale basin). The sole reduction of point source emissions predominantly results in a decrease of chlorophyll a loads in the rivers due to less available ammonium and phosphate as algal food.

The decrease in fertilizer application causes lower nitrate loads in all analyzed river parts, as this nutrient originates mainly from diffuse sources (predominantly from agricultural fields). The reduction is only marginally influenced by climate change. A decrease in fertilization affects NH_4-N only partly, and causes decreased ammonium loads, particularly in the upper part of the Elbe basin. As the changes in NH_4-N and PO_4-P loads are less distinct under the "Fertilization" experiment, chlorophyll a loads are only marginally influenced. The on-average-increasing chlorophyll a trend caused by climate change impacts in the upper Elbe and Saale catchments cannot be reversed by a simple reduction of fertilization in the combined experiments.

An increased nutrient retention and decomposition potential in the soils of the landscape ("Retention") has the highest impact on nutrient loads. Especially the diffuse nitrate nitrogen loads are affected, but also ammonium and phosphate show some reactions, though with different magnitudes for the four analyzed gauges. The diversity in the magnitude of changes for the river parts can be explained by the heterogeneity and distribution of land use patterns and point sources as well as by the diversity in projected climate change within the catchment. As NH_4-N and/or PO_4-P are remarkably reduced under the retention experiment, chlorophyll a shows a decreasing trend due to a lack of nutrients. This reduction is even able to reverse the increasing trend in chlorophyll a caused by climate changes in the upper Elbe and Saale sub-catchments.

Two experiments dealing with a changed land use composition ("Buffer" and "Slope") result in more meadows and less agricultural areas in the sub-catchments and show similar results in the different river parts. Nitrate nitrogen is reduced most in the majority of cases due to less agricultural area with fertilizer application and hence lower total fertilizer loading under the experiments. The highest diversity of changes can be seen under the "Slope" experiment in the upper Elbe and Saale sub-catchments, which are characterized by a high share of mountainous areas, where the share of transformed land use areas is higher than in the lowland sub-catchment of the Havel river. For the latter, the "Slope" experiment has nearly no impact on the model outputs, and the combined changes result only from the climate scenario impacts.

The concentrations of dissolved oxygen are not visibly influenced by the changes in land use or management. The decreasing trend due to increased water temperature is more obvious in the upper part of the Elbe basin (gauge Schöna), probably due to less oxygen production with decreasing chlorophyll a loads in the river.

In general, the shares of cropland and distribution of point sources, as well as the distribution of soils with their specific nutrient retention potentials, are very important factors influencing the nutrient loads coming with the rivers. However, in the model application presented here, it is often difficult to distinguish between the single impacts on nutrient loads caused by certain land use or management changes and the secondary impacts due to altered chlorophyll a concentrations and a resulting change in nutrient uptake in the water body. The in-stream processes include a complex behavior of nutrients with a high number of interactions and feedbacks with the algae population. Chlorophyll a, for example, increases with decreasing NH_4-N availability and *vice versa*, causing increase (or decrease) of PO_4-P due to less (or more) algal uptake. Therefore, the resulting impacts are not only directly caused by land use changes, but are also indirectly caused by the subsequently changed conditions in the river water.

5. Discussion

A comparison of obtained results with the results of previous studies dealing with global change impacts in the Elbe river catchment is sometimes difficult, as different scenarios and downscaling methods were used by different authors. The whole range of model outputs illustrates a high uncertainty in climate change impact assessment.

The majority of published studies for the Elbe basin deal with climate change impacts on the hydrological cycle. The simulated effects of climate change on water cycle and river discharge presented in literature are diverse and differ in intensity and even in direction of change, resulting mainly from the diversity of precipitation change signals projected by different climate scenarios. Studies using 100 realizations of the Statistical Analogue Resampling Scheme (STARS), with a distinct decrease in summer precipitation and a moderate increase of precipitation in winter, project lower river discharge in the Elbe basin [85–88]. However, we have to note that recently the STARS model was critically discussed [89,90] regarding its suitability for producing climate scenarios. Model runs using the Inter-Sectoral Impact Model Intercomparison Project (ISI-MIP) climate scenarios [91] for the Elbe basin also project a decreased discharge on average, but the magnitude of changes is less pronounced [85]. Huang *et al.* [92] report diverse results depending on the driving climate model: the projections driven by the empirical-statistical model WETTREG (WETTerlagen-basierte REGionalisierungsmethode) produce negative trends in flood occurrence, whereas the projections forced by the two dynamical regional climate models REMO (REgional MOdelling) and CCLM (COSMO-Climate Limited-area Modelling) show various results with the prevalence of increasing trends in flood occurrence for this region. Applications of the ENSEMBLES scenarios for assessing future risks of floods and droughts in Germany showed an increasing trend of floods but no significant increase in droughts for the Elbe basin [22]. This is also reflected in our study, where under the same ENSEMBLES climate scenarios' higher discharges are projected on average (compare Figure 7).

There are some studies on the management change impacts on river water quality for often only small parts of the Elbe region (e.g., [93–95]), but only a few publications exist covering water quality issues under climate change. So, Quiel *et al.* [62] used the outputs of a model chain driven by selected realizations of the statistical model STARS as boundary conditions to run the river model QSim for a 700 km reach of the Elbe river. Soluble phosphorus concentrations decrease in all tributaries under all scenarios compared with the reference period for the same scenario. This results in an increased phytoplankton growth along the studied river reach and a shift of the chlorophyll *a* maximum under the dry and medium scenarios, but in a decrease in chlorophyll *a* concentration under the wet scenario conditions [62]. The latter is in accordance with the results for chlorophyll *a* presented in our study (compare Figure 7), as the ENSEMBLES climate projections produce a wetter climate in the future as well.

It seems that the projected nutrient loads presented in literature often correspond to the precipitation change signals, especially when eco-hydrological models using a simple routing of nutrients through the river network are applied. The increased precipitation causes higher nitrogen leaching through soils as well as higher phosphorus erosion rates with surface flow to the river network. Both processes increase nutrient loading to the river waters. Therefore, statistically downscaled climate scenarios with a negative trend in precipitation (e.g., STARS) mostly project decreasing nutrient loads in the Elbe catchment, whereas dynamic climate scenarios (e.g., REMO or the wet years of ISI-MIP scenarios) mostly result in increasing nutrient loads in the river due to positive precipitation change signals (e.g., [96,97]). This simple relationship between precipitation change signals and final nutrient load projections can be disturbed by including in-stream and algal processes in the eco-hydrological models, due to included transformation and ingestion of nutrients in the river network.

Nevertheless, the same conclusion as for river discharge is valid for the water quality impact assessment: a wide range in projections can be found in literature (as well as in our study). However, the diversity in discharge and water quality projections may not necessarily be the result of the application of different model approaches or climate scenario sets. Even with one scenario and one model, a high spatial variability can be observed, and some sub-regional trends can actually be opposite to the overall average trend of a large-scale basin, or local effects can be masked by large-scale aggregation [78,79]. This could also be seen in our study, where changes of model outputs due to climate impact differ in magnitude and intensity or even in the direction of change when comparing several tributaries of the Elbe (Figure 7).

In climate and socio-economic impact assessments in addition to the general (and often large) uncertainty associated with climate scenarios as drivers, there is also uncertainty connected to applied watershed models. The so-called structural and parameterization uncertainty is related to the ability of eco-hydrological models to represent the interrelated processes in landscape, vegetation and river network. The parameterization uncertainty can be especially large in a model with a high number of calibration parameters influencing each other, as in the SWIM model with implemented in-stream processes. Often several calibration parameter combinations exist, delivering the same or very similar model performance, so that it could happen to be "right for the wrong reasons" [98,99]. In general, such uncertainty rises with the rising model complexity, and goes along with a rising need in calibration efforts [68], and this should be taken into account when the model is extended by adding new processes. To overcome the limitations and weaknesses of a single eco-hydrological or climate model approach, it is useful to apply several models with the same input parameter sets (model intercomparison) and ensembles of climate scenarios for a more comprehensive assessment of uncertainties and elicitation of robust outputs [91,100].

The land use change experiments applied in this study do not represent the "full" set of potential future land use scenarios in the Elbe region, which could be elaborated considering options of future socio-economic development. For example, changes in urbanization or forest patterns could also have effects on the environment and the water resources [101,102]. In our study only the effects of single measures connected to nutrient sources and agricultural practices (which are currently considered in the planning of land/water management) on water quantity and quality were tested, also in combination with climate change. This could be regarded as a first step to finding suitable methods for adaptation to climate change impacts. However, for further studies it is recommended to apply a combination of different measures under consideration of the future socio-economic development for a more realistic land use change impact assessment. As climate change can strengthen, revoke or even inverse the land use change impacts, this aspect should always be included in such studies.

6. Summary and Conclusions

The SWIM model supplemented by an in-stream module was successfully calibrated and validated for the entire Elbe river basin, and applied for climate and land use change impact assessment in the region. For that, the commonly used technique was applied, using 19 climate scenario data sets provided by the ENSEMBLES project to drive an eco-hydrological model for 30-year periods in order to evaluate changes in water quantity and quality for the two future periods of 2021–2050 and 2071–2098 in comparison to the reference period of 1971–2000.

The calibration and validation of the extended SWIM for the Elbe region was complicated due to the high number of calibration parameters and the spatial variability within the catchment. Satisfactory model results could be still achieved by applying spatially distributed calibration parameter sets to capture variability in soil type distribution, land use pattern and economic development in the sub-catchments.

The analysis of a potential future climate, as projected by 19 scenarios for the Elbe catchment, indicates increasing trends in temperature and precipitation, but a decreasing trend in solar radiation on average. However, looking at the climate change signals of the 19 scenarios separately, differences can be seen in the intensity and—for precipitation and solar radiation—also in the direction of change signals. The standard deviation of the whole set of climate change signals increases in the second future period.

The results of the climate change impact assessment on water quantity and quality show a high spatial variability within the catchment according to the individual characteristics of the tributaries within the basin. For the entire Elbe catchment, river discharge is projected to increase by 11% and 20% on average for the two future periods. Dissolved oxygen concentration is projected to decrease by 2% and 5%, mainly due to the increased water temperature. The projected changes in nutrient loads do not show the same change direction. While NO_3-N loads slightly decrease on average (−1% and −5%), and NH_4-N shows a distinct decreasing trend (−11% and −24%), PO_4-P loads are expected to increase by 6% and 5% on average. The simulated reaction of nutrient loads to climate change is always influenced by the phytoplankton population, and *vice versa*. The chlorophyll *a* concentrations decrease slightly under the future conditions, by 3% and 4% on average, at the last downstream Elbe gauge.

Five simulation experiments dealing with possible changes in nutrient emissions were applied in the study, also in combination with climate change scenarios. Water discharge was mainly influenced by climate change impacts, and land use change measures had little or no influence on runoff. A reduction of agricultural area or fertilizer application mainly influenced the resulting nitrate nitrogen loads in the Elbe, whereas the reduction of point source emissions had the highest impacts on ammonium nitrogen and phosphate phosphorus loads. The chlorophyll *a* concentrations reacted to a changed food supply in the river, and would be reduced with a reduced nitrogen and phosphorus availability. An increase in nutrient retention and decomposition potential within the catchments would certainly be beneficial to reduce all types of nutrient loads in the river waters.

Nevertheless, the model application in the Elbe basin comes along with a certain degree of structural, parameterization and scenario uncertainty. Due to the lack of more detailed information on the case-specific observations and processes, not all possible methods to reduce uncertainties could be applied in this study, and the climate scenario–related uncertainty is unavoidable. The climate change impact assessment and land use change simulation experiments presented here deliver the first results and rough estimation on probable future developments in the Elbe river basin under climate change. For future research, in order to diminish and better assess (but not to eliminate) uncertainty, it could be recommended to apply two to three eco-hydrological models, as well as a "full" set of socio-economic scenarios, for a more reliable combined climate and land use change impact assessment. It could be also advantageous to additionally include management measures neglected in this model application so far (e.g., reservoirs or different crop types and rotations). These methods would help to identify future risks and threats more realistically, and to virtually test possible adaptation measures, as efforts to cope with the future climate conditions and their impacts are generally needed. Watershed models offer a suitable tool to guide decision-making on water quantity and quality for a sustainable management of water resources to match the requirements of the European Water Framework Directive (WFD).

Acknowledgments: The ENSEMBLES data used in the study were produced in the EU FP6 Integrated Project ENSEMBLES (contract No. 505539). The authors thank Shaochun Huang for delivering the prepared spatial SWIM input data of the entire Elbe river basin. The authors additionally thank Vanessa Wörner for download and preparation of observed water quality data, and Lasse Scheele for climate scenario data download, preparation and evaluation.

Author Contributions: The two authors planned and designed the methods to study climate and management change impacts on water quantity and quality in the Elbe river catchment. Cornelia Hesse set up, adjusted and calibrated the extended SWIM model with included in-stream processes for the Elbe river catchment, ran the climate and land use change scenarios and analyzed the temporal and spatial scenario output data. Cornelia Hesse also prepared the tables and figures for the publication, wrote the text and formatted the paper. Valentina

Krysanova guided and supervised the whole process, discussed results during the modeling study, and edited the manuscript.

Conflicts of Interest: The authors declare no conflict of interest.

References

1. Intergovernmental Panel on Climate Change (IPCC). *Climate Change 2007: The Physical Change Basis. Contribution of Working Group I to the Fourth Assessment Report of the Intergovernmental Panel on Climate Change*; Solomon, S., Qin, D., Manning, M., Chen, Z., Marquis, M., Averyt, K.B., Tignor, M., Miller, H.L., Eds.; Cambridge University Press: Cambridge, UK; New York, NY, USA, 2007; p. 996.

2. Intergovernmental Panel on Climate Change (IPCC). Summary for Policymakers. In *Climate Change 2013: The Physical Science Basis. Contribution of Working Group I to the Fifth Assessment Report of the Intergovernmental Panel on Climate Change*; Stocker, T.F., Qin, D., Plattner, G.-K., Tignor, M., Allen, S.K., Boschung, J., Nauels, A., Xia, Y., Bex, V., Midgley, P.M., Eds.; Cambridge University Press: Cambridge, UK, 2013; pp. 3–29.

3. European Environment Agency (EEA). *The European Environment. State and Outlook 2010. Land Use*; EEA: Copenhagen, Denmark, 2010.

4. Cassardo, C.; Jones, J.A.A. Managing water in a Changing World. *Water* **2011**, *3*, 618–628.

5. Research Institute for Knowledge Systems (RIKS). *Exploration of Land Use Trends under SOER 2010*; RIKS: Maastricht, The Netherlands, 2010.

6. Alcamo, J.; Flörke, M.; Märker, M. Future long-term changes in global water resources driven by socio-economic and climatic changes. *Hydrolog. Sci. J.* **2007**, *52*, 247–275.

7. Christensen, O.B.; Christensen, J.H. Intensification of extreme European summer precipitation in a warmer climate. *Glob. Planet. Chang.* **2004**, *44*, 107–117.

8. Kløve, B.; Ala-Aho, P.; Bertrand, G.; Gurdak, J.J.; Kupfersberger, H.; Kværner, J.; Muotka, T.; Mykrä, H.; Preda, E.; Rossi, P.; *et al.* Climate change impacts on groundwater and dependent ecosystems. *J. Hydrol.* **2014**, *518*, 250–266.

9. Kundzewicz, Z.W. Climate change impacts on the hydrological cycle. *Ecohydrol. Hydrobiol.* **2008**, *8*, 195–203.

10. Wang, X.; Siegert, F.; Zhou, A.; Franke, J. Glacier and glacial lake changes and their relationship in the context of climate change, Central Tibetan Plateau 1972–2010. *Glob. Planet. Chang.* **2013**, *111*, 246–257.

11. Linderholm, H.W. Growing season changes in the last century. *Agric. For. Meteorol.* **2006**, *137*, 1–14.

12. Vittoz, P.; Cherix, D.; Gonseth, Y.; Lubini, V.; Maggini, R.; Zbinden, N.; Zumbach, S. Climate change impacts on biodiversity in Switzerland: A review. *J. Nat. Conserv.* **2013**, *21*, 154–162.

13. Kittel, T.G.F. The Vulnerability of Biodiversity to Rapid Climate Change. In *Climate Vulnerability: Understanding and Addressing Threats to Essential Resources*; Pielke, R.A., Ed.; Elsevier Inc., Academic Press: Oxford, UK, 2013; Volumn 4, Chapter 4; pp. 185–201.

14. Lindner, M.; Fitzgerald, J.B.; Zimmermann, N.E.; Reyer, C.; Delzon, S.; van der Maaten, E.; Schelhaas, M.-J.; Lasch, P.; Eggers, J.; van der Maaten-Theunissen, M.; *et al.* Climate change and European forests: What do we know, what are the uncertainties, and what are the implications for forest management? *J. Environ. Manag.* **2014**, *146*, 69–83.

15. Gabriel, K.M.A.; Endlicher, W.R. Urban and rural mortality rates during heat waves in Berlin and Brandenburg, Germany. *Environ. Pollut.* **2011**, *159*, 2044–2050.

16. Haines, A.; Kovats, R.S.; Campbell-Lendrum, D.; Corvalan, C. Climate change and human health: Impacts, vulnerability and public health. *Public Health* **2006**, *120*, 585–596.

17. Koch, H.; Vögele, S.; Hattermann, F.F.; Huang, S. Sensitivity of electricity generation in Germany to climate change and variability. *Meteorol. Z.* **2015**. in print.

18. Liersch, S.; Cools, J.; Kone, B.; Koch, H.; Diallo, M.; Reinhardt, J.; Fournet, S.; Aich, V.; Hattermann, F.F. Vulnerability of rice production in the Inner Niger Delta to water resources management under climate variability and change. *Environ. Sci. Policy* **2013**, *34*, 18–33.

19. Alcamo, J.; Moreno, J.M.; Nováky, B.; Bindi, M.; Corobov, R.; Devoy, R.J.N.; Giannakopoulos, C.; Martin, E.; Olesen, J.E.; Shvidenko, A. Europe. In *Climate Change 2007: Impacts, Adaptation and Vulnerability; Contribution of Working Group II to the Fourth Assessment Report of the Intergovernmental Panel on Climate Change*; Parry, M.L., Canziani, O.F., Palutikof, J.P., van der Linden, P.J., Hanson, C.E., Eds.; Cambridge University Press: Cambridge, UK, 2007; pp. 541–580.

20. European Environment Agency (EEA). *Climate Change, Impacts and Vulnerability in Europe 2012*; EEA-Report No 12/2012; EEA: Copenhagen, Denmark, 2012; p. 300.

21. Federal Environmental Agency (UBA). *Monitoringbericht 2015 zur Deutschen Anpassungsstrategie an den Klimawandel—Bericht der Interministeriellen Arbeitsgruppe Anpassungsstrategie der Bundesregierung*; UBA: Dessau-Roßlau, Germany, 2015.

22. Huang, S.; Krysanova, V.; Hattermann, F.F. Projections of climate change impacts on floods and droughts in Germany using an ensemble of climate change scenarios. *Reg. Environ. Chang.* **2015**, *15*, 461–473.

23. Kyselý, J.; Beranová, R. Climate-change effects on extreme precipitation in central Europe: Uncertainties of scenarios based on regional climate models. *Theor. Appl. Climatol.* **2009**, *95*.

24. Whitehead, P.G.; Wilby, R.L.; Battarbee, R.W.; Kernan, M.; Wade, A.J. A review of the potential impacts of climate change on surface water quality. *Hydrolog. Sci. J.* **2009**, *54*, 101–123.

25. Kundzewicz, Z.W.; Mata, L.J.; Arnell, N.W.; Döll, P.; Kabat, P.; Jiménez, B.; Miller, K.A.; Oki, T.; Sen, Z.; Shiklomanov, I.A. Freshwater resources and their management. In *Climate Change 2007: Impacts, Adaptation and Vulnerability. Contribution of Working Group II to the Fourth Assessment Report of the Intergovernmental Panel on Climate Change*; Parry, M.L., Canziani, O.F., Palutikof, J.P., van der Linden, P.J., Hanson, C.E., Eds.; Cambridge University Press: Cambridge, UK, 2007; pp. 173–210.

26. Crossmann, J.; Futter, M.N.; Oni, S.K.; Whitehead, P.G.; Jin, L.; Butterfield, D.; Baulch, H.M.; Dillon, P.J. Impacts of climate change on hydrology and water quality: Future proofing management strategies in the Lake Simcoe watershed, Canada. *J. Great Lakes Res.* **2013**, *39*, 19–32.

27. Dunn, S.M.; Brown, I.; Sample, J.; Post, H. Relationship between climate, water resources, land use and diffuse pollution and the significance of uncertainty in climate change. *J. Hydrol.* **2012**, *434–435*.

28. Mimikou, M.A.; Baltas, E.; Varanou, E.; Pantazis, K. Regional impacts of climate change on water resources quantity and quality indicators. *J. Hydrol.* **2000**, *234*, 95–109.

29. Desortová, B.; Punčochář, P. Variability of phytoplankton biomass in a lowland river: Response to climate conditions. *Limnol. Ecol. Manag. Inland Waters* **2011**, *41*, 160–166.

30. Hardenbicker, P.; Rolinski, S.; Weitere, M.; Fischer, H. Contrasting long-term trends and shifts in phytoplankton dynamics in two large rivers. *Int. Rev. Hydrobiol.* **2014**, *99*.

31. Scharfe, M.; Callies, U.; Blöcker, G.; Petersen, W.; Schroeder, F. A simple Lagrangian model to simulate temporal variability of algae in the Elbe River. *Ecol. Model.* **2009**, *220*, 2173–2189.

32. Barclay, J.R.; Walter, M.T. Modeling denitrification in a changing climate. *Sustain. Water Qual. Ecol.* **2015**, *5*, 64–76.

33. Whitehead, P.G.; Wilby, R.L.; Butterfield, D.; Wade, A.J. Impacts of climate change on in-stream nitrogen in a lowland chalk stream: An appraisal of adaptation strategies. *Sci. Total Environ.* **2006**, *365*, 260–273.

34. Macleod, C.J.A.; Falloon, P.D.; Evans, R.; Haygarth, P.M. The effects of climate change on the mobilization of diffuse substances from agricultural systems. *Adv. Agron.* **2012**, *115*, 41–77.

35. Vollenweider, R. *Scientific Fundamentals of the Eutrophication of Lakes and Flowing Waters, with Particular Reference to Nitrogen and Phosphorous as Factors in Eutrophication*; OECD Tech Rep. DAS/CSI/68.27; Organisation for Economic Co-operation and Development (OECD): Paris, France, 1968; p. 159.

36. Anderson, D.M.; Glibert, P.M.; Burkholder, J.M. Harmful algal blooms and eutrophication: Nutrient sources, composition, and consequences. *Estuaries* **2002**, *25*, 704–726.

37. Pieterse, N.M.; Bleuten, W.; Jørgensen, S.E. Contribution of point sources and diffuse sources to nitrogen and phosphorus loads in lowland river tributaries. *J. Hydrol.* **2003**, *271*, 213–225.

38. Schindler, D.W. Recent advances in the understanding and management of eutrophication. *Limnol. Oceanogr.* **2006**, *51*, 356–363.

39. European Environment Agency (EEA). *Land Use Scenarios for Europe: Qualitative and Quantitative Analysis on a European Scale (PRELUDE)*; EEA Technical Report No 9/2007; EEA: Copenhagen, Denmark, 2007.

40. Seitzinger, S.P.; Mayorga, E.; Bouwman, A.F.; Kroeze, C.; Beusen, A.H.W.; Billen, G.; van Drecht, G.; Dumon, E.; Fekete, B.M.; Garnier, J.; *et al.* Global river nutrient export: A scenario analysis of past and future trends. *Glob. Biogeochem. Cy.* **2010**, *24*.

41. Bouraoui, F.; Grizzetti, B. Long term change of nutrient concentrations of rivers discharging in European seas. *Sci. Total Environ.* **2011**, *409*, 4899–4916.

42. De Wit, M.; Behrendt, H.; Bendoricchio, G.; Bleuten, W.; van Gaans, P. *The Contribution of Agriculture to Nutrient Pollution in Three European Rivers, with Reference to the European Nitrates Directive*; European Water Management Online, Official Publication of the European Water Association (EWA): Hennef, Germany, 2002.

43. Grizzetti, B.; Bouraoui, F.; Aloe, A. Changes of nitrogen and phosphorus loads to European seas. *Glob. Chang. Biol.* **2012**, *18*, 769–782.

44. Mehdi, B.; Ludwig, R.; Lehner, B. Evaluating the impacts of climate change and crop land use change on streamflow, nitrates and phosphorus: A modelling study in Bavaria. *J. Hydrol. Reg. Stud.* **2015**, *4*.

45. Huttunen, I.; Lehtonen, H.; Huttunen, M.; Piirainen, V.; Korppoo, M.; Veijalainen, N.; Viitasalo, M.; Vehviläinen, B. Effects of climate change and agricultural adaptation on nutrient loading from Finnish catchments to the Baltic Sea. *Sci. Total Environ.* **2015**, *529*, 168–181.

46. Horn, A.L.; Rueda, F.J.; Hörmann, G.; Fohrer, N. Implementing river water quality modelling issues in mesoscale watershed models for water policy demands—An overview on current concepts, deficits, and future tasks. *Phys. Chem. Earth* **2004**, *29*, 725–737.

47. Daniel, E.B.; Camp, J.V.; LeBoeuf, E.J.; Penrod, J.R.; Dobbins, J.P.; Abkowitz, M.D. Watershed Modeling and its Applications: A State-of-the-Art Review. *Open Hydrol. J.* **2011**, *5*, 26–50.

48. Krysanova, V.; Wechsung, F.; Arnold, J.; Srinivasan, R.; Williams, J. *SWIM (Soil and Water Integrated Model.): User Manual*; PIK Report No. 69; Potsdam Institute for Climate Impact Research (PIK): Potsdam, Germany, 2000; p. 239.

49. Hesse, C.; Krysanova, V.; Päzolt, J.; Hattermann, F.F. Eco-hydrological modelling in a highly regulated lowland catchment to find measures for improving water quality. *Ecol. Model.* **2008**, *218*, 135–148.

50. Hesse, C.; Krysanova, V.; Voß, A. Implementing In-Stream Nutrient Processes in Large-Scale Landscape Modeling for the Impact Assessment on Water Quality. *Environ. Model. Assess.* **2012**, *17*, 589–611.

51. Van der Linden, P., Mitchell, J.F., Eds.; *ENSEMBLES: Climate Change and Its Impacts: Summary of Research and Results from the ENSEMBLES Project*; Met Office Hadley Centre: Exeter, UK, 2009; p. 160.

52. International Commission for the Protection of the Elbe River (IKSE). *Die Elbe und ihr Einzugsgebiet—Ein Geographisch-Hydrologischer und Wasserwirtschaftlicher Überblick*; Schlüter GmbH & Co. KG: Schönebeck (Elbe), Germany, 2005; p. 258.

53. Bronstert, A.; Itzerott, S. Bewirtschaftungsmöglichkeiten im Einzugsgebiet der Havel: Abschlussbericht zum BMBF-Forschungsprojekt. 2006, p. 212. Available online: http://www.havelmanagement.net/Havel-ger/Publikationen/Endberichte/Endbericht_Verbund.pdf (accessed on 28 November 2014).

54. German Working Group on water issues (LAWA). *Beurteilung der Wasserbeschaffenheit von Fließgewässern in der Bundesrepublik Deutschland—Chemische Gewässergüteklassifikation*; Kulturbuchverlag: Berlin, Germany, 1998; pp. 1–35.

55. Koskova, R.; Nemecková, S.; Hesse, C. Using of the soil parametrization based on soil samples databases in rainfall-runoff modelling. In Proceedings of the Adolf Patera Workshop "Extreme Hydrological Events in Catchments"; Jakubíková, A., Broza, V., Szolgay, J., Eds.; Czech Technical University: Prague, Czech Republic, 2007; pp. 241–249. (In Czech).

56. River Basin Community Elbe (FGG-Elbe). *Zusammenfassender Bericht der Flussgebietsgemeinschaft Elbe über die Analysen nach Artikel 5 der Richtlinie 2000/60/EG (A-Bericht)*; FGG-Elbe: Magdeburg, Germany, 2004.

57. International Commission for the Protection of the Elbe River (IKSE). *Bestandsaufnahme von bedeutenden punktuellen kommunalen und industriellen Einleitungen von prioritären Stoffen im Einzugsgebiet der Elbe Appendix 3: Bestandsaufnahme der kommunalen Abwassereinleitungen größer 20000 EGW im Einzugsgebiet der Elbe in der Tschechischen Republik (Stand 1995)*; IKSE: Magdeburg, Germany, 1995.

58. Hussian, M.; Grimvall, A.; Petersen, W. Estimation of the human impact on nutrient loads carried by the Elbe river. *Environ. Monit. Assess.* **2004**, *96*, 15–33.

59. Klöcking, B.; Haberlandt, U. Impact of land use changes on water dynamics—A case study in temperate meso- and macroscale river basins. *Phys. Chem. Earth* **2002**, *27*, 619–629.

60. Lehmann, A.; Rode, M. Long-term behaviour and cross-correlation water quality analysis of the river Elbe, Germany. *Water Res.* **2001**, *35*, 2153–2160.

61. Schneider, P.; Reincke, H. Contaminated Sediments in the Elbe Basin and its Tributary Mulde. In *Uran. Environ.*; Springer: Berlin, Germany, 2006; pp. 655–662.

62. Quiel, K.; Becker, A.; Kirchesch, V.; Schöl, A.; Fischer, H. Influence of global change on Phytoplankton and nutrient cycling in the Elbe river. *Reg. Environ. Chang.* **2011**, *11*, 405–421.

63. Grossmann, M. Economic value of the nutrient retention function of restored floodplain wetlands in the Elbe River basin. *Ecol. Econ.* **2012**, *83*, 108–117.

64. Arnold, J.; Allan, P.; Bernhardt, G. A comprehensive surface-groundwater flow model. *J. Hydrol.* **1993**, *142*, 47–69.

65. Krysanova, V.; Meiner, A.; Roosaare, J.; Vasilyev, A. Simulation modelling of the coastal waters pollution from agricultural watershed. *Ecol. Model.* **1989**, *49*, 7–29.

66. Neitsch, S.L.; Arnold, J.G.; Kiniry, J.R.; Srinivasan, R.; Williams, J.R. *Soil and Water Assessment Tool User's Manual: Version 2000*; TWRI Report TR-192; Texas Water Resources Institute (TWRI): Temple, TX, USA, 2002; p. 412.

67. Hattermann, F.F.; Krysanova, V.; Habeck, A.; Bronstert, A. Integrating wetlands and riparian zones in river basin modelling. *Ecol. Model.* **2006**, *199*, 379–392.

68. Hesse, C.; Krysanova, V.; Vetter, T.; Reinhardt, J. Comparison of several approaches representing terrestrial and in-stream nutrient retention and decomposition in watershed modelling. *Ecol. Model.* **2013**, *269*.

69. Thuringian Regional Office for Agriculture (TLL). *Leitlinie zur effizienten und umwelt-verträglichen Erzeugung von Winterweizen*; 2011; Available online: http://www.tll.de/ainfo/archiv/weiz0508.pdf (accessed on 12 December 2013).

70. Hattermann, F.F.; Krysanova, V.; Hesse, C. Modelling wetland processes in regional applications. *Hydrol. Sci. J.* **2008**, *53*, 1001–1012.

71. Internatiopnal Kommission zum Schutz der Elbe (IKSE). *Informationsdokumente zum Internationalen Messprogramm Elbe 2007, Übersicht der Messstationen und Messstellen*, 2007. Available online: http://www.ikse- mkol.org/index.php?id=212 (accessed on 21 November 2012).

72. Hattermann, F.F.; Wattenbach, M.; Krysanova, V.; Wechsung, F. Runoff simulations on the macroscale with the ecohydrological model SWIM in the Elbe catchment—validation and uncertainty analysis. *Hydrol. Proc.* **2005**, *19*.

73. Huang, S.; Hesse, C.; Krysanova, V.; Hattermann, F. From meso- to macro-scale dynamic water quality modelling for the assessment of land use change scenarios. *Ecol. Model.* **2009**, *220*, 2543–2558.

74. Nash, J.E.; Sutcliffe, J.V. River flow forecasting through conceptual models part I: A discussion of principles. *J. Hydrol.* **1970**, *10*, 282–290.

75. Gudmundsson, L.; Wagener, T.; Tallaksen, L.M.; Engeland, K. Evaluation of nine large-scale hydrological models with respect to the seasonal runoff climatology in Europe. *Water Resour. Res.* **2012**, *48*.

76. Nakicenovic, N., Swart, R., Eds.; *Special Report on Emissions Scenarios. Report of Working Group III of the Intergovernmental Panel on Climate Change*; Cambridge University Press: Cambridge, UK, 2000; p. 570.

77. Gädeke, A.; Hölzel, H.; Koch, H.; Pohle, I.; Grünewald, U. Analysis of uncertainties in the hydrological response of a model-based climate change impact assessment in a subcatchment of the Spree River, Germany. *Hydrol. Process.* **2014**, *28*.

78. Arheimer, B.; Dahné, J.; Donnelly, C. Climate change impact on riverine nutrient load and land-based remedial measures of the Baltic action plan. *Ambio* **2012**, *41*, 600–612.

79. Piniewski, M.; Kardel, I.; Giełczewski, M.; Marcinkowski, P.; Okruszko, T. Climate change and agricultural development: adapting Polish agriculture to reduce future nutrient loads in a coastal watershed. *Ambio* **2014**, *43*.

80. Tebaldi, C.; Knutti, R. The use of the multi-model ensemble in probabilistic climate projections. *Philos. Trans. Roy. Soc.* **2007**, *365*.

81. Kling, H.; Fuchs, M.; Paulin, M. Runoff conditions in the upper Danube basin under an ensemble of climate change scenarios. *J. Hydrol.* **2012**, *424–425*, 264–277.

82. Teutschbein, C.; Seibert, J. Regional Climate Models for Hydrological Impact Studies at the Catchment Scale: A Review of Recent Modeling Strategies. *Geogr. Compass.* **2010**, *4*, 834–860.

83. Ehret, U.; Zehe, E.; Wulfmeyer, V.; Warrach-Sagi, K.; Liebert, J. HESS Opinions "Should we apply bias correction to global and regional climate model data?". *Hydrol. Earth Syst. Sci.* **2012**, *16*.

84. Bundesministerium für Verbraucherschutz, Ernährung und Landwirtschaft (BMVEL). *Gute Fachliche Praxis zur Vorsorge gegen Bodenschadverdichtungen und Bodenerosion*; BMVEL: Bonn, Germany, 2002.

85. Roers, M.; Wechsung, F. Neubewertung der Auswirkungen des Klimawandels auf den Wasserhaushalt im Elbegebiet. *Hydrol. Wasserbewirts* **2015**, *59*, 109–119.

86. Huang, S.; Krysanova, V.; Österle, H.; Hattermann, F.F. Simulation of spatiotemporal dynamics of water fluxes in Germany under climate change. *Hydrol. Process.* **2010**, *24*.

87. Hattermann, F.F.; Post, J.; Krysanova, V.; Conradt, T.; Wechsung, F. Assessment of water availability in a Central-European River Basin (Elbe) under climate change. *Adv. Clim. Chang. Res.* **2008**, *4*, 42–50.

88. Conradt, T.; Koch, H.; Hattermann, F.F.; Wechsung, F. Spatially differentiated management-revised discharge scenarios for an integrated analysis of multi-realisation climate and land use scenarios. *Reg. Environ. Chang.* **2012**, *12*.

89. Wechsung, F.; Wechsung, M. Dryer years and brighter sky—The predictable simulation outcomes for Germany's warmer climate from the weather resampling model STARS. *Int. J. Climatol.* **2014**.

90. Wechsung, F.; Wechsung, M. A methodological critique on using temperature-conditioned resampling for climate projections as in the paper of Gerstengarbe *et al.* (2013) winter storm- and summer thunderstorm-related loss events in Theoretical and Applied Climatology (TAC). *Theor. Appl. Climatol.* **2015**, *1–5*.

91. Warszawski, L.; Frieler, K.; Huber, V.; Piontek, F.; Serdeczny, O.; Schewe, J. The Inter-Sectoral Impact Model Intercomparison Project (ISI-MIP). *Proc. Natl. Acad. Sci. USA* **2013**, *111*.

92. Huang, S.; Hattermann, F.F.; Krysanova, V.; Bronstert, A. Projections of climate change impacts on river flood conditions in Germany by combining three different RCMs with a regional eco-hydrological model. *Clim. Chang.* **2013**, *116*.

93. Kersebaum, K.C.; Steidl, J.; Bauer, O.; Piorr, H.-P. Modelling scenarios to assess the effects of different agricultural management and land use options to reduce diffuse nitrogen pollution into the river Elbe. *Phys. Chem. Earth* **2003**, *28*, 537–545.

94. Meissner, R.; Seeger, J.; Rupp, H. Effects of agricultural land use changes on diffuse pollution of water resources. *Irrig. Drain.* **2002**, *51*.

95. Ullrich, A.; Volk, M. Application of the Soil and Water Assessment Tool (SWAT) to predict the impact of alternative management practices on water quality and quantity. *Agric. Water Manag.* **2009**, *96*.

96. Martinkova, M.; Hesse, C.; Krysanova, V.; Vetter, T.; Hanel, M. Potential impact of climate change on nitrate load from the Jizera catchment (Czech Republic). *Phys. Chem. Earth* **2011**, *36*, 673–683.

97. Roers, M.; Venohr, M.; Wechsung, F.; Müller, E.N. Nährstoffeinträge im Elbegebiet—Modellstudien zu deren zukünftiger Änderung und zur Wirkung von Reduktionsmaßnahmen unter Klimawandel. *Hydrol. Wasserbewirts* Unpublished work. **2015**.

98. Dayton, P.K. Two Cases of Resource Partitioning in an Intertidal Community: Making the Right Prediction for the Wrong Reason. *Am. Nat.* **1973**, *107*, 662–670.

99. Van der Laan, M.; Annandale, J.G.; Bristow, K.L.; Stirzaker, R.J.; du Preez, C.C.; Thorburn, P.J. Modelling nitrogen leaching: Are we getting the right answer for the right reason? *Agric. Water Manag.* **2014**, *133*.

100. Schewe, J.; Heinke, J.; Gerten, D.; Haddeland, I.; Arnell, N.W.; Clark, D.B.; Dankers, R.; Eisner, S.; Fekete, B.; Colón-González, F.J.; *et al.* Multi-model assessment of water scarcity under climate change. *Proc. Natl. Acad. Sci. USA* **2013**, *111*.

101. O'Driscoll, M.; Clinton, S.; Jefferson, A.; Manda, A.; McMillan, S. Urbanization Effects on Watershed Hydrology and In-Stream Processes in the Southern United States. *Water* **2010**, *2*.

102. Wei, X.; Liu, W.; Zhou, P. Quantifying the Relative Contributions of Forest Change and Climatic Variability to Hydrology in Large Watersheds: A Critical Review of Research Methods. *Water* **2013**, *5*.

Assessment of Climate Change Impacts on Water Quality in a Tidal Estuarine System Using a Three-Dimensional Model

Wen-Cheng Liu and Wen-Ting Chan

Abstract: Climate change is one of the key factors affecting the future quality and quantity of water in rivers and tidal estuaries. A coupled three-dimensional hydrodynamic and water quality model has been developed and applied to the Danshuei River estuarine system in northern Taiwan to predict the influences of climate change on water quality. The water quality model considers state variables including nitrogen, phosphorus, organic carbon, and phytoplankton as well as dissolved oxygen, and is driven by a three-dimensional hydrodynamic model. The hydrodynamic water quality model was validated with observational salinity distribution and water quality state variables. According to the analyses of statistical error, predictions of salinity, dissolved oxygen, and nutrients from the model simulation quantitatively agreed with the observed data. The validated model was then applied to predict water quality conditions as a result of projected climate change effects. The simulated results indicated that the dissolved oxygen concentration was projected to significantly decrease whereas nutrients will increase because of climate change. Moreover, the dissolved oxygen concentration was lower than 2 mg/L in the main stream of the Danshuei River estuary and failed to meet the water quality standard. An appropriate strategy for effective water quality management for tidal estuaries is needed given the projected persistent climate trends.

Reprinted from *Water*. Cite as: Liu, W.-C.; Chan, W.-T. Assessment of Climate Change Impacts on Water Quality in a Tidal Estuarine System Using a Three-Dimensional Model. *Water* **2016**, *8*, 60.

1. Introduction

Estuaries are among the world's vital aquatic resources. They provide food resources and a habitat for ecologically and economically important fish and shellfish species, recreational regions, educational and scientific experiences, and other important ecosystem services [1–5]. For example, the Guandu Natural Park in Taipei city, which is located at the confluence of the Danshuei River and the Keelung River, serves as an educational purpose and scientific experience [6]. Ecosystem services are fundamental life-support processes upon which all organisms depend [7]. Two ecosystem services that estuaries provide are water filtration and habit protection. However, adverse impacts on the estuarine ecosystem by environmental perturbations (e.g., anthropogenic nutrient loading, land use change, hydrological modification) have been widely reported [8–10]. The adverse impacts include impaired water quality, habitat loss, and diminished resources [11]. These perturbations result in declining water quality and deleterious changes in ecosystem structure and tropic dynamics [12,13]. The deleterious water quality subsequently produces the problems of odor, aesthetics, human pathogens, and increased public health risk. For example, McKibben *et al.* [14] reported that harmful algal blooms are proliferations of microscopic algae that harm the environment by producing toxins that accumulate in shellfish or fish, or through the accumulation of biomass that in turn affects co-occurring organisms and alters food webs in negative ways. Impacts include human illness and mortality following direct consumption or indirect exposure to toxic shellfish or toxins in the environment.

Climate change occurs naturally, but human population growth and associated land-cover deforestation and burning of fossil fuel have substantially accelerated the increase in greenhouse gases (CO_2, CH_4, N_2O, *etc.*). The elevated concentration of CO_2 and other greenhouse gases from anthropogenic activities have caused warming of the global climate by modifying radiative forcings, and continued changes will result in climate shifts [15–18]. Feng *et al.* [19] used model-projected future surface temperature and precipitation to examine the change/shifts of climate types over the global land area. They concluded that compared to the present-day condition, the boreal winter temperature over the global land area is projected to increase by 3–12 °C by 2071–2100 under a high emission scenario. Strong warming (>8 °C) appears along the Arctic coastal regions, moderate warming (5–7 °C) appears in the mid-latitude of the Northern Hemisphere, while the warming in the tropical and the Southern Hemisphere is relatively smaller (<5 °C). The projected warming in the boreal summer is much weaker (3–6 °C). Xin *et al.* [20] studied climate change projections over East Asia under various representative concentration pathway (RCP) scenarios using simulations conducted with the Beijing Climate Center Climate System Model for the Coupled Model Intercomparison Project phase 5. Under all RCPs, including RCP2.6, RCP4.5, RCP6.0, and RCP8.5, the East Asian climate is found to be warmer and wetter in the 21st century than the present climatology (1986–2005). For 2080–2099, the East Asian mean surface air temperature is higher than for present climatology by 0.98 °C (4.4%) under RCP2.6, 1.89 °C (7.7%) under RCP4.5, 2.47 °C (7.1%) under RCP6.0, and 4.06 °C (9.1%) under RCP8.5.

The impacts of climate change on human health have been widely reported [21–23]. Numerous studies project greater morbidity and mortality from direct exposure, as well as greater health risks due to decreased air quality, water-borne disease, and other infectious diseases [23,24]. Impacts of climate change on river and estuarine systems provide a subject of active research [25–27] because of the importance of water resources for human activities. Potential impacts of climate change

284

on hydrology cover changes in runoff discharge, river flow, and groundwater storage [28]. Impacts on water quality include many factors (physical, including temperature, turbidity; chemical, including pH and concentration; biological, including biodiversity and species abundance across the entire food web from microbial pools and macrophytes up to fishes). With respect to water quality, most climate change impacts can be attributed to changes in either discharge—which controls dilution, flow velocity, and residence time—or water temperature. The impact of climate change on river and estuarine water quality is also heavily dependent on the future evolution of human activities (pollutions, withdrawals, *etc.*), so the direct influence of climate change may end up being relatively small [29].

The impact of climate change on estuaries has been reviewed by Robins *et al.* [30], who reported that potential changes to physical processes include flooding and coastal squeeze, caused by increased sea level rise, changing surge and wave climates, and changing river flow events. Sea level rise will cause a shift towards net sediment accretion, but with reduced transport in UK estuaries. Turbulent mixing that is critical for water quality and coastal ecology is controlled by river flow variability. Therefore, alterations to river flows will change the estuarine fronts, stratification, and mixing. The combination of sea level rise and longer dry periods in summer will cause negative impacts on eutrophication, harmful algal blooms, and hypoxia.

Numerical water quality models are useful in assisting the understanding of biological processes and the assessment of the influences of climate change on water quality conditions in aquatic systems [31–38]. For example, Tu [39] used a GIS-based watershed simulation model, AVGWLF, to simulate the future changes in streamflow and nitrogen load under different climate change and land use change scenarios at a watershed in eastern Massachusetts, USA. AVGWLF simulates daily streamflows and monthly nitrogen loads. As a result, the historical observed daily streamflow and nitrogen loads have to be used for model calibration and validation. The AVGWLF model tracks monthly streamflow and nitrogen load well in both calibration and validation. The coefficient of determination (R^2) and Nash-Sutcliffe coefficient (NS) values in calibration for streamflow in most of the watersheds are higher than 0.7, and for nitrogen load are higher than 0.6. The R^2 and NS values in validation of streamflow and nitrogen loads are even higher than the corresponding values in calibration. The validated model was used to project the impact of different climate scenarios (A1B, B1, and A2) on streamflow ad nitrogen load. The results revealed that the monthly streamflows in late fall and winter increase, whereas those in the summer months decrease, mainly as a result of climate change. Simulated nitrogen loads in late fall and winter months increase greatly, whereas those in spring and summer months have mixed responses affected by both climate and land use changes [39]. Rehana and Mujumdar [40] adopted a water quality model, QUAL2K, to simulate the water quality responses of six climate change scenarios covering different streamflow, air temperature, and water temperature at different stations. The simulated results suggested that all climate change scenarios would cause impairment in water quality. It was found that there was a significant decrease in dissolved oxygen levels owing to the impact of climate change on temperature and flows. For example, Luo *et al.* [41] applied the Soil and Water Assessment Tool (SWAT) to evaluate and enhance the watershed modeling approach in characterizing climate change impacts on water supply and ecosystem stressors. The SWAT was applied to headwater drainage basins in the northern Costal Ranges and Sierra Nevada mountain range in California. SWAT parameters for hydrological simulation were initialized within the ArcSWAT interface. Input data for watershed morphology have a 12-km spatial resolution. Input parameters mainly include the SCS runoff curve number (CN), snowmelt-related parameters, channel hydraulic conductivity, and parameters for groundwater recharge. The model was calibrated with daily streamflow at different selected stations. The calibrated model was then applied to project the effects of climate change. They concluded that the hydrological cycle and water quality of headwater drainage basins in California, especially their seasonality, were very sensitive to projected climate change.

These kinds of numerical models used to resolve one-dimensional and two-dimensional issues cannot well represent the spatial variations in three dimensions. For examples, Wan *et al.* [42] documented the development, calibration, and verification of a three-dimensional water quality model for the St. Lucie Estuary, a small and shallow estuary located on the east coast of south Florida. Modeling results revealed that high algae concentrations in estuaries are likely caused by excessive nutrient and algae supply in freshwater inflows. Cerco and Noel [43] applied the CE-QUAL-ICM (Corps of Engineers Integrated Compartment Water Quality Model) eutrophication model to simulate a 21-year (1985-2005) water quality model of Chesapeake Bay. The most significant finding was the influence of physical processes, notably stratification and associated effects (e.g., anoxic volume), on computed water quality. Li *et al.* [44] developed a three-dimensional hydrodynamic model coupled with a water quality model to determine the environmental capacity of nitrogen and phosphorus in Jiaozhou Bay, China. The model was calibrated based on data collected in 2003. The proposed water quality model effectively reproduced the spatiotemporal variability in nutrient concentration. However, few studies have emphasized the impacts of climate change on estuarine water quality using three-dimensional hydrodynamics and water quality coupling models.

This study aims to apply a coupled three-dimensional hydrodynamic and water quality (SELFE-WQ) model to characterize the water quality conditions in the estuarine system and assess the impacts of climate change scenarios on water quality in the Danshuei River estuarine system in northern Taiwan. The model was validated with observational salinity and water quality state variables. The validated water quality model was then applied to project the water quality conditions in estuarine system responses to climate change scenarios under the low flow condition.

2. Materials and Methods

The Danshuei River, with its tributaries, is the largest river system in northern Taiwan; its watershed encompasses 2726 km^2, with a combined length of 158.7 km (Figure 1). The regional climate is subtropical with the temperature varying between 10 and 35 °C, and the annual precipitation in the region ranges between 1500 mm and 2500 mm, with the majority falling in late spring (May) to early fall (October). The long-term average annual river flow rate is 6.6×10^9 m^3/y. The contributions of freshwater from the three major tributaries are, on average, 27% from the Keelung River, 31% from the Tahan Stream, and 37% from the Hsintien Stream. In addition to the mainstream of the Danshuei River, the lower reaches of the three major tributaries are also affected by tide. The principal tidal constituents of the estuary lean toward semi-diurnal tides, with a mean tidal range of 2.1 m and a spring tidal range of 3.5 m. Seawater intrusion reaches into all three tributaries except during periods of very high river inflows. In general, saltwater intrusion reaches 25–30 km from the Danshuei River mouth. The hydrodynamic characteristics in the system are mainly controlled by tide, river inflow, and the density gradient induced by the mixing of saline and freshwater [45,46]. The average flushing time of the Danshuei River is 2–4 days [47].

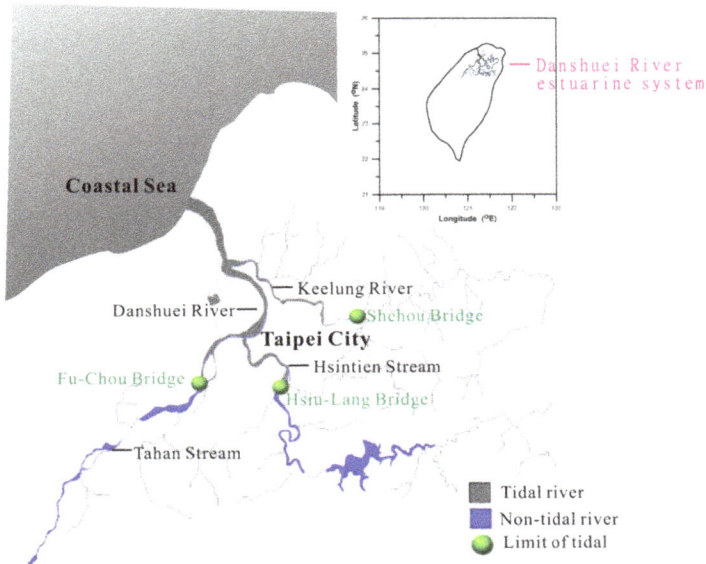

Figure 1. Danshuei River estuarine system and watershed.

287

The Danshuei River flows through the metropolitan area of Taipei, which has a population of approximately 6 million. A huge amount of treated and untreated domestic sewage was discharged into the river system and resulted in low dissolved oxygen and high nutrients. Viable biological activities are observed only in the lowest reach of the estuary, where the pollutant concentrations are reduced as a result of dilution by seawater [48,49].

3. Materials and Methods

3.1. Hydrodynamic Model

The numerical modeling of ocean circulation at scales ranging from estuaries to ocean basins is maturing as a field. Most modern oceanic and estuarine circulation codes solve for some form of the three-dimensional Navier–Stokes equations and can be complemented with conservation equations for a given water volume and salt concentration. In this paper, a three-dimensional, semi-implicit Eulerian-Lagrangian finite element model (SELFE, Zhang and Baptisa [50]) was implemented to simulate the Danshuei River estuarine system and its adjacent coastal sea. SELFE solves the Reynolds stress-averaged Navier–Stokes equations, which use conservation laws for mass, momentum, and salt with hydrostatic and Boussinesq approximations, to determine the free-surface elevation, three-dimensional water velocity, and salinity.

Unlike most 3D models using finite-difference/finite-volume schemes, SELFE is based on a finite-element scheme. No model splitting was used in SELFE, thus eliminating the errors associated with the splitting between internal and external modes [51]. Semi-implicit schemes were applied to all the equations to enhance the stability and maximize the efficiency of the system. An Eulerian-Lagrangian method was used to treat advection in the momentum equation, thus permitting the use of large time steps without compromising on stability. The horizontal space was discretized in the form of an unstructured grid of triangular elements, whereas the hybrid vertical coordinates—partly terrain-following S coordinates and partly Z coordinates—were used in the vertical direction. The wetting and drying algorithm was incorporated into the model. The minimum depth criterion for wetting and drying simulation was set to be 0.05 m.

Because turbulent mixing plays a critical role in determining the stratification in the tidal estuary, several reports have documented the model results of turbulence mixing parameterization. SELFE uses the generic length scale (GLS) turbulence closure of Umlauf and Burchard [52], which has the advantage of encompassing most of the 2.5-equation closure model (K–Ψ). A detailed description of the turbulence closure model, the vertical boundary conditions for the momentum equation, the numerical solution methods, and the numerical stability parameters can be found in Zhang and Baptista [50].

3.2. Water Quality Model

The water quality model used in this study was based on a three-dimensional conventional water quality analysis simulation program called WASP5, originally developed by Ambrose *et al.* [53]. It constitutes a complicated system of four interacting parts: dissolved oxygen, nitrogen cycle, phosphorus cycle, and phytoplankton dynamics. Eight water quality components are included: dissolved oxygen (DO), phytoplankton as carbon (PHYT), carbonaceous biochemical oxygen demand (CBOD), ammonium nitrogen (NH$_4$), nitrate and nitrite nitrogen (NO$_3$), organic nitrogen (ON), ortho-phosphorus or inorganic phosphorus (OP), and organic phosphorus (OP). The conceptual framework for the water quality model is presented in Figure 2.

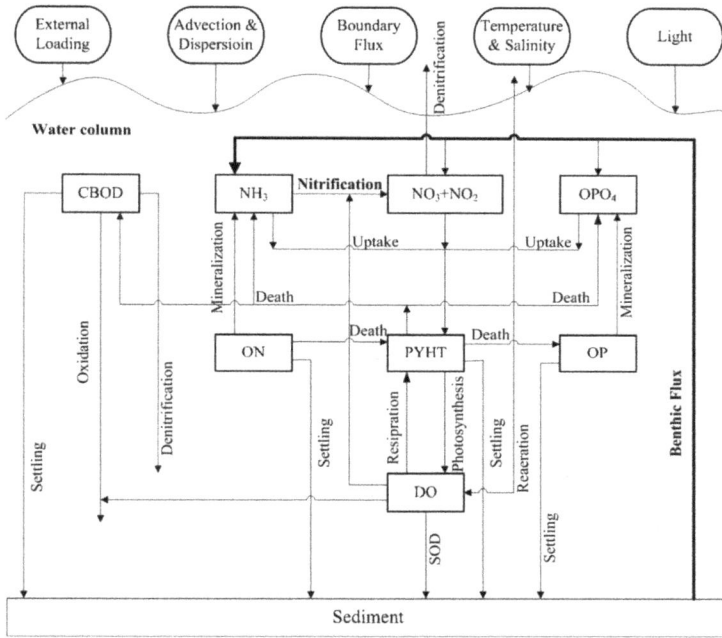

Figure 2. Schematic of water quality model.

A mathematical formulation of the conservation of mass can be written as follows:

$$\frac{\partial C}{\partial t} + \frac{\partial (uC)}{\partial x} + \frac{\partial (vC)}{\partial y} + \frac{\partial (wC)}{\partial z} = \frac{\partial}{\partial x}(A_h \frac{\partial C}{\partial x}) + \frac{\partial}{\partial y}(A_h \frac{\partial C}{\partial y}) + \frac{\partial}{\partial z}(K_v \frac{\partial C}{\partial z}) + S_e + S_i \quad (1)$$

where C is the concentration of water quality components; u, v, and w are the water velocity components corresponding to a Cartesian coordinate system (x, y, z); A_h and

K_v are the coefficients of horizontal viscosity and vertical eddy diffusion, respectively; S_e is the time rate of external additional (withdrawal) across the boundaries; and S_i is the time rate of internal increase/decrease by biogeochemical reaction processes.

Equation (1) gives the distribution of each state variable using the physical parameters determined from the hydrodynamic model. The last two terms, S_e and S_i represent, respectively, the external and internal sources (or sinks), the latter being primarily due to biogeochemical processes.

The present model of DO includes the following processes: source from photosynthesis, reaeration through surface and external loading, and sinks due to decay of CBOD, nitrification, algae respiration, and SOD. The mathematical representation is:

$$S_i = -K_c CBOD - a_{no}\frac{K_{n23}N_2}{K_{h23} + N_2}\frac{DO}{DO + K_{nit}} + a_c a_{co}(PQ \cdot G - \frac{R}{RQ})Chl \qquad (2)$$

$$S_e = (1 - \lambda_1)K_r(DO_s - DO) - \frac{SOD}{\Delta z}\frac{DO}{DO + K_{DO}} + \frac{WDO}{V} \qquad (3)$$

where a_c = ratio of carbon to chlorophyll in phytoplankton (mg C/μg Chl); a_{co} = ratio of oxygen demand to organic carbon recycled = 2.67; a_{no} = ratio of oxygen consumed per unit of ammonia nitrogen nitrified = 4.57; CBOD = concentration of carbonaceous of biochemical oxygen demand (mg/L); Chl = concentration of chlorophyll a (μg/L); DO = concentration of dissolved oxygen (mg/L); DO_s = saturation concentration of DO (mg/L); G = growth rate of phytoplankton (1/day); K_c = first-order decay rate of CBOD (1/day); K_{DO} = half-saturation concentration for benthic flux of CBOD (mg/L); K_{h23} = half-saturation concentration for nitrification (mg/L); K_{n23} = nitrification rate of ammonia nitrogen to nitrite-nitrate nitrogen (mg/L/day); K_{nit} = half-saturation concentration for oxygen limitation of nitrification (mg/L); K_r = reaeration rate (1/day); N_2 = concentration of ammonia nitrogen; PQ = photosynthesis quotient (mole O_2/mole C); R = respiration rate of phytoplankton (1/day); RQ = respiration quotient (mole CO_2/mole O_2); V = layer volume (cm^3); WDO = external loading of DO (mg/day) including point and nonpoint sources; Δz = layer thickness (cm); and $\lambda_1 = 0$ for $k = 1$ (at top layer), $\lambda_1 = 1$ for $2 \leqslant k \leqslant N$, and N is the number of layers.

The sediment oxygen demand (SOD) in Equation (3) is the rate of oxygen consumption exerted by the bottom sediment and the overlay water due to the respiration of the benthic biological communities and the biochemical degradation of organic matter. The SOD is a major component of the dissolved oxygen (DO) budget and a key parameter to be determined through the model validation in the water quality model.

According to previous study implemented by Chen et al. [54], the component of phytoplankton species in the Danshuei River estuary includes diatoms, green algae, and others; therefore, these three major species are taken into account in the model simulation.

3.3. Model Schematization and Implementation

In the present study, the horizontal resolutions, 200 m × 200 m and 40 m × 40 m, of the bathymetric and topographical data in the Taiwan Strait and Danshuei-River estuarine system were obtained from the Ocean Data Bank and Water Resources Agency, Taiwan. The deepest point within the study area is 110 m (below the mean sea level) near the northeast corner of the computational domain (Figure 3). The model mesh for the Danshuei-River estuarine system and its adjacent coastal sea consists of 5119 elements (Figure 3). To meet the accuracy requirements, fine-grid resolution was used locally, and coarse resolution was implemented away from the region of interest. In this computational domain, the mesh size varied from 6000 m in the Taiwan Strait down to 40 m in the upper reach of Danshuei River estuary. The mesh size (40 m) used in the upper reach of Danshuei River estuary would be an appropriate resolution because the bathymetric and topographic data in 40 m × 40 m were only obtained.

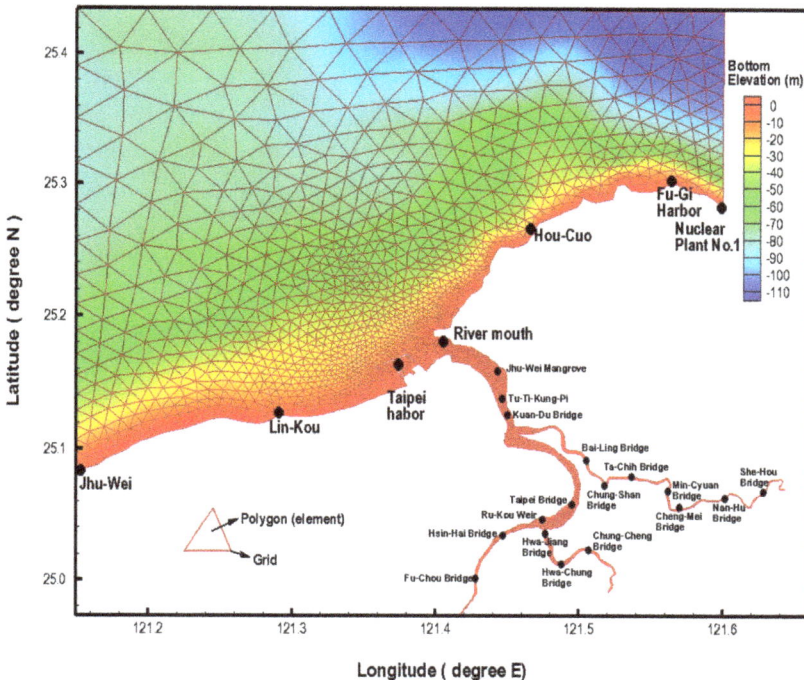

Figure 3. The topography of the Danshuei River estuarine system and its adjacent coastal and unstructured grid for the computational domain.

291

In the vertical direction, ten z-levels and ten evenly spaced S-levels were specified at each horizontal grid, *i.e.*, the thickness of the cell depended on the bottom elevation of each grid. The vertical resolutions in the coastal sea and Danshuei River estuary range from 10–20 m and 0.015–1.2 m, respectively. A 120-s time step was used in our simulations without any signs of numerical instability.

4. Model Validation

4.1. Salinity Distribution

Salinity distributions reflect the combined results of all processes, including density circulation and mixing processes. These processes in turn control the density circulation and modify the mixing processes [45]. In the present study, the salinity distribution along the Danshuei River-Tahan Stream collected by the Water Resources Agency, Taiwan, was used for model validation. Liu *et al.* [55] reported that a five-constituent tide (*i.e.*, M_2, S_2, N_2, K_1, and O_1) is sufficient to represent the tidal components in the Taiwan Strait. The five-constituent tide was adopted in the model simulation as a forcing function at the coastal sea boundaries. The model was run for a two-year simulation. The salinity of the open boundaries in the coastal sea was set to 35 *ppt*. The upstream boundary conditions at the three tributaries (Tahan Stream, Hsintien Stream, and Keelung River) were specified with daily freshwater discharges; therefore, the salinity at the upstream boundaries was set to be 0 *ppt*.

The simulated salinity distribution compared favorably to the salinity measurements along the Danshuei River–Tahan Stream during the flood and ebb tides on 26 November 2010, shown in Figure 4. The measured salinity during the flood and ebb tides means that the salinity was measured at instantaneous flood and ebb tides. Note that the field data of salinity were measured 0.5 m below the water surface and then every 1.0 m below the water surface 0.5 m, and the simulated salinity was presented with the top layer and bottom layer. The measured salinity shown in Figure 4 presents the mean salinity in vertical direction plus/minus one standard deviation. The absolute mean error and root mean square error of the difference between the measured salinities and the computed salinity on 26 November 2010 are 2.71 *ppt* and 3.72 *ppt*, respectively, during the flood tide. The absolute mean error and root mean square error are 0.49 *ppt* and 0.67 *ppt*, respectively, during the ebb tide. It can be seen that the modeling performance for the ebb tide is better than that for the flood tide. This may be the reason that the higher horizontal eddy diffusion is calculated according to 2.5-equation closure model, resulting in salinity diffusion to the upstream region during the flood tide.

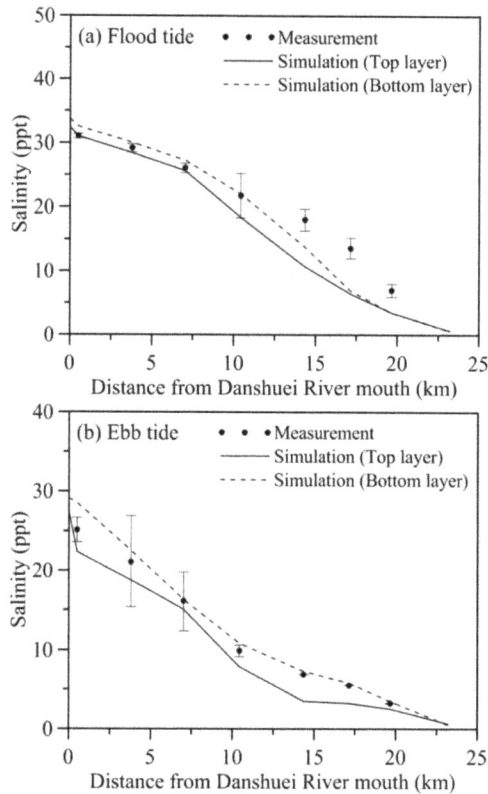

Figure 4. The comparison between the measured and simulated salinities along the Danshuei River–Tahan Stream on 26 November 2010 during (**a**) flood tide and (**b**) ebb tide.

4.2. Water Quality Distribution

Chen *et al.* [56] implemented a comprehensive field sampling and lab analysis program for the Danshuei River to collect the data in 2009 and 2010. They found that adjacent to the metropolitan Taipei City, the spatial trend of the deteriorated water quality is mostly attributed to the wastewaters directly discharged into the river channels. Efforts were made successively to estimate the point source loadings by Montgomery Watson Harza (MWH) [57] adopted in the following water quality simulations. The freshwater discharges in 2010 and 2011 were adopted at the upstream boundaries at the Tahan Stream, the Hsintien Stream, and the Keelung River. The five-constituent tide used to generate the time-series tidal level was employed at the ocean boundaries. Concentrations of water quality state variables ammonium nitrogen, total nitrogen, total phosphorus, carbonaceous biochemical oxygen demand, dissolved oxygen, and chlorophyll *a* at the river boundaries and at the ocean boundaries were established based on the monthly measurement by the Taiwan Environmental Protection Administration (TEPA). The model was conducted with two-year simulation.

Eight measured datasets were collected on 1 March, 1 June, 3 September, 2 December in 2010, 3 March, 1 June, 5 September, and 1 December in 2011 and were used for model validation. The model parameters were initially estimated from the literature [58]. These were adjusted and tuned until a reasonable reproduction of field data at observation stations was obtained. The coefficients adopted for water quality simulations are listed in Table 1. The longitudinal water quality distributions predicted by the water quality model on 3 March 2011 for the Danshuei River–Tahan Stream, the Hsintien Stream, and the Keelung River are shown in Figures 5–7 respectively. Water quality distributions of the dissolved oxygen, carbonaceous biochemical oxygen demand, ammonium nitrogen, and total phosphorus concentrations at the top and bottom layers along the river channels are presented in the figures, together with the observations from monitoring stations. Both the model-predicted and observed dissolved oxygen concentrations along the Danshuei River–Tahan Stream show a decrease from the Danshuei River mouth to Hsin-Hai Bridge and an increase at the Fu-Chou Bridge (Figure 5a). In the lower estuary, the dissolved oxygen concentrations increase toward the river mouth as a result of seawater dilution. It also shows that quite low dissolved oxygen concentrations occur at the Chong-Yang Bridge, Chung-Siao Bridge, and Hsin-Hai Bridge. Carbonaceous biochemical oxygen demand, ammonium nitrogen, and total phosphorus all show the same spatial trends along the river channel from the Tahan Stream to the Danshuei River (Figure 5b–d). The concentrations increase from the Danshuei River mouth to Hsin-Hai Bridge, reach a maximum at the Hsin-Hai Bridge, and then gradually decrease toward the Fu-Chou Bridge. The maximum concentrations of all three occur at the Hsin-Hai Bridge, resulting in low dissolved oxygen. The figure shows that the model generally captured the spatial trends of the observed longitudinal distributions.

The ratio of nitrogen to carbon in different water bodies has been documented in reports [59–61]. The ratio ranges from 0.02–0.25 mg N/mg C. However, we set this ratio to 0.01 mg N/mg C in the model simulation, which is lower than the suggested value. This is the reason that the concentration of ammonium nitrogen (NH_4) in the Danshuei River estuarine system is quite high compared to other estuaries [10,42,62]. If we adopted the higher ratio of nitrogen to carbon in the model, the simulation results of CBOD would be too high and DO would be too low to compare with the measured data.

Table 1. Coefficients used in the water quality model.

Coefficients	Value	Unit
Deoxygenation rate at 20 °C	0.16	day^{-1}
Nitrification rate at 20 °C	0.13	day^{-1}
Phytoplankton respiration rate at 20 °C	0.6	day^{-1}
Denitrification rate at 20 °C	0.09	day^{-1}
Organic nitrogen mineralization at 20 °C	0.075	day^{-1}
Organic phosphorus mineralization at 20 °C	0.22	day^{-1}
Optimum phytoplankton growth rate at 20 °C	2.5	day^{-1}
Optimal temperature for growth of phytoplankton	16	°C
The morality rate of phytoplankton at 20 °C	0.003	day^{-1}
Half-saturation constant for oxygen limitation of carbonaceous deoxygenation	0.5	mg O$_2$ L^{-1}
Half-saturation constant for oxygen limitation of nitrification	0.5	mg O$_2$ L^{-1}
Half-saturation constant for uptake of inorganic nitrogen	25	µg N L^{-1}
Half-saturation constant for uptake of inorganic phosphorus	1	µg P L^{-1}
Half-saturation constant for oxygen limitation of denitrification	0.1	mg O$_2$ L^{-1}
Half-saturation constant of phytoplankton limitation of phosphorus recycle	1	mg C L^{-1}
Sediment oxygen demand at 20 °C	3.5	g/m^2 day
Optimal solar radiation rate	250	langleys/day
Total daily solar radiation	300	langleys/day
Ratio of nitrogen to carbon in phytoplankton	0.25	mg N/mg C
Ratio of phosphorus to carbon in phytoplankton	0.025	mg P/mg C
Ratio of phytoplankton to carbon	0.04	mg Phyt/mg C
Organic carbon (as CBOD) decomposition rate at 20 °C	0.21	day^{-1}
Anaerobic algae decomposition rate at 20 °C	0.01	day^{-1}
Denitrification rate at 20 °C	0.01	day^{-1}
Organic nitrogen decomposition rate at 20 °C	0.01	day^{-1}
Organic phosphorus decomposition rate at 20 °C	0.01	day^{-1}
Benthic NH$_4$ flux	0.04	mg N day^{-1}
Benthic NO$_3$ flux	0.003	mg N day^{-1}
Benthic PO$_4$ flux	0.005	mg P day^{-1}
Ratio of nitrogen to carbon	0.01	mg N/mg C
Ratio of phosphorus to carbon	0.01	mg P/mg C

The longitudinal water quality distributions along the Hsintien Stream are illustrated in Figure 6. Both the observation data and model predictions show that the water quality conditions degrade as the river reach approaches the Hsintien Stream mouth, where it joins the main stream of the Danshuei River; the dissolved oxygen decreases, and the organic carbon, ammonium nitrogen and total phosphorus increase monotonically. The model faithfully represents the observed carbonaceous biochemical oxygen demand, ammonium nitrogen, and total phosphorus along the Hsintien Stream.

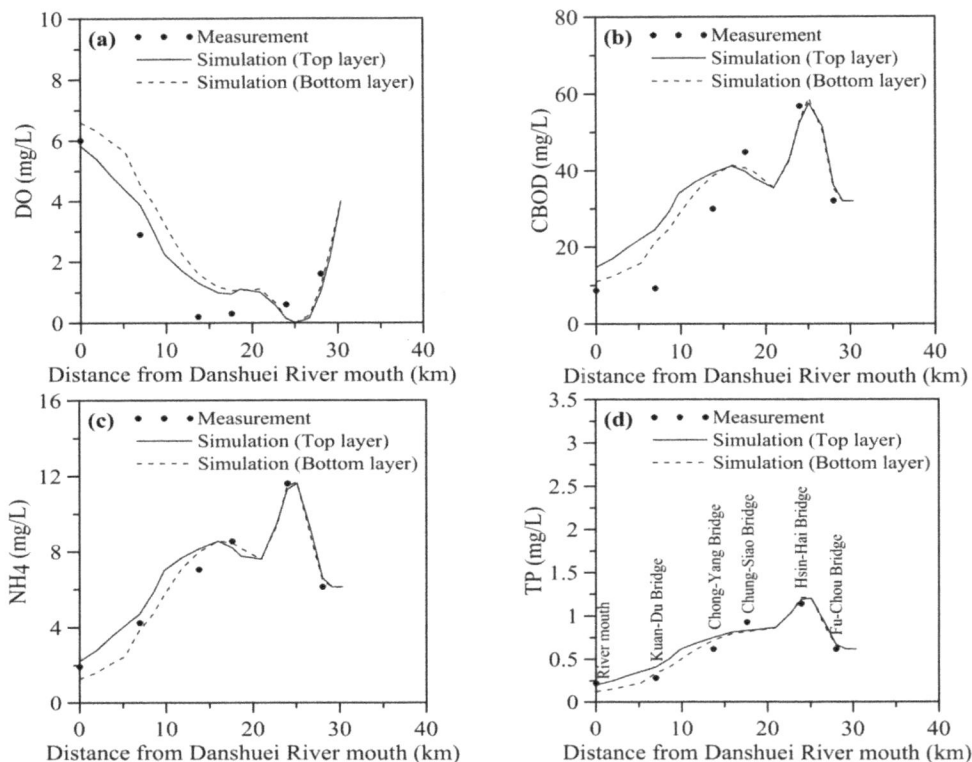

Figure 5. The comparison between the measured and simulated water quality distributions along the Danshuei River to the Tahan Stream on 3 March 2011 (**a**) DO; (**b**) CBOD; (**c**) NH$_4$; and (**d**) TP.

The longitudinal water quality distribution along the Keelung River (Figure 7) shows that significant pollution loadings were discharged into the river section around the Bai-Ling Bridge and the Chung-Shan Bridge, where the lowest dissolved oxygen in the Keelung River was observed. The model can realistically mimic the observed dissolved oxygen. The model was also revealed to match the observed carbonaceous biochemical oxygen demand, ammonium nitrogen, and total phosphorus very well along the Keelung River. Due to the page limitation, the statistical errors, including the absolute mean error and root mean square error on 2 December 2010, 3 March, 1 June, 5 September 2011 are shown only in Tables 2–5.

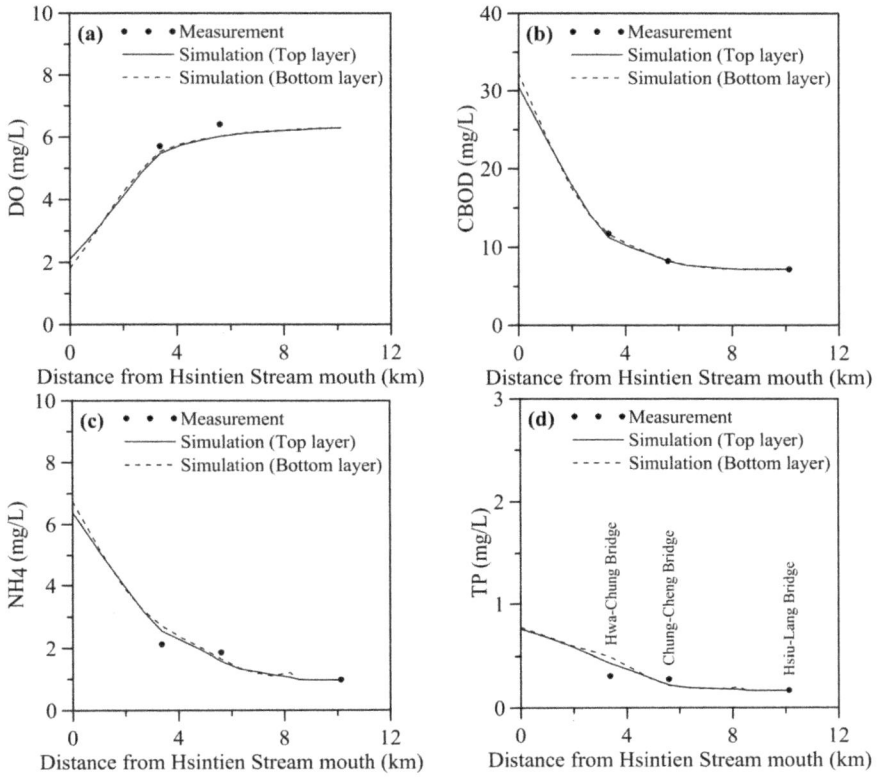

Figure 6. The comparison between the measured and simulated water quality distributions along the Hsintien Stream on 3 March 2011 (**a**) DO; (**b**) CBOD; (**c**) NH$_4$; and (**d**) TP.

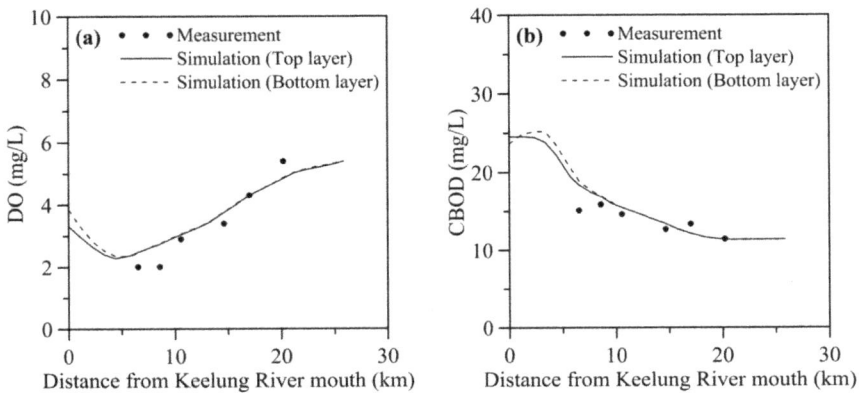

Figure 7. *Cont.*

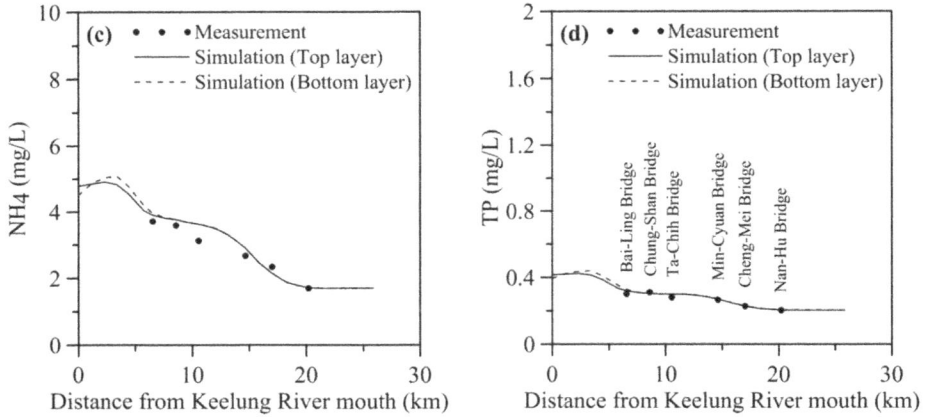

Figure 7. The comparison between the measured and simulated water quality distributions along the Keelung River on 3 March 2011 (**a**) DO; (**b**) CBOD; (**c**) NH$_4$; and (**d**) TP.

Table 2. Statistical error between simulated and measured water quality state variables on 2 December 2010.

Water Quality Variable	Danshuei River–Tahan Stream		Hsintien Stream		Keelung River	
	AME (mg/L)	RMSE (mg/L)	AME (mg/L)	RMSE (mg/L)	AME (mg/L)	RMSE (mg/L)
Dissolved oxygen	0.63	0.88	2.17	3.47	0.91	0.98
Carbonaceous biochemical oxygen demand	2.21	2.79	1.79	2.29	2.02	2.51
Ammonium nitrogen	0.36	0.54	0.52	0.74	0.15	0.19
Total phosphorus	0.07	0.10	0.08	0.12	0.008	0.01

Note that AME $= \dfrac{1}{N}\sum\limits_{i=1}^{N}\left|(C_p)_i - (C_o)_i\right|$, RMSE $= \sqrt{\dfrac{1}{N}\sum\limits_{i=1}^{N}\left[(C_p)_i - (C_o)_i\right]^2}$, where C_p is the predicted water quality concentration; and C_o is the observed water quality concentration.

Table 3. Statistical error between simulated and measured water quality state variables on 3 March 2011.

Water Quality Variable	Danshuei River–Tahan Stream		Hsintien Stream		Keelung River	
	AME (mg/L)	RMSE (mg/L)	AME (mg/L)	RMSE (mg/L)	AME (mg/L)	RMSE (mg/L)
Dissolved oxygen	0.74	0.85	0.86	1.18	0.38	0.45
Carbonaceous biochemical oxygen demand	6.56	7.5	0.11	0.16	1.25	1.67
Ammonium nitrogen	0.35	0.46	0.25	0.32	0.23	0.27
Total phosphorus	0.08	0.08	0.07	0.09	0.01	0.01

Table 4. Statistical error between simulated and measured water quality state variables on 1 June 2011.

Water Quality Variable	Danshuei River–Tahan Stream		Hsintien Stream		Keelung River	
	AME (mg/L)	RMSE (mg/L)	AME (mg/L)	RMSE (mg/L)	AME (mg/L)	RMSE (mg/L)
Dissolved oxygen	0.83	0.97	0.29	0.44	0.68	0.76
Carbonaceous biochemical oxygen demand	2.48	2.84	2.23	2.85	1.60	1.86
Ammonium nitrogen	0.52	0.80	0.44	0.67	0.35	0.42
Total phosphorus	0.99	0.10	0.08	0.14	0.06	0.07

Table 5. Statistical error between simulated and measured water quality state variables on 5 September 2011.

Water Quality Variable	Danshuei River–Tahan Stream		Hsintien Stream		Keelung River	
	AME (mg/L)	RMSE (mg/L)	AME (mg/L)	RMSE (mg/L)	AME (mg/L)	RMSE (mg/L)
Dissolved oxygen	1.41	1.65	0.67	0.83	1.04	1.19
Carbonaceous biochemical oxygen demand	1.10	1.32	006	0.07	0.80	0.91
Ammonium nitrogen	0.47	0.78	0.03	0.04	0.56	0.71
Total phosphorus	0.04	0.06	0.01	0.01	0.02	0.02

5. Model Project Responses to Climate Change Impact

The future climate scenarios frequently used in Taiwan have been based on the Intergovernmental Panel on Climate Change (IPCC) Special Report on Emissions Scenarios (SRES) A1B and A2 scenarios. The Water Resources Agency [58] projected the streamflow in the Danshuei River basin due to the climate change scenarios in year 2039 (*i.e.*, short term). The projected results in streamflow during the dry seasons based on different scenarios are summarized in Table 6. These results indicate that the decreasing rates of streamflows in the Tahan Stream, the Hsintien Stream, and the Keelung River are 45.54%, 4.15%, and 45.65%, respectively, for the A2 scenarios, whereas they are 19.05%, 3.44%, and 24.32%, respectively, for the A1B scenario.

Table 6. The streamflows for the present condition and under climate change scenarios during Q75 low flow.

River	Q_{75} Low Flow under Present Condition (m^3/s)	Decreasing Rate under A2 Scenario (%)	Q_{75} Low Flow under A2 Scenario (m^3/s)	Decreasing Rate under A1B Scenario (%)	Q_{75} Low Flow under A1B Scenario (m^3/s)
Tahan Stream	3.36	45.54	1.83	19.05	2.72
Hsintien Stream	14.23	4.15	13.64	3.44	13.74
Keelung River	3.33	45.65	1.81	24.32	2.52

To perform the model prediction of the water quality in the estuarine system, the five-constituent tide at the ocean boundaries was used to force the model simulation. Concentrations of the water quality state variables ammonium nitrogen, total nitrogen, total phosphorus, carbonaceous biochemical oxygen demand, dissolved oxygen, and chlorophyll *a* at the river boundaries and at the ocean boundaries were established based on mean values calculated from the measured water quality data collected from 2003 to 2013 as observed by TEPA. The river discharges at the tidal limits of the three major tributaries—the Tahan Stream, the Hsintien Stream, and the Keelung River—were conducted using the Q75 low flow condition, where Q75 flow is the flow that is equaled or exceeded 75% of the time. The Q75 river flows at the upstream reaches of the Tahan Stream, the Hsintien Stream, and the Keelung River are 3.36, 14.23, and 3.33 m^3/s, respectively, for the present condition. For the A2 and A1B scenarios, the Q75 river flows at the upstream reaches of the Tahan Stream, the Hsintien Stream, and the Keelung River are presented in Table 6.

The predicted water quality distribution for the present condition and the A2 climate change scenario under Q75 low flow along the Danshuei River to the Tahan Stream, the Hsintien Stream, and the Keelung River, respectively, is shown in Figures 8–10. A comparison of the present condition with the A2 climate change scenario reveals that the dissolved oxygen concentration decreased by a maximum of 1.75 mg/L and that the carbonaceous biochemical oxygen demand, ammonium nitrogen, and total phosphorus increased by a maximum of 6.4, 1.1, and 0.04 mg/L, respectively, in the Danshuei River–Tahan Stream (Figure 8). The dissolved oxygen concentration decreased by a maximum of 0.15 mg/L, and the carbonaceous biochemical oxygen demand, ammonium nitrogen, and total phosphorus increased by a maximum of 0.33, 0.14, and 0.01 mg/L, respectively, in the Hsintien Stream (Figure 9). The dissolved oxygen concentration decreased by a maximum of 1.50 mg/L, and the carbonaceous biochemical oxygen demand, ammonium nitrogen, and total phosphorus increased by a maximum of 0.85, 0.48, and 0.03 mg/L, respectively, in the Keelung River (Figure 10). We found that the dissolved oxygen concentration was lower than 2 mg/L in the Danshuei River-Tahan Stream and did not meet the minimum requirement of TEPA. The maximum rate of dissolved oxygen, carbonaceous biochemical oxygen demand, ammonium nitrogen, and total phosphorus under climate change scenarios A2 and A1B is summarized in Table 7. The maximum rate refers to the maximum values determined by the formula represented by $\dfrac{C_p - C_c}{C_p} \times 100\%$, where C_p is the water quality concentration at the present time and C_c is the water quality concentration under climate change.

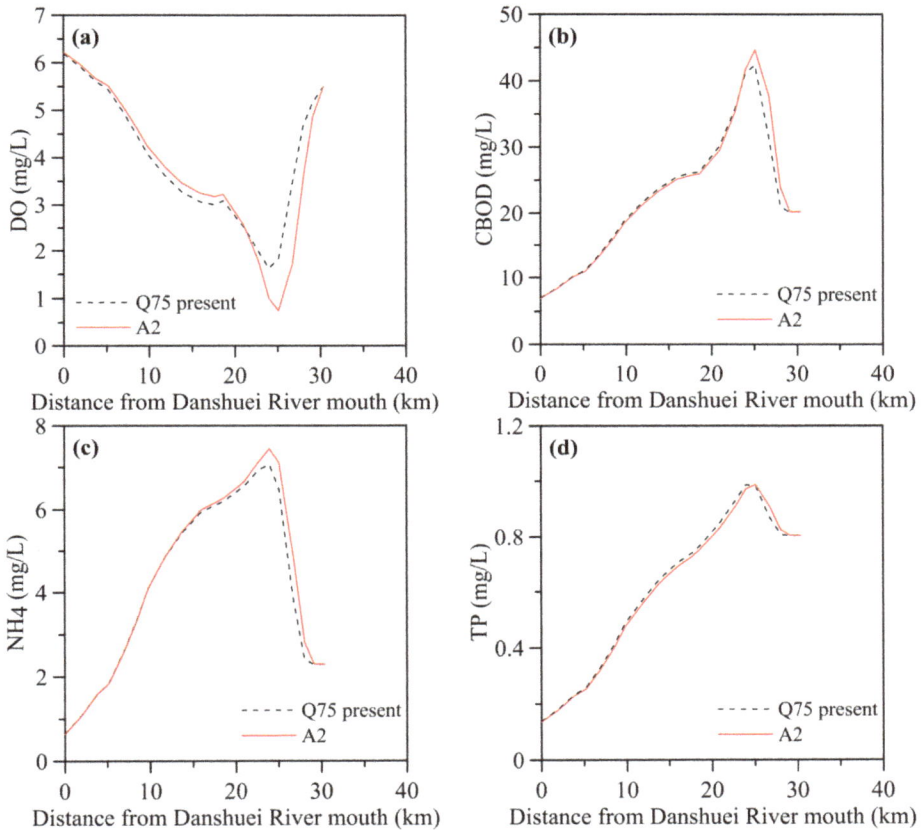

Figure 8. Predicting water quality distributions for present and climate change (A2 scenario) conditions under Q75 low flow along the Danshuei River to the Tahan Stream (**a**) DO; (**b**) CBOD; (**c**) NH4; and (**d**) TP.

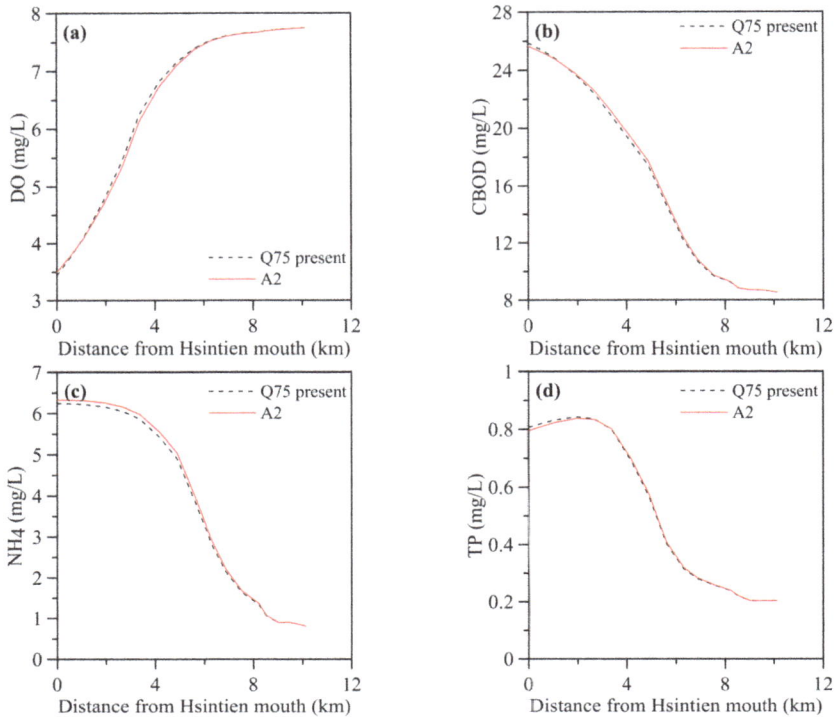

Figure 9. Predicting water quality distributions for present and climate change (A2 scenario) conditions under Q75 low flow along the Hsintien Stream (**a**) DO; (**b**) CBOD; (**c**) NH4; and (**d**) TP.

Table 7. Maximum rate of water quality state variables under climate change scenarios A2 and A1B.

River	Maximum Rate under Climate Change A2 Scenario				Maximum Rate under Climate Change A1B Scenario			
	DO (%)	CBOD (%)	NH$_4$ (%)	TP (%)	DO (%)	CBOD (%)	NH$_4$ (%)	TP (%)
Danshuei River–Tahan Stream	−59.4	20.46	26.9	4.4	−25.8	7.5	9.8	1.8
Hsintien Stream	−2.0	1.9	3.8	1.7	−1.9	1.6	3.0	1.6
Keelung River	−33.7	4.9	13.8	6.2	−14.5	2.3	6.2	3.3

Note: minus and plus represent a decrease and increase, respectively.

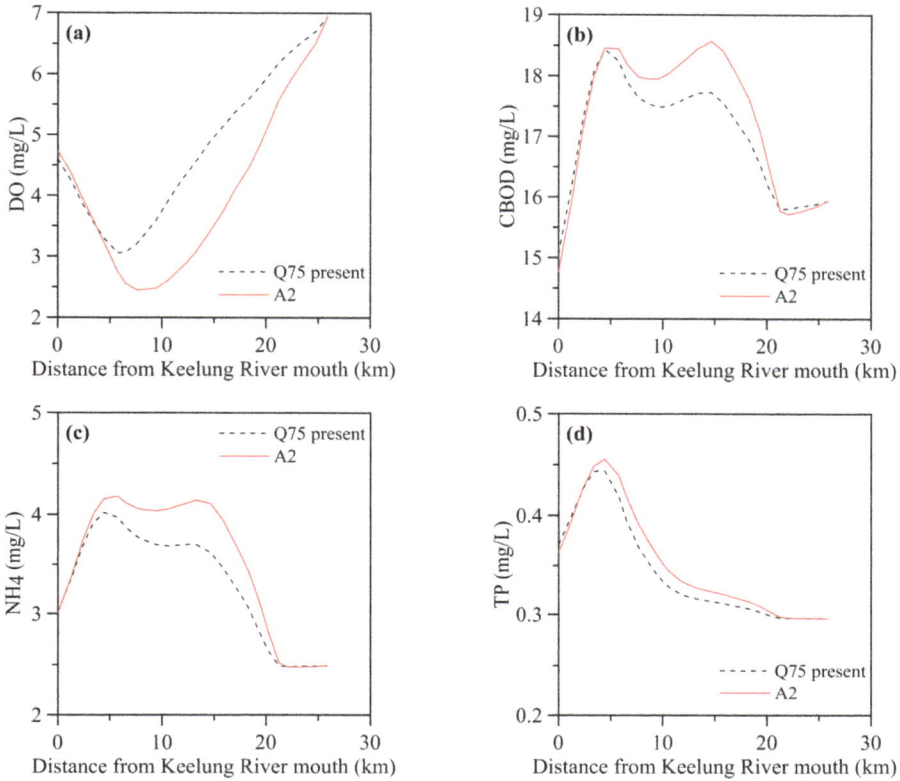

Figure 10. Predicting water quality distribution for present and climate change (A2 scenario) conditions under Q75 low flow along the Keelung River (**a**) DO; (**b**) CBOD; (**c**) NH4; and (**d**) TP.

The vertical distributions of monthly (in August) average salinity, dissolved oxygen, carbonaceous biochemical oxygen demand, ammonium nitrogen, and total phosphorus in the Danshuei River-Tahan Stream under the present condition and A2 scenario, respectively, are shown in Figures 11 and 12. It can be seen that the limit of salt water intrusion for the A2 scenario (Figure 12a) moves further upriver compared with the present condition (Figure 11a). The limit of salt water intrusion for the present condition, A2 scenario, and A1B scenario in the Danshuei River-Tahan Stream, the Hsintien Stream, and the Keelung River is illustrated in Table 8. The differences in the limit of salt water intrusion between the A2 scenario and present condition are 1.52 km, 0.25 km, and 1.33 km, respectively, in the Danshuei River-Tahan Stream, the Hsintien Stream, and the Keelung River. According to Figures 11a and 12a, we can observe the vertical stratification in salinity exhibited in the lower Danshuei River estuary. The dissolved oxygen concentration for the A2 scenario (Figure 12b) in an

303

estuarine system decreases compared to the present condition (Figure 11b), while the concentrations of carbonaceous biochemical oxygen demand, ammonium nitrogen, and total phosphorus (Figure 12c–e) increase (Figure 11c–e). No significant vertical stratification in water quality was found in the Danshuei River estuary.

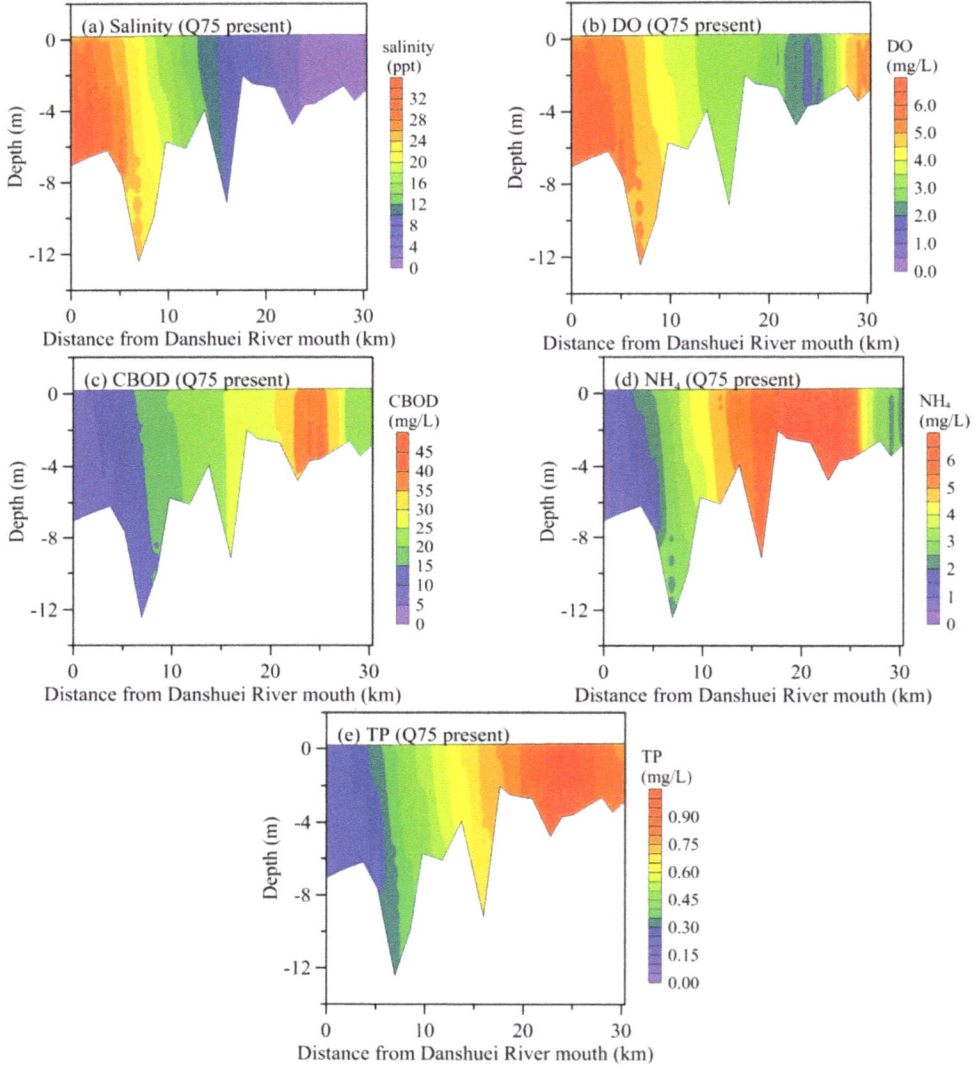

Figure 11. The vertical distribution of the monthly average water quality concentration in the Danshuei River to Tahan Stream under Q75 flow for the present condition (**a**) Salinity; (**b**) DO; (**c**) CBOD; (**d**) NH4; and (**e**) TP.

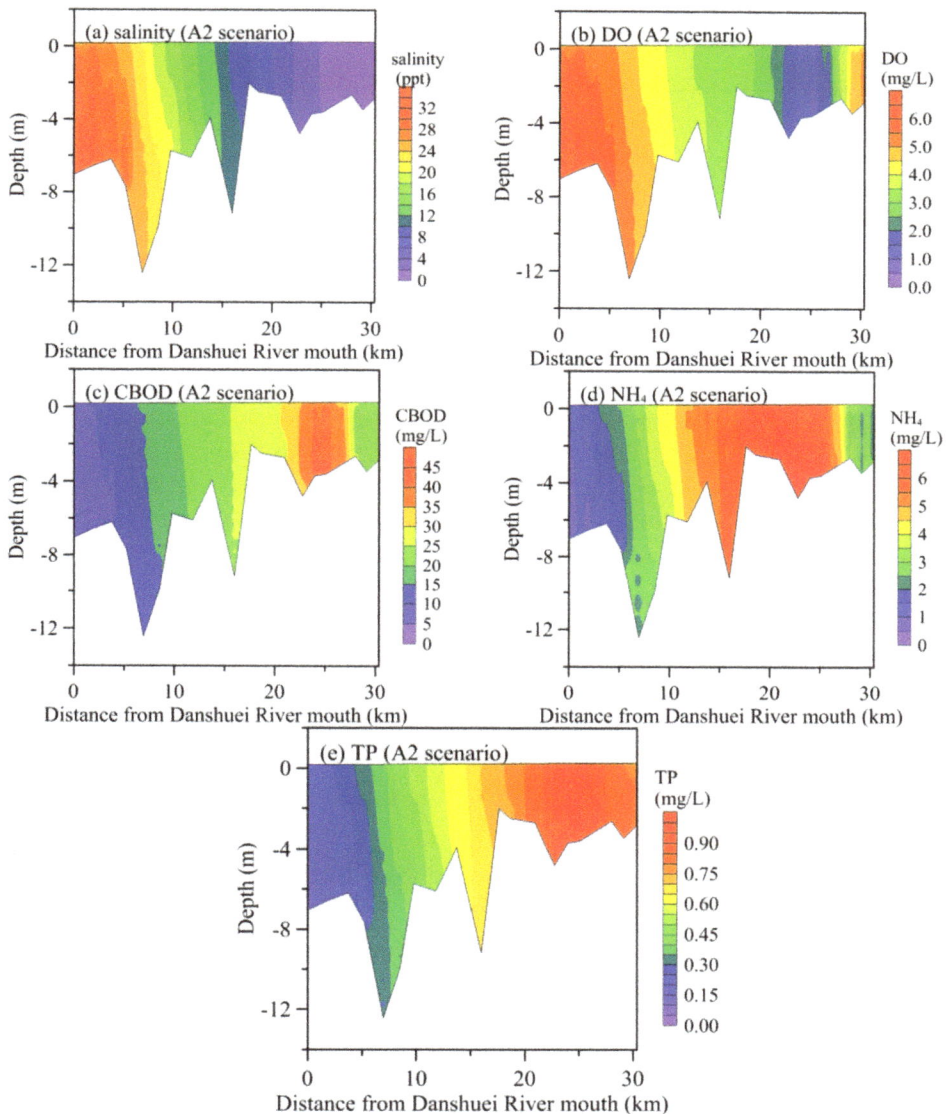

Figure 12. The vertical distribution of the monthly average water quality concentration in the Danshuei River to Tahan Stream under Q75 flow for the A2 Scenario (**a**) Salinity; (**b**) DO; (**c**) CBOD; (**d**) NH4; and (**e**) TP.

Table 8. The limit of salt water intrusion in the Danshuei River estuarine system under different scenarios.

River	Present Condition (km)	A2 Scenario (km)	A1B Scenario (km)
Danshuei River–Tahan Stream	24.67	26.19	25.24
Hsintien Stream	3.06	3.31	3.20
Keelung River	11.29	12.62	11.80

6. Discussion

Water quality studies utilizing a coupled hydrodynamic and water quality model tend to contain some limitations and assumptions. These limitations and assumptions exist in both the data and model. The major data used in this study were climate change scenarios. The climate change scenarios (*i.e.*, A1B and A2) used in this study were obtained from the Water Resources Agency, Taiwan, which projected the streamflow during dry seasons in the Danshuei River basin. We know that climate change is a non-stationary and dynamic problem; however, the streamflow projected from the climate change model and used in this study is a steady-state condition. This could lead to bias in future streamflow estimates that result in more uncertainty in the modeling results of water quality.

As mentioned in the Section "Water Quality Model", there are many parameters in the water quality model. The model was validated with two-year measured data. The model parameters after validation are kept for modeling future conditions without adjustment under future scenarios. However, future climate change might change the parameters. All of these parameters might reduce the accuracy of the modeling results. Nevertheless, after considering the aforementioned limitations and assumptions, the modeling results of water quality are relevant and reliable under the current climate change scenarios. The approaches are useful for assessing the impact of climate change on estuarine water quality.

Some literature has stated that climate change causes the degradation of water quality. For example, Tung *et al.* [59] evaluated the effects of climate change on sustainable water quality management and proposed a systematic assessment procedure including a weather generation model, the streamflow component of GWLF, QUAL2E, and an optimization model. Their studies indicated that streamflows may likely increase in humid seasons and decrease in arid seasons. The reduction of streamflow in arid seasons might further degrade water quality and assimilation capacity. Our study also demonstrated that the dissolved oxygen would decrease as a result of climate change, which reduces the streamflow during dry seasons. Wetz and Yoskowitz [27] reported that drought coupled with burgeoning population growth in coastal watersheds places a serve strain on freshwater supplies and greatly reduces freshwater inflows to estuaries, especially when coincident with seasonal peaks in human freshwater demand. Freshwater contains nutrients and organic matter that upon delivery to the coastal zone, fuels the rich productivity of coastal ecosystems and shapes critical fish habitats through its effects on salinity gradients and stratification. Low freshwater inflow events have the potential to significantly alter the water quality and ecosystem structure. In this study, we found that the dissolved oxygen would decrease and nutrients would increase for the low flow condition as a result of climate change. The decreased dissolved oxygen would result in malodor, fish mortality, and microbial proliferation, which causes the issue of public health.

In future research, future climate scenarios will be performed with a global climate model (GCM model) combined with a rainfall-runoff model to project time-series streamflow, which can be incorporated into the hydrodynamic and water quality model. The impact of sea-level rise on the estuarine water quality can be investigated. A long-term early warning system triggering proper adaptations to reduce climate change effects can also be studied.

7. Conclusions

A coupled three-dimensional hydrodynamic water quality model was applied to predict the water quality conditions in the Danshuei River estuarine system due to the projected effects of climate change. The model was validated against salinity distribution and water quality state variables including dissolved oxygen, carbonaceous biochemical oxygen demand, ammonium nitrogen, and total phosphorus. The simulated results using the three-dimensional hydrodynamic water quality model revealed that the computed salinity and water quality state variables well reproduced the observed data. The overall performance of the model is in qualitative agreement with the available field data.

The validated model was then used to assess the effects of climate change on water quality in the Danshuei River estuarine system during the low flow condition. Two climate change scenarios, A2 and A1B, were considered for model simulation. The simulated results indicated that the dissolved oxygen concentration has significantly decreased and the concentrations of carbonaceous biochemical oxygen demand, ammonium nitrogen, and total phosphorus have obviously increased because of climate change. Moreover, the dissolved oxygen concentration would be lower than 2 mg/L in the main stream of the Danshuei River estuary and would fail to meet the minimum requirement of TEPA. The deleterious water quality would produce other issues related to human pathogens and public health.

The simulated results may vary depending on the estuarine system, climate scenario, water quality model, and parameters considered. Considering the limitations of this study, the results are valid only under current climate change scenarios in the study area. However, the results and methodologies in this study still have implications for future water quality management in the estuarine system for the study area and other regions facing similar stresses from climate change.

Acknowledgments: Acknowledgments: This study was supported in part by the Ministry of Science and Technology (MOST) Taiwan, under grant No. 102-2625-M-239-002. This financial support was greatly appreciated. The authors express their appreciation to the Taiwan Water Resources Agency and Environmental Protection Administration for providing the observed data used in our model validation. The authors sincerely thank three anonymous reviewers for their valuable comments to substantially improve this paper.

Author Contributions: Author Contributions: Wen-Cheng Liu supervised the progress of the MOST project and served as a general editor. Wen-Ting Chan performed the data collection, model establishment, and model simulations and discussed the results with Wen-Cheng Liu. All authors read and approved the final manuscript.

Conflicts of Interest: Conflicts of Interest: The authors declare no conflict of interest.

References

1. Hobbie, J.E. *Estuarine Science—A Synthetic Approach to Research and Practice*; Island Press: Washington, DC, USA, 2000.

2. Pendleton, L.H. *The Economic and Market Value of Coasts and Estuaries: What's at Stake*; Restore America's Estuaries: Arlington, VA, USA, 2008; p. 182.

3. Yoskowitz, D.W.; Santos, C.; Allee, B.; Carollo, C.; Henderson, J.; Jordan, S.; Ritchie, J. Ecosystem Services in the Gulf of Mexico. In Proceedings of the Gulf of Mexico Ecosystem Services Workshop, Bay St. Louis, MS, USA, 16–18 June 2010; Harte Research Institute for Gulf of Mexico Studies: Corpus Christi, TX, USA, 2010; p. 16.

4. Skerratt, J.; Wild-Allen, K.; Rizwi, F.; Whitehead, J.; Coughanowr, C. Use of a high resolution 3D fully coupled hydrodynamic, sediment and biogeochemical model to understand estuarine nutrient dynamics under various water quality scenarios. *Ocean Coast. Manag.* **2013**, *83*, 52–66.

5. Yoskowitz, D.W.; Werner, S.R.; Carollo, C.; Santos, C.; Washburn, T.; Isaksen, G.H. Gulf of Mexico offshore ecosystem services: Relative evaluation by stakeholders. *Mar. Policy* **2015**. in press.

6. Lee, R.W. Research of Construct Wetland Environment Model for Guandu Natural Park. Master's Thesis, National Taiwan Ocean University, Keelung, Taiwan, 2012.

7. Daily, G.C. *Introduction: What are Ecosystem Services? Nature's Services: Societal Dependence on Natural Ecosystems*; Island Press: Washington, DC, USA, 1997.

8. Chau, K.W. An unsteady three-dimensional eutrophication model in Tolo harbour, Hong Kong. *Mar. Pollut. Bull.* **2005**, *51*, 1078–1084.

9. McGlathery, K.J.; Sundback, K.; Anderson, I.C. Eutrophication in shallow coastal bay and lagoons: The role of plants in the coastal filter. *Mar. Ecol. Prog. Ser.* **2007**, *348*, 1–18.

10. Wild-Allen, K.; Skerratt, J.; Whitehead, J.; Rizwi, F.; Parslow, J. Mechanisms driving estuarine water quality: A 3D biogeochemical model for informed management. *Estuar. Coast. Shelf Sci.* **2013**, *135*, 33–45.

11. Kennish, M.J. Environmental threats and environmental future of estuaries. *Environ. Conserv.* **2002**, *29*, 78–107.

12. Paerl, H.W.; Valdes, L.M.; Peierls, B.L.; Adolf, J.E.; Harding, L.W. Anthropogenic and climatic influences on the eutrophication of large estuarine ecosystem. *Limnol. Oceanogr.* **2006**, *51*, 448–462.

13. Rabalais, N.N.; Turner, R.E.; Diaz, R.J.; Justic, D. Global change and eutrophication of coastal waters. *ICES J. Mar. Sci.* **2009**, *66*, 1528–1537.

14. McKibben, S.M.; Watkins-Brandt, K.S.; Wood, A.M.; Hunter, M.; Forster, Z.; Hopkins, A.; Du, X.; Eberhart, B.T.; Peterson, W.T.; White, A.E. Monitoring Oregon coastal harmful algae: Observations and implications for a harmful algal bloom-monitoring project. *Harmful Algae* **2015**, *50*, 32–44.

15. Stone, M.C.; Hotchkiss, R.H.; Hubbard, C.M.; Fontaine, T.A.; Mearns, L.O.; Arnold, J.G. Impacts of climate change on Missouri River Basin water yield. *J. Am. Water Resour. Assoc.* **2001**, *37*, 1119–1129.

16. Chang, H.; Knight, C.G.; Staneva, M.P.; Kostov, D. Water resource impacts of climate change in southwestern Bulgaria. *Geojournal* **2002**, *57*, 159–168.

17. Moss, R.H.; Edmonds, J.A.; Hibbard, K.A.; Manning, M.R.; Ross, S.K.; Van Vuuren, D.P.; Carter, T.R.; Emori, S.; Kainuma, M.; Kram, T.; *et al.* The next generation of scenarios for climate change research and assessment. *Nature* **2010**, *463*, 747–756.

18. Trenberth, K.E. Framing the way to relate climate extremes to chimate change. *Clim. Change* **2012**, *115*, 283–290.

19. Feng, S.; Hu, Q.; Hung, W.; Ho, C.H.; Li, R.; Tang, Z. Projected climate regime shift under future global warming from multi-model, multi-scenario CMIP5 simulations. *Glob. Planet Change* **2014**, *112*, 41–52.

20. Xin, X.; Zhang, L.; Zhang, J.; Wu, T.; Fang, Y. Climate change projection over East Asia with BCC_CSM1.1 climate model under RCP scenarios. *J. Meteorl. Soc. Jpn.* **2013**, *91*, 413–429.

21. Braks, M.; van Ginkel, R.; Wint, W.; Sedda, L.; Sprong, H. Climate change and public health policy: Translating the science. *Int. J. Environ. Res. Public Health* **2014**, *11*, 13–29.

22. Kendrovski, V.; Spasenovska, M.; Menne, B. The public health impacts of climate change in the former Yugoslav Republic of Mecedonia. *Int. J. Environ. Res. Public Health* **2014**, *11*, 5975–5988.

23. Petkova, E.P.; Ebi, K.L.; Culp, D.; Redlener, I. Climate change and health on the U.S. Gulf Coast: Public health adaption is needed to address future risks. *Int. J. Environ. Res. Public Health* **2015**, *12*, 9342–9356.

24. Luber, G.; Prudent, N. Climate change and human health. *Trans. Am. Clin. Climatol. Assoc.* **2009**, *120*, 113–117.

25. Whitehead, P.; Wade, A.; Butterfield, D. Potential impacts of climate change on water quality and ecology in six UK rivers. *Hydrol. Res.* **2009**, *40*, 113–122.

26. Wu, Y.; Liu, S.; Gallant, A.L. Predicting impacts of increased CO_2 and climate change on the water cycle and water quality in the semiarid James River Basin of the Midwestern USA. *Sci. Total Environ.* **2012**, *430*, 150–160.

27. Wetz, M.S.; Yoskowitz, D.W. An "extreme" future for estuaries? Effects of extreme climate events on estuarine water quality and ecology. *Mar. Pollut. Bull.* **2013**, *69*, 7–18.

28. Middelkoop, H.; Daamen, K.; Gellens, D.; Grabs, W.; Kwadijk, J.C.J.; Lang, H.; Parmet, B.W.; Schadler, B.; Schulla, J.; Wilke, K. Impact of climate change on hydrological regimes and water resources management in the Rhine Basin. *Clim. Change* **2001**, *49*, 105–128.

29. Ducharne, A.; Baubion, C.; Beaudoin, N.; Benoit, M.; Billen, G.; Brisson, N.; Garnier, J.; Kieken, H.; Lebonvallet, S.; Ledoux, E.; *et al.* Long term prospective of the Seine River system: Confronting climate and direct anthropogenic changes. *Sci. Total Environ.* **2007**, *375*, 292–311.

30. Robins, P.E.; Skov, M.W.; Lewis, M.J.; Gimenez, L.; Davies, A.G.; Malham, S.K.; Neill, S.P.; McDonald, J.E.; Whitton, T.A.; Jackson, S.E.; *et al.* Impact of climate change on UK estuaries: A review of past trends and potential projections. *Estuar. Coast. Shelf Sci.* **2016**, *169*, 119–135.

31. Garcia, A.; Juanes, J.A.; Alvarez, C.; Revilla, J.A.; Medina, R. Assessment of the response of a shallow macrotidal estuary to changes in hydrological and wastewater inputs through numerical modeling. *Ecol. Model.* **2010**, *221*, 1194–1208.

32. Zhou, N.; Westrich, B.; Jiang, S.; Wang, Y. A coupling simulation based on hydrodynamics and water quality model of the Pearl River Delta, China. *J. Hydrol.* **2010**, *396*, 267–276.

33. Piliwal, R.; Patra, R.R. Applicability of MIKE 21 to assess temporal and spatial variation in water quality of an estuary under the impact of effluent from an industrial estate. *Water Sci. Technol.* **2011**, *63*, 1932–1943.

34. Long, T.Y.; Wu, L.; Meng, G.H.; Guo, W.H. Numerical simulation for impacts of hydrodynamic conditions on algae growth in Chongqing Section of Jialing River, China. *Ecol. Model.* **2011**, *222*, 112–119.

35. Gao, G.; Falconer, R.A.; Lin, B. Modelling effects of a tidal barrage on water quality indicator distribution in the Severn Estuary. *Front. Environ. Sci. Eng.* **2013**, *7*, 211–218.

36. Hanfeng, Y.E.; Guo, S.; Li, F.; Li, G. Water quality evaluation in tidal river reaches of Liaohe River estuary, China using a revised QUAL2K model. *Chin. Geogr. Sci.* **2013**, *23*, 301–311.

37. Sun, J.; Lin, B.; Jiang, G.; Li, K.; Tao, J. Modelling study on environmental indicators in an estuary. *Proc. Inst. Civil Eng. Water Manag.* **2014**, *167*, 141–151.

38. Seo, D.; Song, Y. Application of three-dimensional hydrodynamics and water quality model of the Youngsan River, Korea. *Desalin. Water Treat.* **2015**, *54*, 3712–3720.

39. Tu, J. Combined impact of climate and land use changes in streamflow and water quality in eastern Massachusetts, USA. *J. Hydrol.* **2009**, *379*, 268–293.

40. Rehana, S.; Mujumdar, P.P. River water quality response under hypothetical climate change scenarios in Tunga-Bhadra river, India. *Hydrol. Process.* **2011**, *25*, 3372–3386.

41. Luo, Y.; Ficklin, D.L.; Liu, X.; Zhang, M. Assessment of climate change impacts on hydrology and water quality with a watershed modeling approach. *Sci. Total Environ.* **2013**, *450*, 72–82.

42. Wan, Y.; Ji, Z.G.; Shen, J.; Hu, G.; Sun, D. Three dimensional water quality modeling of a shallow subtropical estuary. *Mar. Environ. Res.* **2012**, *82*, 76–86.

43. Cerco, C.F.; Noel, M.R. Twenty-year simulation of Chesapeake Bay water quality using the CE-QUAL-ICM eutrophication model. *J. Am. Water Resour. Assoc.* **2013**, *49*, 1119–1133.

44. Li, K.; Zhang, L.; Li, Y.; Zhang, L.; Wang, X. A three-dimensional water quality model to determine the environmental capacity of nitrogen and phosphorus in Jiaozhou Bay, China. *Mar. Pollut. Bull.* **2015**, *91*, 306–316.

45. Hsu, M.H.; Kuo, A.Y.; Kuo, J.T.; Liu, W.C. Procedure to calibrate and verify numerical models of estuarine hydrodynamics. *J. Hydraul. Eng.* **1999**, *125*, 166–182.

46. Liu, W.C.; Chen, W.B.; Hsu, M.H. Influence of discharge reductions on salt water intrusion and residual circulation in Danshuei River Estuary. *J. Mar. Sci. Technol.* **2011**, *19*, 596–606.

47. Wang, C.F.; Hsu, M.H.; Kuo, A.Y. Residence time of Danshuei River estuary, Taiwan. *Estuar. Coast. Shelf Sci.* **2004**, *60*, 381–393.

48. Liu, W.C.; Liu, S.Y.; Hsu, M.H.; Kuo, A.Y. Water quality modeling to determine minimum instream flow for fish survival in tidal rivers. *J. Environ. Manag.* **2005**, *76*, 293–308.

49. Wang, C.F.; Hsu, M.H.; Liu, W.C.; Hwang, J.S.; Wu, J.T.; Kuo, A.Y. Simulation of water quality and plankton dynamics in the Danshuei River Estuary, Taiwan. *J. Environ. Sci. Health A* **2007**, *42*, 933–953.

50. Zhang, Y.L.; Baptista, A.M. SELFE: A semi-implicit Eulerian-Lagrangian finite-element model for cross-scale ocean circulation. *Ocean Model.* **2008**, *21*, 71–96.

51. Shchepetkin, A.F.; McWilliams, J.C. A method for computing horizontal pressure-gradient force in an oceanic model with a nonaligned vertical coordinate. *J. Geophys. Res.* **2003**, *108*.

52. Umlauf, L.; Buchard, H. A generic length-scale equation for geophysical turbulence models. *J. Mar. Res.* **2003**, *61*, 235–265.

53. Ambrose, R.B., Jr.; Wool, T.A.; Martin, J.L. *The Water Quality Analysis Simulation Program, WASP5, Part A: Model Documentation*; U.S. Environmental Protection Agency: Athens, GA, USA, 1993; p. 202.

54. Chen, W.B.; Liu, W.C.; Hsu, M.H. Water quality modeling in a tidal estuarine system using a three-dimensional model. *Environ. Eng. Sci.* **2011**, *28*, 443–459.

55. Liu, W.C.; Chen, W.B.; Kuo, J.T. Modeling residence time response to freshwater discharge in a mesotidal estuary, Taiwan. *J. Mar. Syst.* **2008**, *74*, 295–314.

56. Chen, C.H.; Lung, W.S.; Yang, C.H.; Lin, C.F. Spatially variable deoxygenation in the Danshui River: Improvement in model calibration. *Water Environ. Res.* **2013**, *85*, 2243–2253.

57. Montgomery Watson Harza (MWH). *Evaluation of Hydrodynamics and Water Quality Monitoring and Management in the Danshuei River Basin of New Taipei City*; Report to New Taipei City Government; Montgomery Watson Harza (MWH): Broomfield, CO, USA, 2011.

58. Bowie, G.L.; Mills, W.B.; Porcella, D.B.; Campbell, J.R.; Pagenkopf, J.R.; Rupp, G.L.; Johnson, K.M.; Chan, P.W.H.; Gherini, S.A. *Rates, Constants, and Kinetics Formulations in Surface Water Quality Modeling*, 2nd ed.; EPA/600/3–85/040; Environmental Research Laboratory, Office of Research and Development, U.S. Environmental Protection Agency: Athens, GA, USA, 1985; p. 455.

59. Gray, K.R.; Biddlestone, A.J. Composing-process parameters. *Chem. Eng.* **1973**, *2*, 71–76.

60. Meyers, P.A.; Ryoshi, I. Lacustrine organic goechemistry—An overview of indicators of organic matter sources and diagenesis in lake sediments. *Org. Geochem.* **1993**, *20*, 867–900.

61. Emerson, S.; Hedges, J. Sediment diagenesis and benthic flux. In *Treatise on Geochemistry*, 2nd ed.; Henry, E., Heinrich, D.H., Karl, K.T., Eds.; Elsevier: Amsterdam, The Netherlands, 2013; pp. 293–319.

62. Kim, T.; Sheng, Y.P.; Park, K. Modeling water quality and hypoxia dynamics in Upper Charlotte Harbor, Florida, U.S.A. during 2000. *Estuar. Coast. Shelf Sci.* **2010**, *90*, 250–263.

63. Water Resources Agency. *Strengthening Sustainable Water Resources Utilization and Adaptive Capability to Climate Change*; Final Report; Water Resources Agency: Nantun, Taiwan; Taichung, Taiwan, 2008. (In Chinese)

64. Tung, C.P.; Lee, T.C.; Liao, W.T.; Chen, Y.J. Climate change impact assessment for sustainable water quality management. *Terr. Atmos. Ocean. Sci.* **2012**, *23*, 565–576.

Estimating the Risk of River Flow under Climate Change in the Tsengwen River Basin

Hsiao-Ping Wei, Keh-Chia Yeh, Jun-Jih Liou, Yung-Ming Chen and
Chao-Tzuen Cheng

Abstract: This study evaluated the overflow risk of the Tsengwen River under
a climate change scenario by using bias-corrected dynamic downscaled data as
inputs for a SOBEK model (Deltares, the Netherlands). The results showed that
the simulated river flow rate at Yufeng Bridge (upstream), Erxi Bridge (midstream),
and XinZong (1) (downstream) stations are at risk of exceeding the management
plan's flow rate for three projection periods (1979–2003, 2015–2039, 2075–2099). After
validation with the geomorphic and hydrological data collected in this study, the
frequency at which the flow rate exceeded the design flood was 2 in 88 events in
the base period (1979–2003), 6 in 82 events in the near future (2015–2039), and 10 in
81 events at the end of the century (2075–2099).

Reprinted from *Water*. Cite as: Wei, H.-P.; Yeh, K.-C.; Liou, J.-J.; Chen, Y.-M.;
Cheng, C.-T. Estimating the Risk of River Flow under Climate Change in the
Tsengwen River Basin. *Water* **2016**, *8*, 81.

1. Introduction

Extreme typhoon precipitation events frequently result in socioeconomic
impacts and loss of human life. Increased incidences of extreme rainfall events
indicate one of the common features signaling climate change worldwide. The
International Panel on Climate Change [1] reported that on average, precipitation has
increased globally by approximately 8%. According to Liu *et al.* [2], scientists have
contended that the increase in global temperature over the past decade has prompted
an increase in extreme precipitation events and a decrease in moderate and mild
precipitation events. The 2010 Taiwan Climate Change Projection and Information
Platform (TCCIP) Project Report II included statistical data regarding the frequency
of extreme typhoon precipitation events in Taiwan from 1970 to 2009. The statistical
results indicated that prior to 2000, the frequency of extreme typhoon precipitation
events was approximately once every 2 years; however, this frequency increased to
at least once a year after 2000 [3]. Because of river flow changes caused by extreme
rainfall, discharge control structures (culverts, flap gates, weirs, and sluice gates) in
river basins are at a high risk of destruction.

The main scientific tool used in long-term climate simulations is the general circulation model (GCM), the main purpose of which is to project global climate characteristics and trends. However, GCM projections (e.g., rainfall, temperature, and humidity) cannot provide adequate and effective information for simulating small areas. In past decades, scientists have developed downscaling methods to increase the spatial resolution, providing more information for correcting the error margin from the GCM simulations and presenting the influence of topographic distribution in local areas. Currently, high-resolution climate data can be obtained through high-resolution GCM, dynamical downscaling, and statistical downscaling. Although numerous recent studies have attempted to increase the spatial resolution of the output from the GCM, for example, by using statistical and dynamical downscaling, the results typically yield only certain points of information that are inadequate for resolving the climate characteristics of small areas with complex terrain such as in Taiwan.

The present study used the Tsengwen River as the study area. High-resolution dynamical downscaling data were used to simulate changes in the hourly flow rate of typhoon events. Based on the selection criteria [4], the number of extreme typhoon events selected from the base period (1979–2003), near future (2015–2039), and end of the century (2075–2099) were 88, 82, and 81, respectively. The high-resolution dynamical downscaling data were used as the input for a SOBEK river channel routing model to simulate changes in the river flow rate under climate change. Results were further compared with the design flow rate, as well as recorded river water levels of the most severe typhoon events in history, to evaluate the risk of river flooding under climate change.

2. Literature Review

The GCM is the main tool for simulating future climate conditions; however, it has a relatively low resolution (approximately 200–500 km) [5], which is inadequate for detailed assessments of land surface processes and climate change effects at local to regional scales, particularly in regions with varied topography [6–8]. Chen *et al.* [9] observed that the GCM has been widely applied in simulating future climate scenarios; however, GCM data have a relatively low spatial resolution and cannot be used for detailed discussions on climate scenarios for small areas. Present-day regional climate models (RCM) are most often used for simulating the climate of more local spatial regions. Over the past few decades, dynamical downscaling has mainly been performed using high-resolution GCM or RCM data, with a spatial resolution less than 100 km. Recently, a high-resolution atmospheric GCM model with a resolution of approximately 20 km was developed by the Meteorological Research Institute (MRI) of Japan (hereafter, MRI-AGCM) [10] to include explicitly simulated extreme weather events, such as tropical storms and meso-scale systems, in long-term climate simulations. Although the MRI-AGCM showed marked improvements in simulating extreme precipitation events, the details of local rainfall over complex terrain may still be difficult to simulate. However, a 20-km resolution remains insufficient for describing the local weather and climate characteristics in some areas of Taiwan because of the complex terrain.

In recent years, hydrologic and hydraulic models such as Hydrologic Engineering Centers River Analysis System, Mike-11, SOBEK, and the integrated flood analysis system (IFAS) have been applied to predict potential disasters by using future climate data. Linde et al. [11] used a SOBEK model to simulate low-probability flood-peak events in the Rhine basin. The results showed a basin-wide increase of 8%–17% in the peak discharge of the Rhine basin in 2050 for probabilities between 1/10 and 1/1250. Kimura et al. [12] applied the IFAS to simulate the peak discharges in Tsengwen reservoir watershed in Taiwan from extreme rainfall events (TP1–10) during three periods: the present (1979–2003), near future (2015–2039), and future (2075–2099). The peak discharges during the future climate change period were higher than those during the present climate change period. Lenderink [13] discussed the discharge of the Rhine during future climate change by investigating two periods: the present (1960–1989) and future (2070–2099). A Rhineflow method was employed to simulate discharges for the UK Met Office RCM HadRM3H [14–16]. The mean discharge in the present (1960–1989) and future (2070–2099) climate change periods increased by approximately 30% in winter and decreased by approximately 40% in summer. This model estimated the effect of climate change on river discharges. Climate data such as temperature, precipitation, and evapotranspiration were used as inputs for the hydrologic and hydraulic models of the river basin; the outputs were for typical river discharge structures [13,17].

Previous studies have rarely focused on hydrological changes in Taiwan because of the low resolution of GCM data. Taiwan can currently generate its own high-resolution data for future climate scenarios, which were employed in the present study for hydraulic and hydrologic routing to project future flow rates under climate change. This study directly compared the river flow and water level determined through hydraulic and hydrologic routing. In addition, the risk of flood protection facilities under climate change was evaluated.

3. Research Methodology

This study was aimed at quantifying the effects of climate change in the Tsengwen catchment area. A flowchart of the research process employed in this study is shown in Figure 1. The first stage focused on introducing related climate change data. For the second stage, the major focus was developing a hydrodynamic model, including its calibration and validation. In the final stage, river discharge changes and river bank overtopping-frequency results were evaluated.

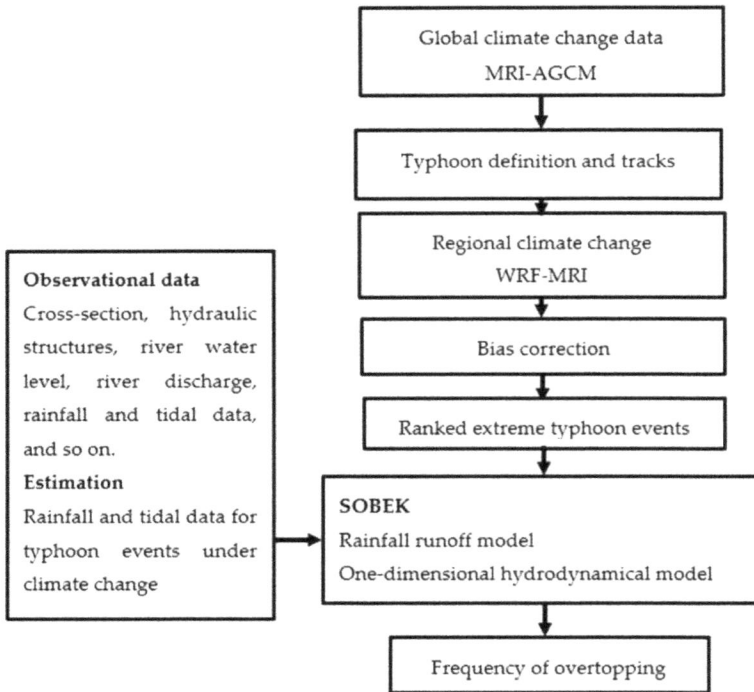

Figure 1. Conceptual scheme of the evaluation of the effects of climate change on river flow and water level.

3.1. SOBEK Model

The SOBEK model, which is in the SOBEK modeling suite developed by Deltares (formerly WL Delft Hydraulics), the Netherlands, integrates the commercial hydrologic and hydrodynamic programs of urban drainage systems along with river and regional drainages. The present study used the SOBEK channel-flow (CF) module along with the rainfall-runoff (RR) module for river channel simulations. The estimated RR volume was calculated as the lateral inflow (node) that converges in the main stream when calculating the unsteady flow of the river channel [18].

3.1.1. Rainfall Runoff

The SOBEK model incorporates the Sacramento RR model for simulating the process of rainfall forming runoff, including evaporation, infiltration, subsurface runoff, and underground water. The concept is to convert effective rainfall at the surface through a unit hydrograph into surface runoff, and to then add soil surface moisture, intermediate flow, and ground water discharge (base flow) to obtain the total runoff [19]. The Sacramento model defines a mathematical equation that accounts for each process in the transformation of rainfall into outflow toward a river. The concept of the Sacramento model and its parameters are shown in Figure 2. According to the Sacramento model, the soil column is divided into two soil zones: upper and lower [18]. The model has 17 parameters, the values of which must be specified [18]. Table 1 lists all the Sacramento parameters [19] and parameter ranges used in this study [20].

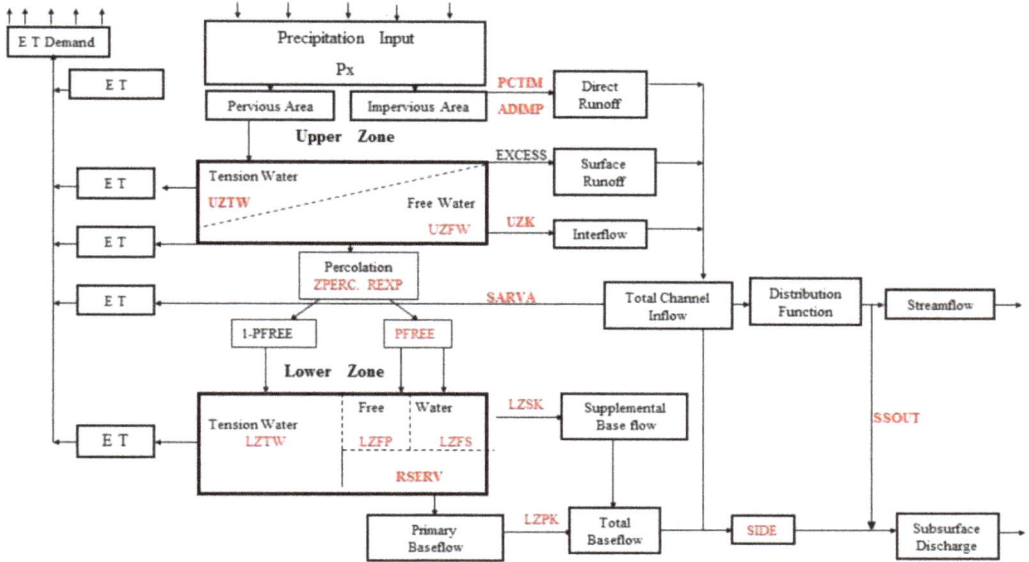

Figure 2. Conceptualization of the Sacramento model and parameters.

3.1.2. River Hydraulics

River flood routing is based on the dynamic wave transfer theory for one-dimensional (1D) varied flow; that is, de Saint Venant's gradually varied flow equation for describing water flow in rivers. This study used the nonlinear implicit difference method for calculating the depth and flow rate for each period. Water depth and flow rate at each cross-section point where main and branch streams converge were determined on the basis of conditions that the main and branch streams have the same water level, and inflow equals outflow. Equations of continuity (1) and motion (2) were considered for flood routing on the basis of de St. Venant's 1D gradually varied flow equation, which is the dynamic wave model. River simulations included the simulation of bridges, reservoirs, and cross-river structures such as weirs, culverts, orifices, and pump stations.

317

Table 1. Sacramento model parameters and their allowable ranges.

Parameters	Description	Allowable Range
UZTWM	Capacity of the upper tension water zone (mm)	250–300
UZFWM	Capacity of the upper free water zone (mm)	240–300
UZK	Upper zone lateral drainage rate (fraction of contents per day)	0.2
PCTIM	Permanent impervious fraction of the segment contiguous with stream channels	0.02
ADIMP	Additional impervious fraction when all tension water requirements are met	0.3–0.5
SARVA	Fraction of the segment covered by streams, lakes, and riparian vegetation	0.01
ZPERC	Proportional increase in the percolation under saturated to dry conditions in the lower zone	10–20
REXP	Exponent in the percolation equation, for determining the rate at which percolation demand changes from dry to wet conditions	1.5–2.5
LZTW	Capacity of the lower zone tension water storage (mm)	210–330
LZFPM	Capacity of the lower zone primary free water storage (mm)	230–450
LZFSM	Capacity of the lower zone supplemental free water storage (mm)	200–340
LZPK	Drainage rate of the lower zone primary free water storage (fraction of contents per day)	0.004–0.04
LZSK	Drainage rate of the lower zone supplemental free water storage (fraction of contents per day)	0.06–0.14
PFREE	Fraction of percolated water that drains directly to the lower zone free water storage	0.2
RSERV	Fraction of the lower zone free water storage that is unavailable for transpiration purposes	0.3
SIDE	Ratio of the unobserved to observed base flow	0
SSOUT	Fixed rate of discharge lost during the total CF (mm/t)	0

$$\frac{\partial A_f}{\partial t} + \frac{\partial Q}{\partial x} = q_{lat} \tag{1}$$

$$\frac{\partial Q}{\partial t} + \frac{\partial}{\partial x}\left(\frac{Q^2}{A_f}\right) + gA_f\frac{\partial h}{\partial x} + \frac{gQ\,|Q|}{C^2 R A_f} - w_f\frac{\tau_{wind}}{\rho_w} = 0 \tag{2}$$

where Q is the discharge (m^3/s), h is the water depth (m), R is the hydraulic radius (m), q_{lat} is the lateral discharge per unit length (m^2/s), A_f is the wetted area (m^2), w_f is the flow width (m), τ_{wind} is the wind shear stress (N = m^2), ρ_w is the density of water (kg/m^3), t denotes time (s), x refers to distance (m), and g denotes acceleration due to gravity (m/s^2) ($\approx$9.81).

When the SOBEK CF module processes the equation of motion, the influence of wind shear is considered, and wind force and direction are set to a fixed value or time sequence. Moreover, the SOBEK model can account for the influence of wind on the water level, which is not considered in the urban drainage and flood model. When considering the lateral inflow of a unit length of a river, including culverts, pumps, and weirs, the flow rate can be computed using the stage–discharge relationship of the hydraulic structures.

3.2. Indicators for Model Error Analysis

To validate the simulation model, the simulated and observed water level values were compared, and three statistical indices, namely, the coefficient of efficiency (CE), error of peak water level (EL_P), and error of the time to peak (E_{TP}), were calculated. The three indices are computed as follows:

$$CE = 1 - \frac{\sum_{i=1}^{n} (L_{obs} - L_{est})^2}{\sum_{i=1}^{n} (L_{obs} - \overline{L}_{obs})^2} \tag{3}$$

$$EL_P = \frac{L_{est} - L_{obs}}{L_{obs}} \tag{4}$$

$$E_{TP} = T_{est} - T_{obs} \tag{5}$$

where L_{est} denotes the estimated flood discharge (cm), L_{obs} represents the observed flood discharge (cm), and $\overline{L}_{obs}$ is the mean value of the observed flood discharge (cm); L_{est} and L_{obs} are the observed and estimated peak water levels of the flood, respectively; and T_{est} and T_{obs} denote the estimated and observed time to peak discharges, respectively.

3.3. Study Area

We selected the Tsengwen River basin as the study area, which covers an area of approximately 1176.7 km^2. The Tsengwen river basin is complex; the mountains are over 3000 m high, and the valley is narrower than 20 km. The average annual rainfall received by the drainage basin is 2643 mm. The Tsengwen River basin comprises the Tsengwen, Nanhua, and Wu Shantou Reservoirs. The Tsengwen Reservoir is located upstream of the Tsengwen Creek, and is the largest reservoir in Taiwan and the major source of water supply for downstream irrigation systems in Chiayi and Tainan Counties. The Tsengwen Reservoir has a large net water storage capacity (approximately 0.5 billion m^3). The mean annual inflow to the reservoir is approximately 1.1 billion m^3 [21]. The Tsengwen River Basin includes the Tsengwen River main stream, Cailiao River, Guantian River, and Houjue Creek. The location of the Tsengwen River Basin is shown in Figure 3.

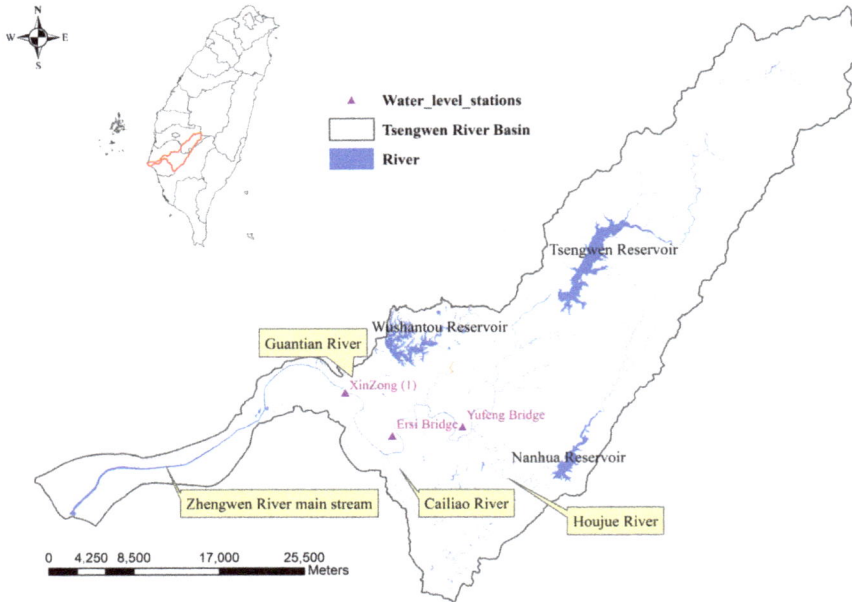

Figure 3. Tsengwen River Basin.

3.4. Hydrologic and Geomorphic Data

Hydrologic and geomorphic data must be collected before simulating the river flow rate and water depth for different scenarios. River cross-sections, hydraulic structures, rainfall in future climate, land use, river flow, and water level data are the basic data.

Rainfall data include observations from historical typhoon events and simulated rainfall data of extreme typhoon events under future climate change. The water level data include those of gauging stations along the river and tidal stations near the estuary (120°06′43″ E, 23°01′25″ N). Hourly water level data from current gauging stations were collected to validate hydraulic routing. Tide levels at the estuary were considered downstream boundary conditions in the model. The finite volume coastal ocean model (FVCOM) was employed to project changes in the astronomical tide at the estuary under a future climate change scenario. Chen and Liu [22] provided a detailed description of the FVCOM structure and parameters. River cross-sectional data of 2010 were provided by the projects of the WRA's Sixth River Management Office and Water Resources Planning Institute, and these include data of the cross-sections of the Tsengwen main stream, Cailiao River, Guantian River, and Houku Creek. The reservoir data include data of the Tsengwen, Nanhua, and Wushantou Reservoirs. The SOBEK model is based on reservoir operations [23–25], in which settings for the reservoir include reservoir area, volume, and contributing

320

area, and settings for the dam include spillway, water gate, power plant discharge, and emergency spillway. Discharge functions were set according to reservoir operation rules.

This study used the atmospheric general circulation model MRI-AGCM; the climate during three periods over a total of 75 years was simulated: base period (1979–2003), near future (2015–2039), and end of the century (2075–2099), developed by Japan Meteorological Agency (JMA) and Meteorological Research Institute (MRI); and ECHAM5, the climate model developed by the German research institute MPI, for climate projections. Simulation results were used as the initial field and boundary conditions for dynamical downscaling in the WRF modeling system, which was developed by the U.S. National Center for Atmospheric Research (please refer to the report of TCCIP (2010) for details).

This study adopted the high-resolution MRI-AGCM (20 km) to define typhoon events. MRI-AGCM revises the definition of typhoon provided by Vitart *et al.* [4], and uses the conditions of 850-hpa vorticity, sea-level pressure, presence or absence of warm-core structure near the typhoon center, and maximum local thickness to detect typhoons. Moreover, the wind speed at the bottom layer of typhoons must reach at least 17 m/s for 1.5 days or more. The process of selecting typhoon events can be divided into two steps: screening typhoon events and tracing typhoon routes [4]. The number of typhoon events determined using the aforementioned definition for the three periods and MRI-AGCM are 88, 82, and 81.

We ranked extreme typhoon rainfall events from each of the three 25-year periods based on the total rainfall over 24 h in the Tsengwen River basin. Although the TCCIP (2/3) reported that projections must be revised, this study bias-corrected the rainfall data by using the cumulative distribution function model [26] for the extreme typhoon rainfall events during the three periods. Figure 4 shows the average rainfall of the TOP1–20 events during the base period, near future, and end of the century. Moreover, we observed that the rainfall of the typhoon events at the end of the century (2075–2099) is higher than that of the base period and near future. Table 2 shows the statistical values for the TOP1–20 extreme events during the three periods. The Central Weather Bureau of Taiwan defines 24-h accumulated rainfall of 250 mm as extremely torrential rain. After observing rainfall characteristics that resulted in floods in Taiwan, Yu *et al.* [27] defined 3-h accumulated rainfall of 130 mm as short-duration disastrous rain. Table 1 shows that the TOP1–2 events in the base period, TOP1–5 in the future and the TOP1–12 events at the end of the century are extremely torrential rain events. The TOP2 event in the near future and the TOP1-3, TOP5, and TOP6 events at the end of the century are short-duration disastrous rainfall events.

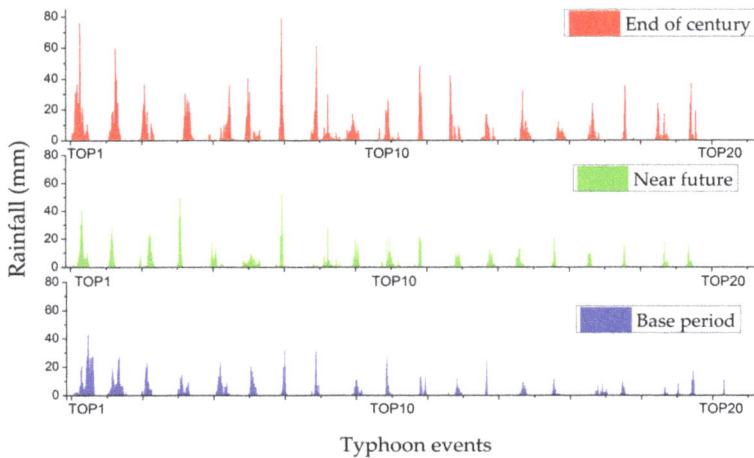

Figure 4. Rainfall hyetograph for the TOP1-20 extreme typhoon rainfall events during the three periods.

Table 2. Precipitation analysis of extreme events.

Typhoon Events	Base Period (1979–2003)				Near Future (2015–2039)				End of This Century (2075–2099)			
No.	(1)	(2)	(3)	(4)	(1)	(2)	(3)	(4)	(1)	(2)	(3)	(4)
Top1	851.4	108.6	491.3	120	548.2	111.7	420.3	90	1027.8	191.7	722.1	48
Top2	505.9	63	268.4	90	370	132.3	367.8	78	738.2	160.6	572.5	66
Top3	298.6	57.1	247.3	90	304.9	68.8	296.3	66	551.1	200.4	549.5	36
Top4	295.7	58.7	223.1	60	344.3	64.2	295.8	60	534.6	84.8	433.2	132
Top5	248.8	55.7	222	66	288.3	117.0	280.9	67	677.0	154.3	430.1	48
Top6	194.8	73.5	192.6	78	227	39.6	197.2	90	407.5	135.9	404.9	66
Top7	174.8	75.7	164.4	42	197.7	47.0	183.3	42	503.0	94.2	394.7	102
Top8	149.1	68	147.6	42	171.9	55.5	162.7	72	475.7	91.2	366.6	48
Top9	242.5	36.3	141.8	48	167.5	40.3	158.2	72	484.8	97.2	359.9	42
Top10	132.4	117.1	131.8	96	218.1	28.2	149.9	48	344.9	63.9	283.4	48
Top11	153.4	31.0	123.0	54	156.9	33.7	145.9	48	338.3	92.0	268.2	54
Top12	112.9	106.0	112.9	48	295.2	27.1	145.7	150	328.0	82.2	250.8	42
Top13	104.3	89.6	98.7	72	147.9	62.7	143.3	72	258.0	61.9	219.6	66
Top14	95.6	95.6	95.6	108	225.4	32.7	133.4	42	259.7	79.0	203.6	42
Top15	92.1	90.6	92.1	84	122.2	56.7	121.8	66	201.1	94.6	195.4	54
Top16	88.6	88.6	88.6	24	120.8	28.7	120.2	84	239.0	43.6	191.8	72
Top17	85.7	77.7	85.7	30	112.2	37.3	106.8	60	306.8	42.7	186.1	60
Top18	91.8	52.7	85.3	54	110.9	37.3	106.8	78	290.1	57.3	173.7	30
Top19	75.2	75.2	75.2	48	98.5	42.6	98.1	30	190.8	28.6	147.1	48
Top20	67.5	34.8	37.0	138	76.5	23.6	74.8	42	68.7	21.5	66.8	78
Typhoon Morakot	1007.5	144.3	636.2	72	–	–	–	–	–	–	–	–

(1) Total precipitation (mm); (2) Maximum of 3 h; (3) Maximum of 24 h; (4) Total duration (h).

322

In 2009, typhoon Morakot induced long-duration continuous rainfall, and the total rainfall received was approximately equal to the rainfall received during the most extreme typhoon event at the end of the century. However, the maximum rainfall induced by typhoon Morakot after continuously raining for 3 and 24 h was lower than the rainfall induced by the top few extreme typhoon events at the end of the century. In other words, rainfall distribution during extreme typhoon events during the future climate will be high over a short period.

Because the weather research and forecasting (WRF) climate data are grid data, this study collected the rainfall data from WRF grid points of the WRA's rainfall stations nearby, and used the data as input for the SOBEK model. The location of rainfall stations and WRF grid points are shown in Figure 5. Because of the historical rainfall data length and data acquisition constraints, this study selected 11 rainfall stations: MUZHA, TSOCHEN, BEILIAO, CHIKULAOS, BEILIAOS, SHANHUA, YUTEN, NANXI, WANGYEGONG, ZHENGWEN, and BIAOHU.

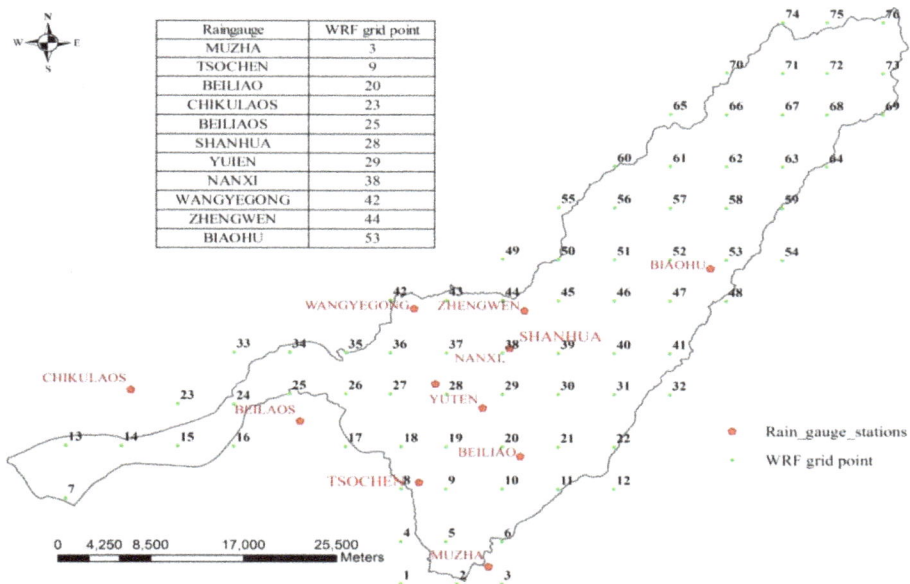

Raingauge	WRF grid point
MUZHA	3
TSOCHEN	9
BEILIAO	20
CHIKULAOS	23
BEILIAOS	25
SHANHUA	28
YUTEN	29
NANXI	38
WANGYEGONG	42
ZHENGWEN	44
BIAOHU	53

Figure 5. Map of the Tsengwen River Basin rainfall stations and WRF grid points.

4. Case Analysis

This study used 88 extreme typhoon rainfall events for the base period, 81 for the near future, and 82 for the end of the century periods. The data were used as inputs for the SOBEK routing model, which is used for simulating changes in the river flow rate during future climate change.

4.1. River Hydraulic Structure Impact Assessment

Common hydraulic structures in rivers include weirs, piers, dams, embankments, and groundsills. When a river channel requires hydraulic structures, which can be for various purposes (flood disaster prevention or hydraulic design), the flow rate and flood stage of the river channel must first be estimated to protect the hydraulic structures as well as the lives and assets of residents.

Conventional river flood prevention plans incorporate the concept of a return period when considering risk [28]; the design standard of river flood prevention facilities in Taiwan considers return periods of 50, 100, or 200 years. Hydrological data used for return period analysis are obtained through statistical analysis of historical data (20–60 years). During flood prevention facility planning and designing, the flow rate is projected on the basis of the hydrological data of the return period and geomorphic data of the river channel, along with a safety factor to reduce the uncertainty. The projected flood stage is calculated using a hydraulic model test or 1D hydraulic model based on the river's physical characteristics. Table 3 shows the design flow rate and flood stage at XinZong (1), Erxi Bridge, and Yufeng Bridge, as well as the highest water level observed in the past.

Table 3. Design discharge and water level.

Gauge Station	Return Period (Years)	Design Discharge (cm)	Design Stage (m)	* Historical Maximum Stage (m)
XinZong (1)		9890	15.71	18.36
Erxi Bridge	100	8740	21.37	23.56
Yufeng Bridge		6900	46.06	46.98

* occurred during typhoon Morakot.

4.2. Model Calibration and Validation

This study used the data of the rainfall for Typhoon Kalmaegi (2008) and Typhoon Morakot (2009) to calibrate the SOBEK model parameters and rainfall of 0610 torrential rains to validate the model parameters. Figures 6–8 compare the water levels measured at the XinZong Bridge No. 1 station in the Tsengwen River basin by using the SOBEK model. The figures show that the SOBEK simulations match the measured water levels.

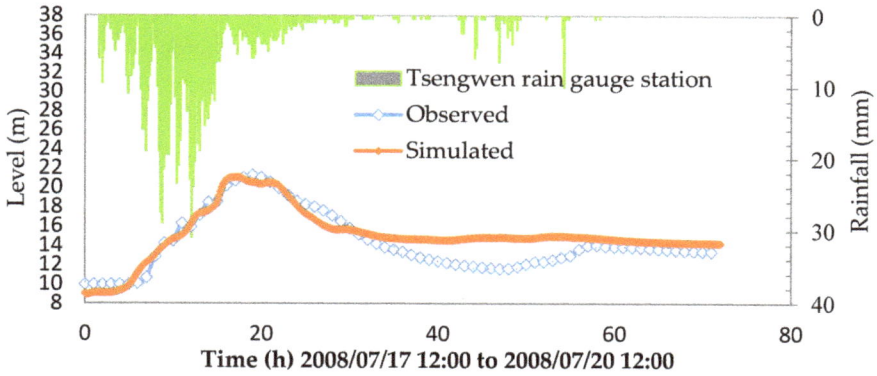

Figure 6. Comparison between the estimated and observed results of the water level at the XinZong (1) water level station during Typhoon Kalmaegi.

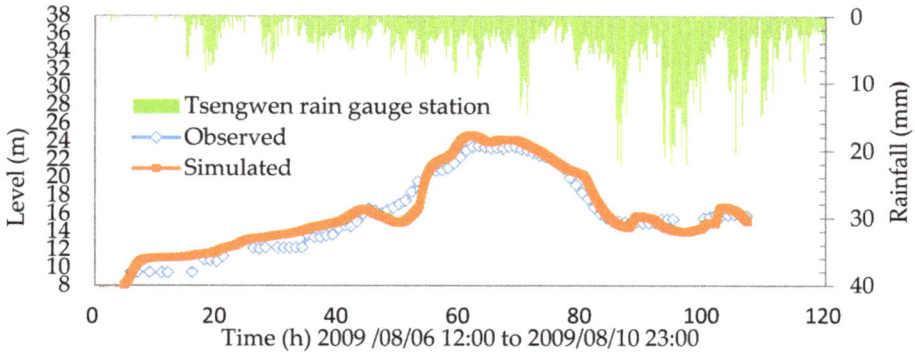

Figure 7. Comparison between the estimated and observed results of the water level at the XinZong (1) water level station during Typhoon Morakot.

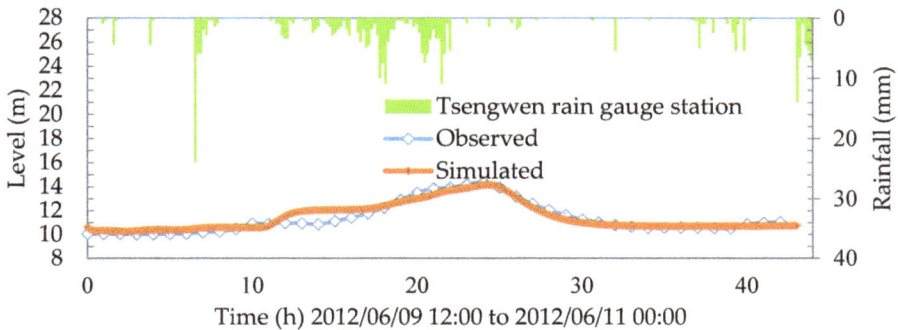

Figure 8. Comparison between the estimated and observed results of the water level at the XinZong (1) water level station for 0610 extreme rain.

To discuss the performance of the model, this study used CE, EL_p, and E_{TP} as a basis for the model validation. A CE approximating 1 indicates that the routing model has a higher goodness of fit, an EL_p greater than 0 indicates that the peak water level projected by the model is higher than the observed peak water level, an EL_p less than 0 indicates that the peak water level projected by the model is lower than the observed peak water level, and a lower E_{TP} indicates that the model is more accurate when projecting the time to peak. The results in Table 4 show that the simulated water level approximated the observed water level.

Table 4. Calibrated and verified results.

Item	Typhoon Events	CE	EL_P	E_{TP} (h)
Calibrated	Kalmaegi(2008)	0.8	−1.31	−2
	Morakot(2009)	0.9	4.75	−1
Verified	0610 Extreme rain (2009)	0.9	−1.66	0

4.3. Simulation Results

Figures 9 and 10 show the simulated discharge and water level hydrograph for the TOP1–20 extreme typhoon events during the base period, near future, and end of the century at the XinZong (1), Erxi Bridge, and Yufeng Bridge. According to these figures, the peak discharge and water level for the end of the century is higher than those for the base period and near future.

The design flood stages at XinZong (1), Erxi Bridge, and Yufeng Bridge gauging stations are 15.71, 21.37, and 46.06 m, respectively. Table 5 shows the water levels that exceeded the design values for extreme typhoon events in the base period, near future, and end of the century at XinZong (1), Erxi Bridge, and Yufeng Bridge. The simulated water levels at these three gauging stations exceeded the design values for the water levels in the three periods. In the base period, the peak flows at XinZong 1, Erxi Bridge, and Yufeng Bridge exceeded the management plan flow rate in 2 of 88, 3 of 88, and 1 of 88 events, respectively. For the near future, the corresponding peak flow rates exceeded the design discharge in 6 of 82, 6 of 82, and 1 of 82 events, and at the end of the century, the corresponding flow rates exceeded the flow rate in 10 of 81, 12 of 81, and 8 of 81 events. At the end of the century, extreme peak flow events were forecasted to increase in both frequency and intensity. The simulation results show that the upstream area of the Tsengwen River is already at risk of flooding at the end of the century.

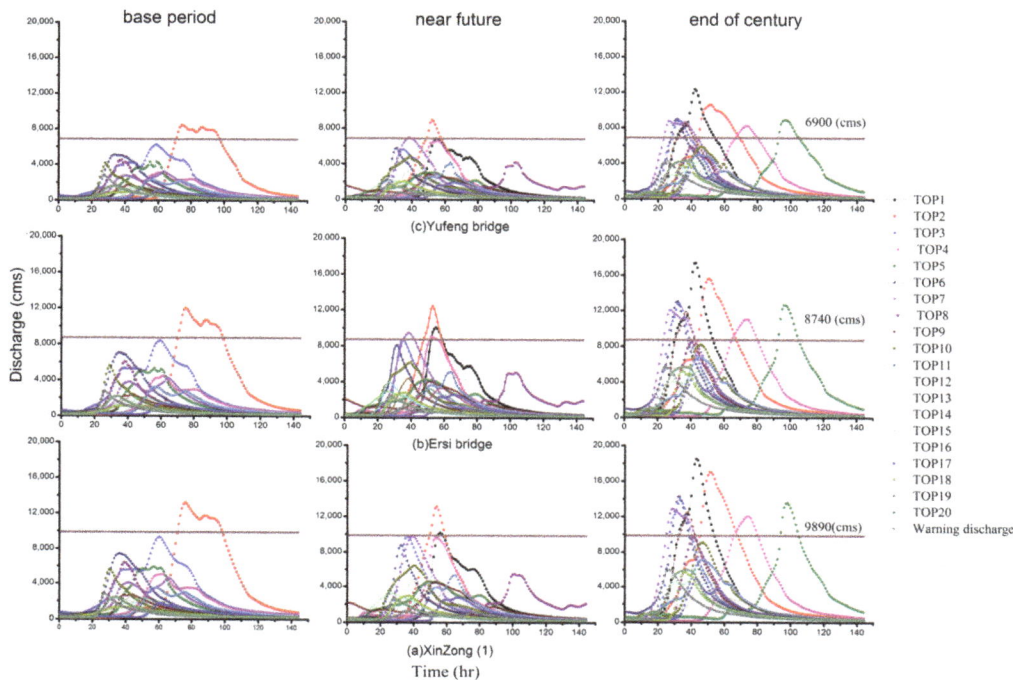

Figure 9. Discharge hydrographs for the TOP1-20 extreme typhoon events during the three periods at (**a**) XinZong (1); (**b**) Erxi Bridge; and (**c**) Yufeng Bridge.

Table 5. Water levels exceeding the design stage during extreme typhoon events.

Water Level Station	Design Water Level (m)		
	Base Period (88)	Future (82)	End of Century (81)
XinZong (1)	2	6	10
Erxi Bridge	3	6	12
Yufeng Bridge	1	1	8

The highest water levels measured at Erxi Bridge during the TOP1 and TOP2 extreme typhoon events at the end of the century were 24.55 and 24.32 m, respectively, which are higher than the highest water level of 23.56 m during typhoon Morakot. This simulation result indicates that a severe flood could reoccur under climate change.

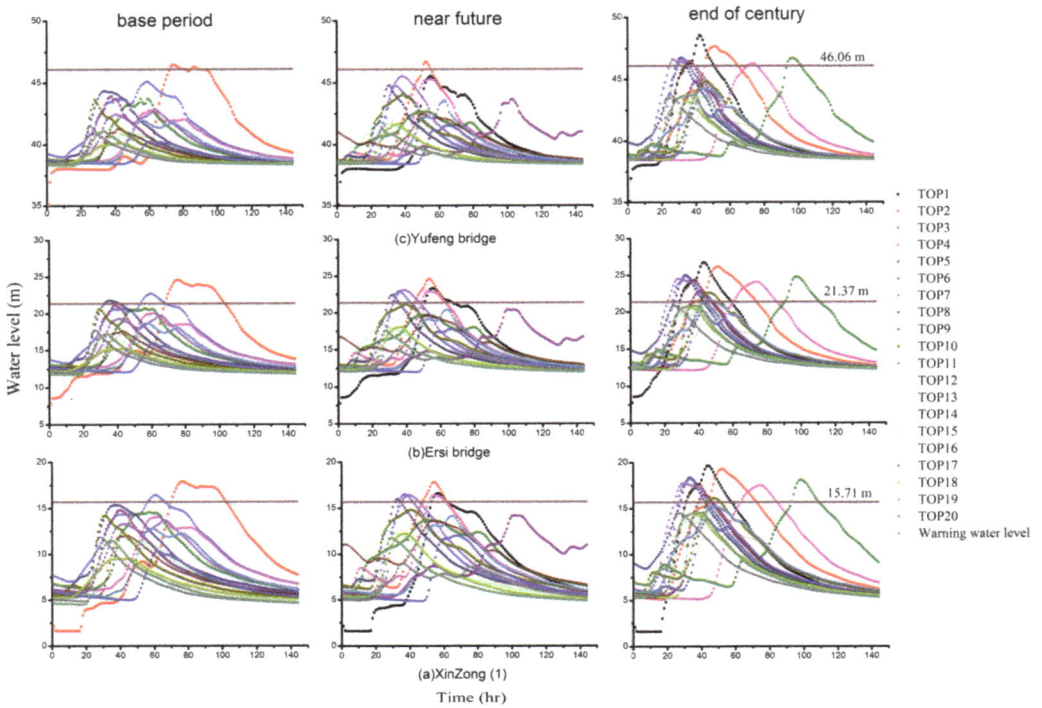

Figure 10. Water level hydrographs for the TOP1–20 extreme typhoon events during three periods at (**a**) XinZong (1); (**b**) Erxi Bridge; and (**c**) Yufeng Bridge.

5. Conclusions and Recommendations

This study used dynamic downscaling data produced by the TCCIP project for river flow rate simulation, and the results highlight the risk of overflow in the Tsengwen River in the future under a climate change scenario.

In 2009, Typhoon Morakot induced continuous rainfall over a long period, and the total rainfall received was lower than that received during the most extreme typhoon events forecasted for the end of the century. Furthermore, the maximum total rainfall received in 3 and 24 h during the top extreme typhoon events at the end of the century was higher than that received at those times during typhoon Morakot, indicating that extreme typhoon events under future climate change will induce strong rainfall over a short period.

Based on the flow rate simulation results, the flow rate at Yufeng Bridge (upstream,) Erxi Bridge (midstream), and XinZong. (1) (downstream) will potentially exceed the management plan at the end of the century. At XinZong (1), the number of times that the flow rate exceeded the management plan rate was 2 in 88 events in the base period, 6 in 82 events in the near future, and 10 in 81 events at the end of the century; that for the end of the century was 5-fold higher than that of the near future and 3-fold higher than that of the base period. At the end of the century, extreme peak flow events will increase in frequency and intensity. Simulation results show that the peak flow rate at the end of the century will be higher than that during Typhoon Morakot. Therefore, a severe flood could reoccur in the future.

In this study, the river cross-section was assumed to be the same when simulating flow rates for future climate change. In future, we will consider the influence of erosion and land use change on the river cross-section when carrying out simulations for future climate change.

Acknowledgments: The authors thank Akio Kitoh of the Meteorological Research Institute, Japan, for providing the MRI data. The authors also thank the Taiwan Central Weather Bureau and Water Resources Agency for providing the observational data. This project was funded by the National Science Council of Taiwan. The authors thank Yuan-Fong Su and Wei-Bo Chen for allowing use of their data in our research.

Author Contributions: Hsiao-Ping Wei executed the model simulations and discussed the results with Keh-Chia Yeh, Jun-Jih Liou, Yung-Ming Chen, and Chao-Tzuen Cheng.

Conflicts of Interest: The authors declare no conflict of interest.

References

1. Intergovernmental Panel on Climate Change (IPCC). *Climate Change 2007: The Physical Science Basis. Contribution of Working Group I to the Fourth Assessment Report of the Intergovermental. Panel on Climate Change*; Cambridge University Press: Cambridge, UK, 2007.

2. Liu, K.F.; Li, H.C.; Hsu, Y.C. Debris flow hazard assessment with numerical Simulation. *Nat. Hazards* **2009**, *49*, 137–161.

3. National Science and Technology Center for Disaster Reduction (NCDR): Introduction of Taiwan Climate Change Projection and Information Platform Project (TCCIP). *NCDR Rep.* **2010**, *7*, 2010–2054. (In Chinese)

4. Vitart, F.; Anderson, J.L.; Stern, W.F. Simulation of the interannual variability of tropical storm frequency in an ensemble of GCM integrations. *J. Clim.* **1997**, *10*, 745–760.

5. Lenderink, G.; Buishand, A.; Deursen, W.V. Estimates of future discharges of the river Rhine using two scenario methodologies: Direct *versus* delta approach. *Hydrol. Earth Syst. Sci.* **2007**, *11*, 1145–1159.

6. Wilby, R.L.; Wigley, T.M.L.; Conway, D.; Jones, P.D.; Hewitson, B.C.; Main, J.; Wilks, D.S. Statistical downscaling of general circulation model output: A Comparison of Methods. *Water. Resour. Res.* **1998**, *34*, 2995–3008.

7. Wilby, R.L.; Charles, S.P.; Zorita, E.; Timbal, B.; Whetton, P.; Mearns, L.O. *IPCC Task Group on Data and Scenario Support for Impact and Climate Analysis (TGICA). Guidelines for Use of Climate Scenarios Developed from Statistical Downscaling Methods*; IPCC Data Distribution Centre: Geneva, Switzerland, 2004.

8. Tabor, K.; Williams, J.W. Globally downscaled climate projections for assessing the conservation impacts of climate change. *Ecol. Appl.* **2010**, *20*, 554–565.

9. Chen, H.; Xu, C.Y.; Guo, S. Comparison and evaluation of multiple GCMs, statistical downscaling and hydrological models in the study of climate change impacts on runoff. *J. Hydrol.* **2012**, *434–435*, 36–45.

10. Mizuta, R.; Yoshimura, H.; Murakami, H.; Matsueda, M.; Endo, H.; Ose, T.; Kamiguchi, K.; Hosaka, M.; Sugi, M.; Yukimoto, S.; *et al.* Climate simulations using the improved MRI-AGCM with 20-km grid. *J. Meteor. Soc.* **2012**, *90A*.

11. Linde, A.H.T.; Aerts, J.C.J.H.; Bakker, A.M.R.; Kwadijk, J.C.J. Simulating low probability peak discharges for the Rhine basin using resampled climate modeling data. *Water. Resour. Res.* **2010**, *46*, 1–19.

12. Kimura, N.; Chiang, S.; Wei, H.P.; Su, Y.F.; Chu, J.L.; Cheng, C.T.; Liou, J.J.; Chen, Y.M.; Lin, L.Y. Tsengwen reservoir watershed hydrological flood simulation under global climate change using the 20 km mesh meteorological research institute atmospheric general circulation model (MRI-AGCM). *Terr. Atmos. Ocean. Sci.* **2014**, *25*, 449–461.

13. Lenderink, G.; Ulden, A.V.B.; van den Hurk, B.; Keller, F. A study on combining global and regional climate model results for generating climate scenarios of temperature and precipitation for the Netherlands. *Clim. Dyn.* **2007**, *29*, 157–176.

14. Jones, R.; Murphy, J.; Hassell, D.; Taylor, R. Ensemble mean changes in a simulation of the European climate of 2071–2100: Using the New Hadley Centre Regional Climate Modelling System HadAM3H/HadRM3H. Available online: http://prudence.dmi.dk/public/publications/hadley_200208.pdf (accessed on 6 October 2005).

15. Jones, C.G.; Willen, U.; Ullerstig, A.; Hansson, U. The Rossby Centre Regional Atmospheric Climate Model (RCA). Part I: Model Climatology and Performance for the Present Climate over Europe. *Ambio* **2004**, *33*, 199–210.

16. Giorgi, F.; Bi, X.; Pal, J. Means, trends and interannual variability in a regional climate change experiment over Europe. PartI: Present-Day Climate (1961–1990). *Clim. Dyn.* **2004**, *22*, 733–756.

17. Linde, A.H.T.; Aerts, J.C.J.H.; Kwadijk, J.C.J. Effectiveness of flood management strategies on peak discharges in the Rhine basin. *J. Flood Risk Manag.* **2010**, *3*, 248–269.

18. *Delft Hydraulics SOBEK User Manuals*; Deltares: Delft, The Netherlands, 2007.

19. Burnash, R.J.C. The NWS River Forecast System—Catchment model. In *Computer Models of Watershed Hydrology*; Water Resources Publications: Littleton, CO, USA, 1995.

20. Republic of China Ministry of Economic Affairs. *Water Resources Agency (WRA) Calibration of Flood Forecasting Model and Review and Establishment of Warning Stages*; Water Resources Agency report (2/2). Republic of China Ministry of Economic Affairs: Taipei, Taiwan, 2014. (In Chinese)

21. Republic of China Ministry of Economic Affairs. *Water Resources Agency (WRA) Operation Directions for Nanhua Reservoir*; (2011 updated version). Republic of China Ministry of Economic Affairs: Taipei, Taiwan, 2004. (In Chinese)

22. Chen, W.B.; Liu, W.C. Modeling Flood Inundation Induced by River Flow and Storm Surges over a River Basin. *Water* **2014**, *6*, 3182–3199.

23. Republic of China Ministry of Economic Affairs. *Water Resources Agency (WRA) Operation Directions for Tsengwen Reservoir*; (2014 updated version). Republic of China Ministry of Economic Affairs: Taipei, Taiwan, 2002. (In Chinese)

24. Republic of China Ministry of Economic Affairs. *Water Resources Agency (WRA) Operation Directions for Wusanto Reservoir*; (2011 updated version). Republic of China Ministry of Economic Affairs: Taipei, Taiwan, 2008. (In Chinese)

25. Tang, Y.; Reed, P.M.; Wagener, T. How effective and efficient are multiobjective evolutionary algorithms at hydrologic model calibration? *Hydrol. Earth Syst.* **2006**, *10*, 289–307.

26. Su, Y.F.; Cheng, C.T.; Liu, J.J.; Chen, Y.M. Bias correction of MRI-WRF dynamic downscaling datasets. *Terr. Atmos. Ocean. Sci.* **2015**. Submitted.

27. Yu, Y.C.; Lee, T.J.; Kung, C.Y. Disaster warning and scenario analysis in Typhoons and Heavy Rainfall events. *NCDR Rep.* **2014**, *2014*, 1–15. (In Chinese)

28. Johnson, F.; Sharma, A. Accounting for interannual variability: A Comparison of Options for Water Resources Climate Change Impact Assessments. *Water Resour. Res.* **2011**, *47*.

Optimal Choice of Soil Hydraulic Parameters for Simulating the Unsaturated Flow: A Case Study on the Island of Miyakojima, Japan

Ken Okamoto, Kazuhito Sakai, Shinya Nakamura, Hiroyuki Cho, Tamotsu Nakandakari and Shota Ootani

Abstract: We examined the influence of input soil hydraulic parameters on HYDRUS-1D simulations of evapotranspiration and volumetric water contents (VWCs) in the unsaturated zone of a sugarcane field on the island of Miyakojima, Japan. We first optimized the parameters for root water uptake and examined the influence of soil hydraulic parameters (water retention curve and hydraulic conductivity) on simulations of evapotranspiration. We then compared VWCs simulated using measured soil hydraulic parameters with those using pedotransfer estimates obtained with the ROSETTA software package. Our results confirm that it is important to always use soil hydraulic parameters based on measured data, if available, when simulating evapotranspiration and unsaturated water flow processes, rather than pedotransfer functions.

Reprinted from *Water*. Cite as: Okamoto, K.; Sakai, K.; Nakamura, S.; Cho, H.; Nakandakari, T.; Ootani, S. Optimal Choice of Soil Hydraulic Parameters for Simulating the Unsaturated Flow: A Case Study on the Island of Miyakojima, Japan. *Water* **2015**, *7*, 5676–5688.

1. Introduction

There are no rivers on Miyakojima, a semitropical island in southernmost Japan, because of its flat topography and the high permeability of the limestone that forms the island. Local residents there are dependent on groundwater for almost all of their domestic water use. In the Okinawa region, temperatures are predicted to rise in response to climate change, while annual rainfall is expected to decrease [1], with resultant depletion of groundwater resources becoming a concern. The main land use on Miyakojima is sugarcane farming; thus, understanding both water movement in the unsaturated zone of the farmland soil and the total water budget is important.

The HYDRUS-1D software package [2] has often been used for analyses of these type of problems [3,4]. HYDRUS-1D provides versatile numerical modeling of the movement of moisture, solutes, and heat in soil. One option in the code is to estimate soil hydraulic properties by using pedo-transfer functions (PTFs). Since it is difficult to measure soil hydraulic parameters, PTFs that estimate them from readily measurable soil characteristics, such as particle size distribution and bulk density provide a very attractive tool for numerical analyses.

The accuracy of HYDRUS-1D simulations has been analytically verified [5]. Most studies to evaluate the performance of HYDRUS-1D were simulations of the transfer of heat and moisture in semiarid and humid regions. For example, Saito *et al.* [6] reported that HYDRUS-1D was useful for predicting the transfer of heat and moisture in sandy soils, and Kato *et al.* [7] reported that it was useful for predicting soil temperature and moisture in volcanic soils.

Other studies have used vadose zone models and PTFs. Steinzer *et al.* [8] simulated evapotranspiration, infiltration, and VWCs distribution in lysimeter experiments on sandy and clay soils in Germany. Wang *et al.* [9] simulated groundwater recharge in sand and sandy loam soil by using HYDRUS-1D. They found that recharge in their soils was strongly dependent on the parameter n of the van Genuchten model [10] and uncertainties in the simulated recharge were affected by uncertainties in the n estimated from PTFs.

Thus, although HYDRUS-1D is known to be useful for simulations of water movement in the unsaturated zone, the code must be tested in a particular area before practical applications; for example, for the development of a water management plan. Wang *et al.* [11] simulated groundwater recharge at four sites in the continental United States with different climate conditions using HYDRUS-1D along with datasets for sand and loamy sand. They showed that the distribution patterns of mean annual groundwater recharge varied considerably across the sites, mainly depending on soil texture and climatic conditions.

To date, the use of HYDRUS-1D in the island of Miyakojima has not yet been tested. Although the necessary weather data are readily available from the Japan Meteorological Agency, the collection of soil data is more difficult. Most of the soil data collected and analyzed in past investigations for land development projects consists only of particle size distributions and bulk density; water retention curves were not always determined. Accurate simulation results are dependent on the quality of the soil hydraulic parameters used as input to the simulation. Okamoto *et al.* [12] reported that the retention curve of Shimajiri mahji soil (dark-red soil [13], which was classified as a Cambisol [14]) estimated using the ROSETTA software package [15–17] was considerably different from that derived from measured data. However, they did not comment on the influence of this difference on the simulated water budget.

To validate the use of HYDRUS-1D on Miyakojima, we examined the influence of input soil hydraulic parameters on HYDRUS-1D simulations of evapotranspiration and VWCs. We first optimized the parameters for root water uptake and examined the influence of soil parameters on simulations of evapotranspiration (details are shown in the later section). We then compared VWCs simulated using measured soil hydraulic parameters with those simulated using parameters derived by using ROSETTA software.

2. Materials and Methods

2.1. Study Site

Our study site is in a sugarcane field at Saratake in the Shirakawada groundwater basin, which is one of several fault-bounded groundwater basins on Miyakojima (Figure 1). We measured precipitation (CTK-15PC, Climatec, Inc., Tokyo, Japan), temperature (CVS-HMP-155D, Climatec, Inc., Tokyo, Japan), wind speed (CPR010C, Climatec, Inc., Tokyo, Japan), net radiation (CHF-NR01, Climatec, Inc., Tokyo, Japan), evapotranspiration (CS7500, Campbell Scientific, Logan, UT, USA and SAT-540, SONIC Co., Tokyo, Japan), and VWCs (EC5, Decagon Devices, Pullman, WA, USA). Evapotranspiration was measured by the eddy covariance method [18]. Time-domain reflectometry soil moisture sensors were installed at depths of 15, 30, 50, and 70 cm. Data were collected every 30 min. Measurement started on 29 August 2009 and ended on 31 December 2009. Cultivation started on 21 February 2009 and ended on 20 January 2010.

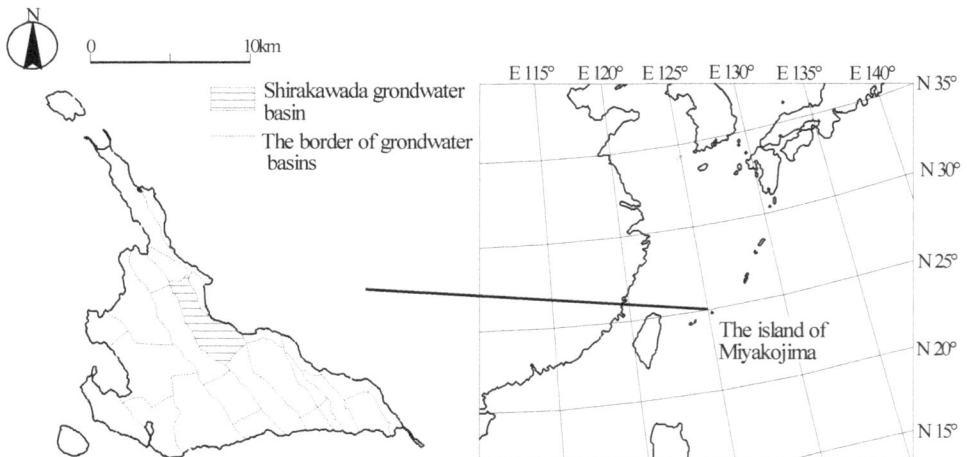

Figure 1. Groundwater basins on the island of Miyakojima.

The surface soil at the study site is a Jahgaru soil (gray soil [13,19]), which extends to a depth of 120 cm at our sampling site. Jahgaru soil derived from Shimajiri-mudstone is classified as Calcaric Regosols [14]. We sampled both disturbed and undisturbed soil at depths of 15, 30, 50, 70, and 100 cm. Undisturbed soil samples were collected using a 100 cm^3 soil core sampler (inside diameter 5 cm, height 5.1 cm).

We measured soil particle size distributions (%), bulk densities (g·cm^{-3}), hydraulic conductivities (cm·day^{-1}), and water retention curves in the laboratory. Soil particle size distributions were obtained using sieve analysis for particle sizes greater than 75 μm and by hydrometer analysis for particle sizes smaller than 75 μm, according to Japanese Industrial Standards (JIS A1202). We measured bulk densities by using the 100 cm^3 soil core sampler, hydraulic conductivities by the constant-head method, and water retention curves by the three methods shown in Table 1.

Table 1. Experimental procedures used for VWC measurements at selected soil suction ranges.

Method	Log$_{10}$ [Suction (cm)]								
	1.0	1.5	2.0	2.5	3.0	3.5	4.0	4.5	5.0
Hanging water column	———————————								
Pressure plate				———————					
Psychrometer *							———————		

Note: * WP4-T, Decagon Devices, Pullman, WA, USA.

2.2. HYDRUS-1D

2.2.1. Overview of HYDRUS-1D

One-dimensional water flow in soil is described by the Richards equation as follows:

$$\frac{\partial \theta}{\partial t} = \frac{\partial}{\partial z}\left(K\frac{\partial h}{\partial z}\right) + \frac{\partial K}{\partial z} - S, \tag{1}$$

where θ is volumetric water content (cm^3·cm^{-3}), t is time (s), z is the spatial coordinate, assumed positive upward (cm), h is pressure head (cm), S is a sink term for water uptake by plant roots (cm^3·cm^{-3}·s^{-1}), and K is unsaturated hydraulic conductivity (cm·s^{-1}). We applied the van Genuchten-Mualem equation [10,20] (VG hereafter) to estimate the water retention curve and hydraulic conductivity of the soil by using Equations (2) and (3), respectively:

$$S_e = \frac{\theta - \theta_r}{\theta_s - \theta_r} = \left(1 + |\alpha h|^n\right)^{-m}, \tag{2}$$

$$K(h) = K_s S_e{}^l \left[1 - S_e\left(1 - S_e^{\frac{1}{m}}\right)^m\right]^2, \tag{3}$$

where θ_r is residual water content ($cm^3 \cdot cm^{-3}$), θ_s is saturated water content ($cm^3 \cdot cm^{-3}$), and S_e is normalized water content. α (cm^{-1}), n, m ($= 1 - 1/n$) and l ($= 0.5$) are empirical parameters, and K_s is saturated hydraulic conductivity ($cm\ s^{-1}$).

We simulated five soil layers (0–20, 20–40, 40–60, 60–80, and 80–120 cm) that correspond to the depth ranges of the soil samples we collected. Atmospheric conditions of daily precipitation and potential evaporation were used as upper boundary conditions. Potential transpiration was used to calculate water uptake by plant roots. The lower boundary condition was free drainage. Initial VWC was the VWC measured in the field.

2.2.2. Crop Model in HYDRUS-1D

We used the method of van Genuchten *et al.* [21] to simulate water uptake by plant roots:

$$S(h) = \alpha(h) S_p, \tag{4}$$

$$\alpha(h) = \frac{1}{1 + \left(\frac{h}{h_{50}}\right)^p}, \tag{5}$$

where $\alpha(h)$ is the water stress response function, S_p is potential water uptake by plants roots, h_{50} is the soil suction at which water uptake by roots is reduced by 50%, and p is an empirical component that is usually assumed to be 3 [22]. We used $p = 3$ to optimize h_{50}.

To determine daily root length, we used a logistic root growth function in HYDRUS-1D [2] and the results of a previous study [23]. We used the leaf area index (*LAI*) growth model of Larsbo and Jarvis [24]. The *LAI* growth model from crop emergence day (D_{min}) to the day of maximum *LAI* (D_{max}) is described by Equation (6), and that from D_{max} to the day of harvest (D_{harv}) by Equation (7):

$$LAI = LAI_{min} + (LAI_{max} - LAI_{min}) \left(\frac{D^* - D_{min}}{D_{max} - D_{min}}\right)^{x_1}, \tag{6}$$

$$LAI = LAI_{harv} + (LAI_{max} - LAI_{harv}) \left(\frac{D_{harv} - D^*}{D_{harv} - D_{max}}\right)^{x_2}, \tag{7}$$

where LAI_{min} is *LAI* at D_{min}, LAI_{max} is *LAI* at D_{max}, LAI_{harv} is *LAI* at D_{harv}, D^* is the number of days after planting, and x_1 and x_2 are empirical components. We derived daily values of *LAI* from the results of a previous study of sugarcane [25] (Figure 2).

Potential evapotranspiration (ET_p) was calculated using the Penman Equation (8) [26,27]:

$$ET_p = \frac{\Delta}{\Delta + \gamma} \cdot \frac{S}{L} + \frac{\gamma}{\Delta + \gamma} \cdot u_2 (e_{sa} - e_a), \tag{8}$$

where ET_p is the reference evapotranspiration (mm·day^{-1}), Δ is the slope of the saturation vapor pressure curve (kPa·°C^{-1}), γ is the psychometric constant (kPa·°C^{-1}), S is the net radiation (MJ·m^{-2}·day^{-1}), L is the latent heat of evaporation (MJ·kg^{-1}) is, u_2 is wind speed at 2 m height (m·s^{-1}), ($e_{sa} - e_a$) is the saturation vapor pressure deficit (kPa). ET_p was partitioned into potential evaporation (E_p) and potential transpiration (T_p) using Campbell's equation [28]:

$$T_p = ET_p \left[1 - \exp\left(-8.2LAI\right)\right] \tag{9}$$

$$E_p = ET_p - T_p \tag{10}$$

Total precipitation during the observation period was 675.5 mm (Figure 3). Since the days after planting during the observation period varied from 189 to 313 and LAI from 2.4 to 3.3 (from Equations (6) and (7)), E_p was low throughout the observation period.

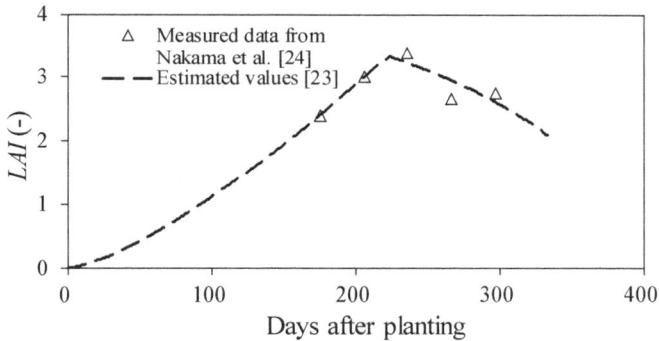

Figure 2. Change of LAI after planting of sugarcane.

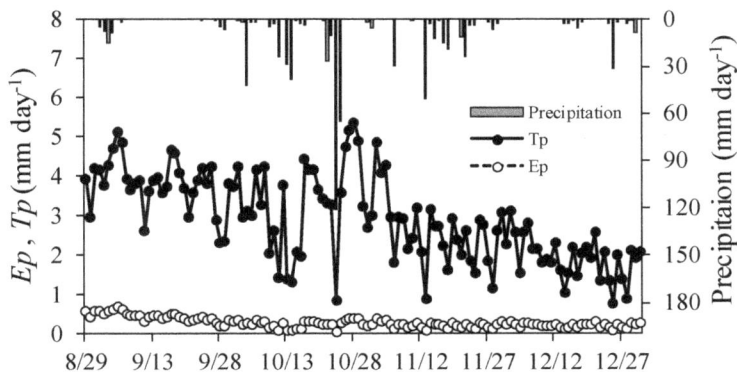

Figure 3. Daily precipitation, potential evaporation (Ep) and potential transpiration (Tp) from 29 August to 31 December 2009.

2.3. Comparison of Measured Soil Hydraulic Parameters with Those Estimated Using ROSETTA

We determined the parameters for the VG equation by using our analyses of soil samples and by application of the ROSETTA module in HYDRUS-1D. We determined VG parameters so as to minimize the difference between measured values and estimated values according to the nonlinear least-squares method using the solver of an EXCEL add-in. ROSETTA can use combinations of soil texture, particle size distribution, bulk density, and one or two points on the water retention curve to determine the parameters for the VG equation and for hydraulic conductivity. In this study, we used particle size distribution and bulk density, since these properties were readily available.

Measured particle size distribution, bulk density and hydraulic conductivity are shown in Table 2. For all samples, the sand content was less than 20% and the soil texture was silt. There were no clear differences in bulk density among soil layers. The standard deviation of hydraulic conductivity was large for all layers, which we attributed to cracks formed in response to shrinkage during drying of the soil samples. We, therefore, used the geometric mean of hydraulic conductivity in the simulations.

Table 2. Measured particle distribution, bulk density, and hydraulic conductivity of all layers.

z (cm)	Sand (%)	Silt (%)	Clay (%)	Bulk Density $(g \cdot cm^{-3})$	Saturated Conductivity, K_s Geometric mean $\pm$ SD (cm$\cdot$day^{-1})
15	13.1	69.6	17.3	1.319	24.7 $\pm$ 97.8
30	11.1	71.2	17.7	1.278	24.3 $\pm$ 152.6
50	7.5	75.3	17.2	1.268	21.8 $\pm$ 90.0
70	8.7	74.8	16.5	1.305	116.3 $\pm$ 343.8
100	10.4	70.3	19.4	1.159	21.9 $\pm$ 99.8

Note: Particle size distribution was derived from one sample and hydraulic conductivity from four samples.

In our application of HYDRUS-1D, we considered four combinations of input soil hydraulic parameters (Cases 1 to 4; Table 3).

Table 3. Combinations of parameters input to HYDRUS-1D.

Soil Hydraulic Parameter	Case 1	Case 2	Case 3	Case 4
Retention curve	Measured	ROSETTA	ROSETTA	Measured
Hydraulic conductivity	Measured	ROSETTA	Measured	ROSETTA

2.4. Simulation of Evapotranspiration and Volumetric Water Contents

To optimize the value of h_{50}, we needed to minimize the difference between measured and simulated evapotranspiration for each of Cases 1 to 4. To achieve this, for each case we ran HYDRUS-1D using h_{50} values from 100 to 1000 cm at intervals of 100 cm. Then, we chose the optimum value of h_{50} as the value with the lowest root-mean-square error (RMSE) for total evapotranspiration calculated at 10-day intervals.

To examine the influence of input soil hydraulic parameters on simulated VWCs movement, we calculated the RMSE between measured VWCs and simulated VWCs (optimized h_{50}) for each of the four soil layers for Cases 1 to 4.

3. Results and Discussion

3.1. Comparison of Measured Soil Hydraulic Parameters with Those Estimated by ROSETTA

In general, ROSETTA underestimated measured VWCs, particularly for soil suctions greater than 1000 cm (Figure 4). Okamoto *et al.* [12] reported similar results from their application of ROSETTA to Shimajiri mahji soil. Since Jahgaru soil has a poorly developed structure [19] it often does not drain well [18]. Therefore, the measured VWC values at high suction tended to be larger than that estimated by ROSETTA (Figure 4). Schaap and Leij [29] reported that the use of PTFs to estimate soil hydraulic parameters might depend strongly on the data used for calibration. The characteristics of Jahgaru soil might be different from the soil data used to develop on the parameters used in ROSETTA.

Measured hydraulic conductivities were 21.8–116.3 cm d^{-1}, whereas those estimated by ROSETTA were 30.3–53.5 cm·d^{-1} (Table 4). We attributed the difference in these results to the influence of cracks and aggregations of soil in the sugarcane field.

Table 4. Parameters used in simulations.

z (cm)	θ_r (cm^3·cm^{-3}) Fitted	θ_r ROS	θ_s (cm^3·cm^{-3}) Fitted	θ_s ROS	α (cm^{-1}) Fitted	α ROS	n Fitted	n ROS	K_s (cm·d^{-1}) Measured	K_s ROS
15	0	0.070	0.493	0.441	0.0037	0.0049	1.193	1.685	24.7	30.3
30	0	0.073	0.493	0.456	0.0148	0.0049	1.140	1.683	24.3	35.0
50	0	0.074	0.449	0.469	0.0117	0.0051	1.125	1.672	21.8	35.2
70	0	0.072	0.462	0.456	0.0107	0.0050	1.147	1.677	116.3	32.3
100	0	0.078	0.556	0.492	0.0082	0.0049	1.166	1.675	21.9	53.5

Note: ROS = from ROSETTA.

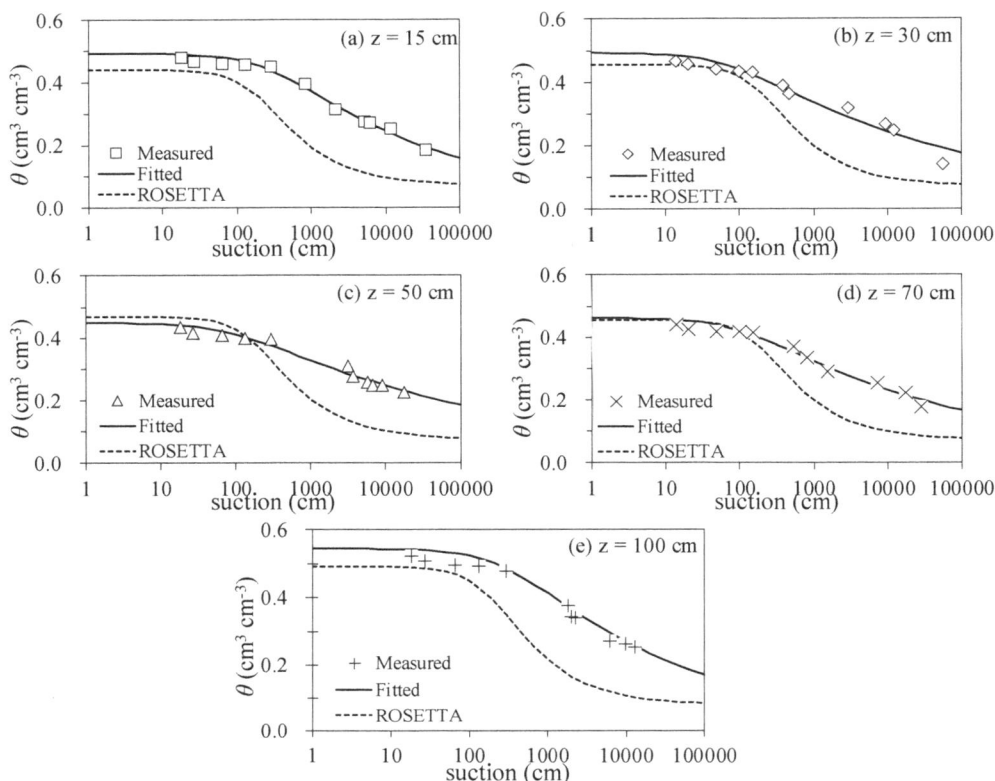

Figure 4. Soil water retention curves from measured data and as estimated by ROSETTA. (**a**) z = 15 cm; (**b**) z = 30 cm; (**c**) z = 50 cm; (**d**) z = 70 cm; (**e**) z = 100 cm.

3.2. Simulation of Evapotranspiration and Volumetric Water Contents

3.2.1. Optimization of h_{50} and Simulation of Evapotranspiration

The simulations run to optimize h_{50} (Figure 5) show that total evapotranspiration (TET) increased with increasing h_{50} for all four cases considered. The optimal h_{50} (smallest RMSE) was 600 cm for Case 1, 400 cm for Case 2, 300 cm for Case 3 and 700 cm for Case 4. For each of the four cases, the TET estimated with the optimized h_{50} was almost the same as the measured TET (204.5 mm).

The simulation results for the pairs of cases with the same retention curve (Cases 1 and 4, Cases 2 and 3) were similar (compare Figures 5 and 6). The suction required to deplete VWCs during normal growth has been reported to be about 1000 cm [30]; therefore, we considered that Case 1 (600 cm) and Case 4 (700 cm) provided the more realistic values of h_{50} for application in HYDRUS-1D, even though the RMSEs of Cases 2 and 3 were lower. The common factor for Cases 1 and 4 was the use of the measured retention curve.

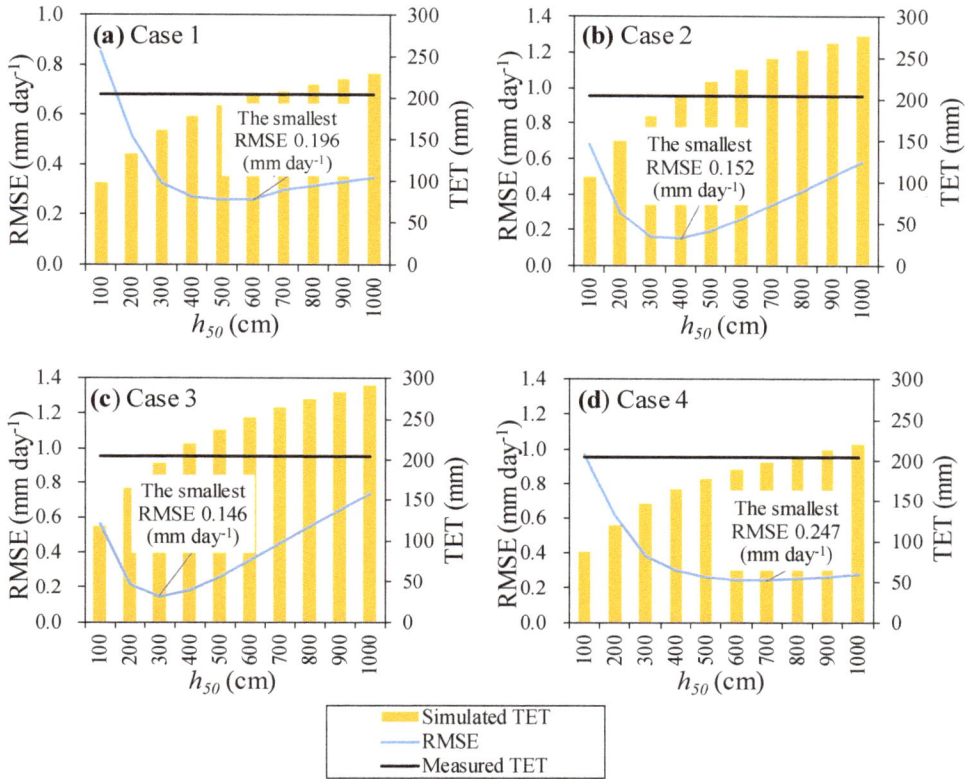

Figure 5. Relationships of the value of h_{50} to RMSE and TET for each Case. (**a**) Case1; (**b**) Case 2; (**c**) Case 3; (**d**) Case 4.

Thus, comparison of our simulation results for evapotranspiration did not clearly indicate which soil hydraulic parameters were better for application in HYDRUS-1D.

3.2.2. Influence of Soil Hydraulic Parameters on Volumetric Water Contents Simulation

The time series of simulated VWCs (Figure 7) and RMSEs between measured and simulated VWCs (Table 5) show that the simulated VWC values for Cases 2 and 3 (retention parameters estimated using ROSETTA) were lower than measured values, whereas for Cases 1 and 4 (retention parameters calculated from measured data) simulated and measured VWCs agreed well.

341

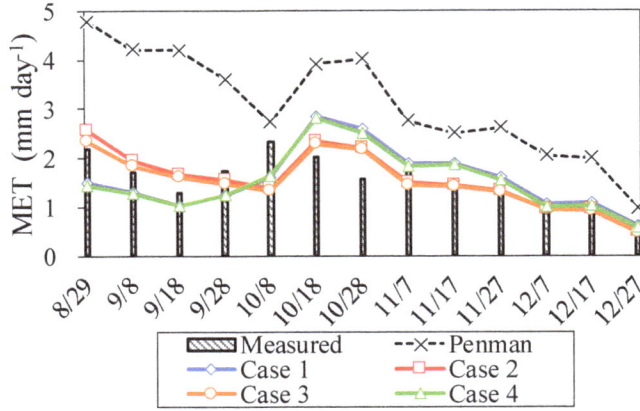

Figure 6. Comparison of simulated, measured and potential evapotranspiration at ten-day intervals (MET) from 29 August to 31 December 2009.

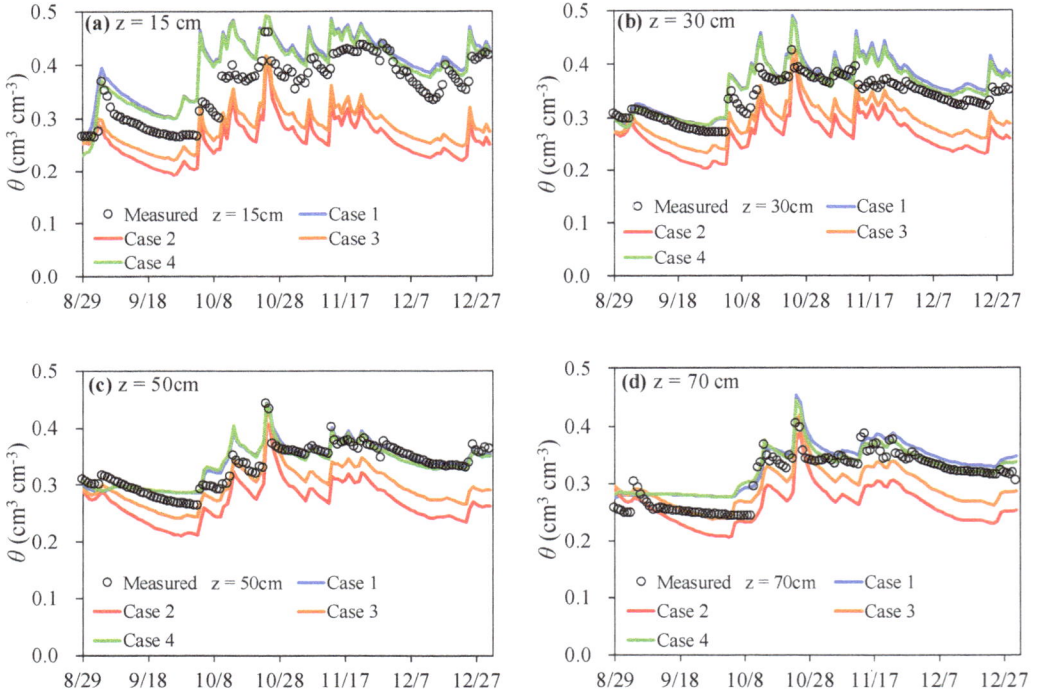

Figure 7. Comparison of simulated and measured daily VWCsat 4 depths from 29 August to 31 December 2009. (**a**) z = 15 cm; (**b**) z = 30 cm; (**c**) z = 50 cm; (**d**) z = 70 cm.

Table 5. RMSE between measured and simulated VWCs.

z (cm)	RMSE ($cm^3 \cdot cm^{-3}$)			
	Case 1	Case 2	Case 3	Case 4
15	0.052	0.101	0.083	0.043
30	0.035	0.067	0.047	0.028
50	0.017	0.060	0.040	0.017
70	0.026	0.051	0.031	0.021

When considering the water budget, water recharge (Re) is calculated using

$$R_e = P - ET - \Delta SW, \tag{11}$$

where P is precipitation, ET is evapotranspiration and ΔSW is the change of VWCs. To estimate water recharge from this equation, accurate simulations of VWCs and evapotranspiration are needed. The results of our simulations of evapotranspiration for Cases 1 to 4 (Figure 6) did not differ greatly (Section 3.2.1). However, our simulations of VWCs indicated that Cases 1 and 4 provided better results than Cases 2 and 3. These results indicate the importance of using soil hydraulic parameters based on retention curves derived from measured data to calculate the water budget for the island of Miyakojima, and likely for many or most other applications.

4. Conclusions

We drew the following main conclusions about the optimum application of HYDRUS-1D in our study area.

1) ROSETTA software underestimated measured VWCs.
2) Optimized values of h_{50} were dependent on the parameters defined by the retention curve. Simulated and measured total evapotranspiration rates agreed well for all four cases considered. Since, for normal growth the amount of suction required to deplete VWCs is about 1000 cm, we consider the h_{50} values we obtained that were closest to 1000 cm to be the more realistic. Thus, our HYDRUS-1D simulations using the measured soil hydraulic parameters provided better results than those based on parameters estimated by ROSETTA.
3) VWCs simulated by HYDRUS-1D using parameters estimated by ROSETTA were lower than the measured values, whereas those using measured parameters agreed well with measured values.

Our study confirmed that it is important to use soil hydraulic parameters derived from measured retention data on the island of Miyakojima, rather than estimates obtained with pedotransfer function.

343

Acknowledgments: The authors gratefully acknowledges the constructive comments and suggestions from the anonymous reviewers.

Author Contributions: Ken Okamoto designed and performed the experiments, analyzed and interpreted the data and wrote the manuscript. Kazuhito Sakai designed the study and interpreted the results. Shinya Nakamura, Hiroyuki Cho, Tamotsu Nakandakari and Shota Ootani interpreted the data. All authors read and approved the final manuscript.

Conflicts of Interest: The authors declare no conflict of interest.

References

1. Climate Change and Its Impacts in Japan. Available online: https://www.env.go.jp/en/earth/cc/report_impacts.pdf (accessed on 4 September 2015).
2. Ficklin, D.L.; Luedeling, E.; Zhang, M. Sensitivity of groundwater recharge under irrigated agriculture to changes in climate, CO_2 concentrations and canopy structure. *Agric. Water Manag.* **2010**, *97*.
3. Leterme, B.; Mallants, D.; Jacques, D. Sensitivity of groundwater recharge using climatic analogues and HYDRUS-1D. *Hydrol. Earth Syst. Sci.* **2012**, *16*.
4. Šimůnek, J.; Šejna, M.; Saito, H.; van Genuchten, M.T. *The HYDRUS-1D Software Package for Simulating the One-Dimensional Movement of Water, Heat, and Multiple Solutes in Variably-Saturated Media*; Department of Environmental Science, University of California Riverside: Riverside, CA, USA, 2013.
5. Zlotnik, V.A.; Wang, T.; Nieber, J.L.; Šimunek, J. Verification of numerical solutions of the Richards equation using a traveling wave solution. *Adv. Water Resour.* **2007**, *30*, 1973–1980.
6. Saito, H.; Šimůnek, J.; Mohanty, B.P. Numerical analysis of coupled water, vapor, and heat transport in the vadose zone. *Vadose Zone J.* **2006**, *5*.
7. Kato, C.; Nishimura, T.; Imoto, H.; Miyazaki, T. Predicting soil moisture and temperature of andisoils under a monsoon climate in Japan. *Vadose Zone J.* **2011**, *10*, 541–551.
8. Stenitzer, E.; Diestel, H.; Zenker, T.; Schwartengräber, R. Assessment of capillary rise from shallow groundwater by the simulation model SIMWASER using either estimated pedotransfer functions or measured hydraulic parameters. *Water Resour. Manag.* **2007**, *21*, 1567–1584.
9. Wang, T.; Zlotnik, V.A.; Šimunek, J.; Schaap, M.G. Using pedotransfer functions in vadose zone models for estimating groundwater recharge in semiarid regions. *Water Resour. Res.* **2009**, *45*.
10. Van Genuchten, M.T. A closed-form equation for predicting the hydraulic conductivity of unsaturated soils. *Soil Sci. Soc. Am. J.* **1980**, *44*, 892–898.
11. Wang, T.; Franz, T.E.; Zlotnik, V.A. Controls of soil hydraulic characteristics on modeling groundwater recharge under different climatic conditions. *J. Hydrol.* **2015**, *521*.
12. Okamoto, K.; Sakai, K.; Cho, H.; Nakamura, S.; Nakandakari, T. Influences of a bulk density for hydraulic conductivity and water retention curve of Shimajiri maji soil, and examination of PTFs' applicability. *J. Rainwater Catchment Syst.* **2015**, *21*, 1–6. (In Japanese with English Abstract)

13. Miyamura, N.; Iha, S.; Gima, Y.; Toyota, K. Factors limiting organic matter decomposition and the nutrient supply of soils on the Daito Ilands. *Soil Microorg.* **2011**, *65*, 119–124.

14. Classification of cultivated soils in Japan. Available online: http://soilgc.job.affrc.go.jp/Document/Classification.pdf (accessed on 4 September 2015).

15. Schaap, M.G.; Leiji, F.J.; van Genuchten, M.T. Neural network analysis for hierarchical prediction of soil water retention and saturated hydraulic conductivity. *Soil Sci. Soc. Am. J.* **1998**, *62*, 847–855.

16. Schaap, M.G.; Leiji, F.J. Improved prediction of unsaturated hydraulic conductivity with the Mualem-van Genuchten model. *Soil Sci. Soc. Am. J.* **2000**, *64*, 843–851.

17. Schaap, M.G.; Leiji, F.J.; Van Genuchten, M.T. Rosetta: A computer program for estimating soil hydraulic parameters with hierarchical pedotransfer function. *J. Hydrol.* **2001**, *251*, 163–176.

18. Kosugi, Y.; Katsuyama, M. Evapotranspiration over a Japanese cypress forest. II. Comparison of the eddy covariance and water budget methods. *J. Hydrol.* **2007**, *334*, 305–311.

19. Jayasinghe, G.Y.; Tokashiki, Y.; Kitou, M.; Kinjyo, K. Effect of synthetic soil aggregates as a soil ameliorant to enhance properties of problematic gray ("Jahgaru") soils in Okinawa, Japan. *Commun. Soil Sci. Plant Anal.* **2010**, *41*, 649–664.

20. Mualem, Y. A new model predicting the hydraulic conductivity of unsaturated porous media. *Water Resour. Res.* **1976**, *12*, 513–522.

21. Van Genuchtenm, M.T. *A Numerical Model for Water and Solute Movement in and Below the Root Zone*; U.S. Salinity Laboratory, USDA, ARS, Riverside: CA, USA, 1987.

22. Van Genuchten, M.T.; Gupta, S.K. A reassessment of the crop tolerance response function. *Indian Soc. Soil. Sci.* **1993**, *41*, 730–737.

23. Smith, D.M.; Inman-Bamber, N.G.; Thorburn, P.J. Growth and function of the sugarcane root system. *Field Crops Res.* **2005**, *92*, 169–183.

24. Larsbo, M.; Jarvis, N. *MACRO 5.0: A Model of Water Flow and Solute Transport in Macroporous Soil*; Department of Soil Sciences, Swedish University of Agricultural Sciences: Uppsala, Sweden, 2003.

25. Nakama, M.; Nose, A.; Miyazato, K.; Murayama, S. *The Science Bulletin of the Faculty of Agriculture, University of the Ryukyus*, 1st ed.; Faculty of Agriculture, University of the Ryukyus: Naha-shi, Japan, 1987; pp. 187–198.

26. Penman, H.L. Natural evaporation from open water, bare soil and grass. *Proc. R. Soc. Lond. A* **1948**, *193*, 120–145.

27. Miura, T.; Okuno, R. Detailed description of calculation of potential evapotranspiration using the Penman equation. *Trans. Jpn. Soc. Irrig. Drain Reclam Eng.* **1993**, *164*, 163.

28. Campbell, G.S. *Soil Physics with Basics, Transport Models for Soil-Plant Systems*, 1st ed.; Elsever: New York, UN, USA, 1985; p. 144.

29. Schaap, M.G.; Leiji, F.J. Database-related accuracy and uncertainty of pedotransfer functions. *Soil Sci.* **1998**, *163*, 765–779.

30. Yanagawa, A.; Fujimaki, H. Tolerance of canola to drought and salinity stresses in terms of root water uptake model parameters. *J. Hydrol. Hydromech.* **2013**, *61*, 73–80.

Farmers' Perceptions about Adaptation Practices to Climate Change and Barriers to Adaptation: A Micro-Level Study in Ghana

Francis Ndamani and Tsunemi Watanabe

Abstract: This study analyzed the farmer-perceived importance of adaptation practices to climate change and examined the barriers that impede adaptation. Perceptions about causes and effects of long-term changes in climatic variables were also investigated. A total of 100 farmer-households were randomly selected from four communities in the Lawra district of Ghana. Data was collected using semi-structured questionnaires and focus group discussions (FGDs). The results showed that 87% of respondents perceived a decrease in rainfall amount, while 82% perceived an increase in temperature over the past 10 years. The study revealed that adaptation was largely in response to dry spells and droughts (93.2%) rather than floods. About 67% of respondents have adjusted their farming activities in response to climate change. Empirical results of the weighted average index analysis showed that farmers ranked improved crop varieties and irrigation as the most important adaptation measures. It also revealed that farmers lacked the capacity to implement the highly ranked adaptation practices. The problem confrontation index analysis showed that unpredictable weather, high cost of farm inputs, limited access to weather information, and lack of water resources were the most critical barriers to adaptation. This analysis of adaptation practices and constraints at farmer level will help facilitate government policy formulation and implementation.

Reprinted from *Water*. Cite as: Ndamani, F.; Watanabe, T. Farmers' Perceptions about Adaptation Practices to Climate Change and Barriers to Adaptation: A Micro-Level Study in Ghana. *Water* **2015**, *7*, 4593–4604.

1. Introduction

Climate change prediction models have indicated that the Sudan and Guinea Savanna zones of Ghana will continue to experience increasing temperature and decreasing precipitation trends [1]. This confirms previous findings that between 2030 and 2039 the rainy season might start in June or even later in Northern Ghana [2]. It is also projected that the standard deviation for the onset of the rainy season will increase [3], which suggests that not only will it shift but also it will become even more "erratic" [4]. The implications are that Northern Ghana would witness more extreme weather conditions such as droughts, dry spells, and floods. This situation will eventually affect agriculture, the environment, and human livelihoods. In particular, it is anticipated that adverse impacts on the agricultural sector will exacerbate the incidence of rural poverty [5]. Adaptation practices are therefore needed to help agrarian communities better face extreme weather conditions associated with climate variations [6].

Adaptations are adjustments or interventions that take place to manage the losses or take advantage of the opportunities presented by a changing climate. Adaptive capacity has been defined as the ability of a system to adjust to climate change (including climate variability and extremes), to moderate potential damages, to take advantage of opportunities, or to cope with the consequences [7]. Adaptation practices are therefore pre-emptive in nature. They are designed to mitigate potential adverse effects and take advantage of the potential benefits of an envisaged change in climatic variables.

Several studies in Ghana have reported adaptation practices in agriculture, including crop diversification, change of planting date, hybrid varieties, and soil moisture conservation techniques [8,9] In Uganda, income diversification, digging of drainage channels, and the use of drought-tolerant varieties have been reported [10]. In addition, mixed farming, mixed cropping, tree planting, use of different crop varieties, changing planting and harvesting dates, increased use of irrigation, increased use of water and soil conservation techniques, and diversifying from farm to non-farm activities have also been reported in Nigeria and in South Africa [11,12].

Globally, many studies have been used to understand farmers' perceptions about climate change and its associated effects on agriculture. Although perceptions are not necessarily consistent with reality, they must be considered to address socioeconomic challenges [13]. Perception has been defined as the process by which organisms interpret and organize sensation to produce a meaningful experience of the world [14]; and that a person's perceptions are based on experiences with natural and other environmental factors that vary in the extent to which such perceptions are enabled [15]. Previous studies have shown that the way in which people experience climate shocks varies across different social groups, geographic locations, and seasons of the year, with men, women, and children all experiencing different levels of hardship and opportunity in the face of climate change [16].

Discussions of adaptation practices and barriers to adoption need to be informed by empirical data from farmers. Adaptation practices in agriculture are generally location-specific [17]; hence, it is crucial to understand farmers' perceptions about the risks they face. To ensure farmers' readiness for extreme weather events and collaboratively learn about the evolution of weather patterns, efforts to focus on farmers and their current activities, knowledge, and perceptions are essential [18,19]. Farmers' willingness to accept and use prescribed measures could be enhanced if their perceptions and understanding are considered in designing such measures. By contrast, current models used in predictions of climate change and adaptation practices are at a global scale and need to be downscaled to accommodate realities at the community level [9].

In the Lawra district of Ghana, agriculture production is the dominant source of food and household incomes for the vast majority of rural households. Agriculture production is largely rain-fed. Farmers' dependence on an annual mono-modal rainfall pattern coupled with farm resource constraints make agriculture very vulnerable to the impacts of climate change. Results of previous studies have revealed a negative correlation between seasonal rainfall and volume of staple crops (*i.e.*, sorghum, millet, and groundnut) produced annually in the Lawra district over the past 20 years [20]. This study explored farmers' perceptions regarding long-term changes in climatic variables and the associated effects on farming. It also identified and prioritized adaptation practices based on farmers' perceived importance. Constraints on the use of adaptation measures were also identified and ranked. This study will help government policy decisions about suitable adaptation practices that are applicable and most preferred by farmers. It will also ensure that critical barriers to adoption are effectively addressed.

2. Materials and Methods

2.1. Survey Design and Study Area

This study is based on a cross-sectional survey data from farming households across four communities (*i.e.*, Brifo-chaa, Methuo, Kalsagri, and Oribili) in the Lawra district of Ghana, located at longitude 10°30' N and latitude 2°35' W. The district lies within the Guinea Savanna Zone, with mean annual rainfall ranging from 900 to 1200 mm. It has two seasons: the dry season (November–April) and the rainy season (May–October). The vegetation is guinea savanna grassland characterized by shrubs and medium-sized trees, such as shea-tree, dawadawa, baobab, and acacia. The soils are mainly laterite soils developed from birimian and granite rocks. These soils are shallow sandy loam with medium coarse quartz stones. Recurrent droughts, dry spells, and floods tend to have adverse effects on crop production. The major crops produced include maize, sorghum, millet, and groundnut. Crop production activities take place within the rainy season. Eighty percent of the district's total population of 100,929 is engaged in rain-fed subsistence agriculture [21] The district was chosen because, based on historical data from the Ghana Meteorological Agency, it is more prone to extreme weather conditions. According to the [21], Lawra is the poorest district in the upper west region of Ghana.

A total of 100 farming households were randomly selected for the interviews. Semi-structured questionnaires were used to investigate farmers' perceived changes in temperature and rainfall, causes and effects of climate change, and adaptation practices being used by farmers. Four focus group discussions (FGDs) were conducted to double check the survey data. The household survey and FGDs were conducted between February and November 2014 with the assistance of three regional and four district agricultural officers. The selection of communities was based on the accessibility and knowledge of agricultural officers.

2.2. Statistical Analysis

Data were entered and analyzed using statistical package for the social sciences (SPSS). Frequencies, percentages, and means are the basic descriptive statistical tools used to represent farmers' perceptions about long-term changes in climatic variables and the associated causes (Table 1). In determining farmers' perceived importance of adaptation practices, respondents were requested to score selected practices based on a 0–3 scale, where 0 is the least important practice and 3 is the most important practice. The adaptation practices were then ranked using the weighted average index (WAI):

$$\text{WAI} = \frac{\sum F_i W_i}{\sum F_i} \tag{1}$$

where F = frequency of response; W = weight of each score; and i = score (3 = highly important; 2 = moderately important; 1 = less important; 0 = not important).

Table 1. Description of data variables.

Variables	Mean	Standard Deviation
Age (15–34 years = 1; 35–54 years = 2; above 55 years = 3)	2.26	0.73
Education level (literate = 1; illiterate = 0)	0.21	0.41
Farm size (continuous)	4.95	1.70
Household size (continuous)	8.20	5.12
Family labor (continuous)	4.65	3.04
Annual farm income—Ghana cedi (continuous)	949.42	1909.55
Annual off-farm income—Ghana cedi (continuous)	797.20	2459.92
Farmer's adaptation (adapted = 1; not adapted = 0)	0.67	0.47

Previous studies have also applied the weighted average index (WAI) to assess farmers' perceived important adaptation strategies in Bangladesh and barriers of adaptation to climate change in Nepal [22,23].

To identify the critical constraints that hinder farmers from using adaptation practices, a ranking was conducted using the Problem Confrontation Index (PCI). Respondents were asked to grade their perceived barriers based on a 0–3 Likert scale (*i.e.*, ranging from "not a problem" to "highly problematic"). The PCI value was estimated using the formula below:

$$\text{PCI} = P_n \times 0 + P_1 \times 1 + P_m \times 2 + P_h \times 3 \tag{2}$$

where:

PCI = Problem Confrontation Index;
P_n = Number of respondents who graded the constraint as no problem;
P_l = Number of respondents who graded the constraint as low;
P_m = Number of respondents who graded the constraint as moderate;
P_h = Number of respondents who graded the constraint as high.

3. Results

3.1. Farmers' Perceptions of Long-Term Temperature and Rainfall Changes

The majority of farmers (82%) perceived an increase in temperature over the past 10 years (Figure 1). About 9% of respondents perceived no change, 6% perceived a decreasing change in temperature, and 3% did not know if there was a long-term change in temperature (Figure 1). Similar results were obtained from the focus group discussion. Generally, farmers believe that the increasing temperature trend was associated with the changes in precipitation. A total of 87% of respondents claimed that the rainfall amount has been decreasing over the past 10 years, 6% perceived no change in precipitation, and 7% gave other responses. Results obtained from FGDs proved that this perception was unanimous among farmers.

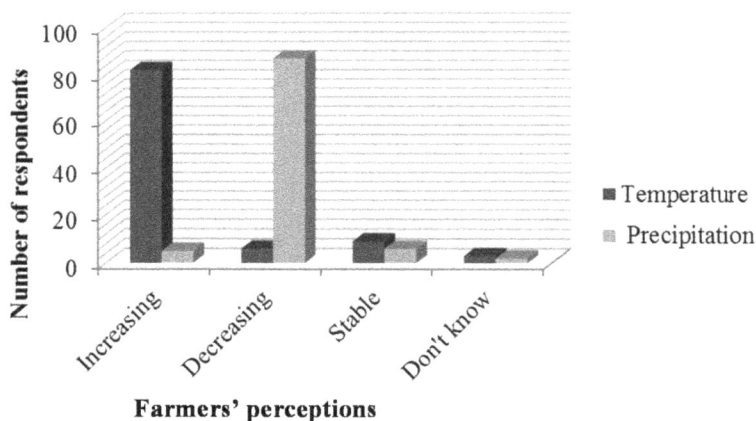

Figure 1. Farmers' perceptions of long-term changes in temperature and precipitation in the Lawra district of Ghana.

To verify farmers' perceptions regarding the precipitation trend, available historical annual rainfall data from 1980 to 2012 were obtained from the upper west regional weather station of the Ghana Meteorological Agency. The results indicated high variability rather than a clear decreasing trend in precipitation (Figure 2).

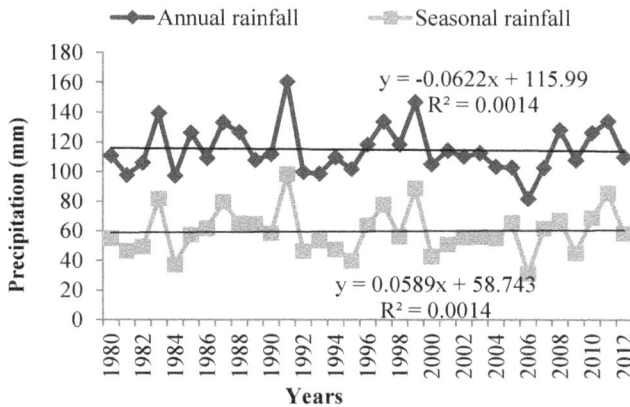

Figure 2. Annual and seasonal rainfall (mm) in the Lawra district of Ghana (The rainfall data was collected directly from the office of the meteorological agency: Ghana Meteorological Agency, 2014).

In addition, the results of discussions with the district agricultural officers confirmed the results of statistical analysis on the long-term trend in precipitation. Thus, the difference between farmers' perceptions and statistical results is due to the fact that farmers' responses are based solely on recall. The high illiteracy rate among farmers in Lawra district hinders their ability to keep formal records, and so accurately recalling long-term trends of rainfall could be difficult.

3.2. Farmers' Perceived Causes of Climatic Variability on Agriculture

Most farmers attributed climate change to human-related causes such as bush fires (51%) and deforestation (14%). While deforestation is largely perceived as being for the purposes of fuel wood, charcoal, and farm expansion, bush fires are believed to be caused by the 'negligence' of hunters and cigarette smokers. About 9.3% of respondents also claimed that traditional gods and ancestral spirits were responsible for the perceived changes in rainfall and temperature trends. During the FGDs, farmers indicated that the gods/ancestors were angry because many taboos have been broken by people (e.g., destroying sacred groves or woods, catching of sacred fish, *etc.*). Additionally, 23.3% of respondents claimed that climate change is caused by many factors, while 2.4% could not give any cause (Table 2).

351

Table 2. Farmers' perceptions about the causes of climate change in the Lawra district of Ghana (*Number of respondents* = 100).

Cause Variable	Percentage of Respondents
Deforestation	14.0
Bush fires	51.0
More than one cause	23.3
Gods/ancestral spirits	9.3
Do not know	2.4
Total	100

3.3. Farmers' Adaptation to Climate Change

The results revealed that farmers' adaptations are largely in response to dry spells (73%). However, 20% and 7% of respondents indicated that they used adaptation options in response to droughts and floods, respectively. Although an overwhelming majority of farmers recognized climate change, 33% of respondents still do not use any adaptation practices (Table 3).

Table 3. Proportion of farmers by adaptation classification and reasons for adaptation in the Lawra district of Ghana (*Number of respondents* = 100).

Variable	Percentage of Respondents
a. Adaptation classification	
Adapted	67
Not adapted	33
b. Reasons for adaptation	
Reduce effects of flood	7
Reduce effects of drought	20
Reduce effects of dry spell	73

3.4. Farmer-Perceived Importance of Adaptation Practices

The ranking of adaptation practices based on farmers' perceived importance is presented in Table 4. Among the seven adaptation practices, improved crop varieties and irrigation practice ranked first and second with a WAI of 2.15 and 2.09, respectively. The increasing incidence of drought and dry spells makes drought-tolerant crop varieties and irrigation preferable to farmers. On the other hand, income-generating activities and agroforestry practice were ranked the least important with a WAI of 0.77 and 0.74, respectively. Results of FGDs showed that farmers considered trading and agroforestry as capital-intensive activities. Crop diversification, farm diversification, and change of planting date were ranked as moderately important.

Table 4. Farmers' ranking of adaptation practices in the Lawra district of Ghana (Number of respondents = 100).

Adaptation Practice	Frequency by Each Level of Importance				WAI	Rank
	Highly Important	Moderately Important	Less Important	Not Important		
Improved crop varieties	35	48	14	3	2.15	1
Irrigation	30	51	17	8	2.09	2
Crop diversification	14	76	8	2	2.02	3
Farm diversification	7	67	23	3	1.78	4
Change of planting date	10	44	26	20	1.44	5
Income generating activities	3	20	28	49	0.77	6
Agroforestry practice	0	9	56	35	0.74	7

The results of actual adaptation measures being implemented by farmers are presented in Figure 3. The majority of farmers use crop diversification practices such as mixed cropping (41%) and crop rotation (10%). About 12% of the respondents use improved crop varieties (*i.e.*, drought-tolerant and early maturing varieties), while 23% adopted change of planting date. Other identified adaptation practices being implemented are off-farm jobs (6%), composting and mulching (3%), reduction in farm size (3%), and dry season gardening (2%).

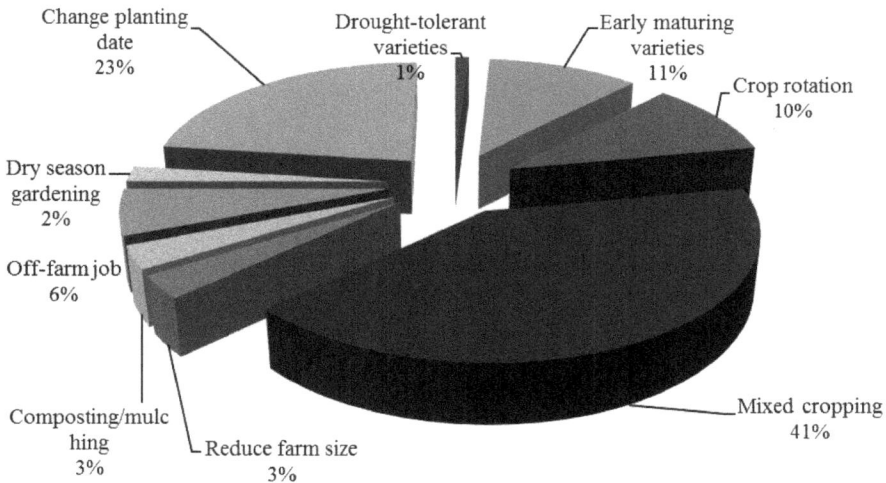

Figure 3. Actual adaptations being used by farmers in the Lawra district of Ghana.

3.5. Perceived Constraints to Adaptation to Climate Change

Results on barriers to use of adaptation practices are presented in Table 5. With a PCI value of 215, unpredictable weather was ranked the most critical impediment to use of adaptation options. High cost of farm inputs, lack of access to timely weather information, and lack of water resources were ranked the second, third, and fourth most pressing problems, respectively. The FGDs showed that farmers' main source of weather information was colleagues who visited peri-urban towns during market days. The focus group discussions also confirmed that farmers considered lack of credit facilities, agricultural subsidies, and poor soil fertility as moderate constraints, while lack of access to agricultural extension officers, agricultural markets, farm labor, and farm size are the lowest constraints.

Table 5. Problems affecting implementation of adaptation practices in the Lawra district of Ghana (Number of respondents = 100).

Constraints to Adoption	Degree of Constraint				PCI	Rank
	High	Moderate	Low	No Problem		
Unpredictable weather	35	48	14	3	215	1
High cost of farm inputs	14	76	8	2	202	2
Lack of access to timely weather information	7	67	23	3	178	3
Lack of access to water resources (e.g., dams)	10	44	26	20	144	4
Lack of access to credit facilities	2	32	11	55	81	5
Lack of access to agricultural subsidies	3	20	28	49	77	6
Poor soil fertility	0	9	56	35	74	7
Limited access to agricultural extension officers	3	19	7	71	54	8
Limited access to agricultural markets	0	0	24	76	24	9
Inadequate farm labor	0	1	19	80	21	10
Limited farm size	0	0	13	87	13	11

4. Discussion

Generally, farmers are aware of climate change, since more than 80% of the surveyed respondents have perceived long-term changes in temperature and precipitation trends. In Sub-Saharan Africa, similar findings have been reported in the Sekyedumase and Wa West districts of Ghana [8,13], Uganda [10], and Senegal [24]. Other studies have also shown that, in the last 100 years, there has been an average global temperature increase of 0.74 °C [25].

Climate change model predictions for the Guinea Savannah Zone of Ghana revealed that the increasing temperature and decreasing precipitation trends will continue [1]. This implies that agricultural stakeholders should identify relevant and applicable adaptation practices to mitigate the effects of the impending change in climatic variables.

The study showed that farmers' perceptions about the causes of climate change are mostly centered on human factors (*i.e.*, deforestation and bushfires) and gods/ancestral curses. Similar findings have been reported in the Wa West district of Ghana [13] and in Northern Nigeria [5].

This study showed that some farmers are already adjusting their farming activities in response to droughts, dry spells, and floods. The FGDs revealed that increased access to agricultural extension officers has impacted positively (67%) on farmers' implementation of adaptation options. Similar findings were reported in Bangladesh, where more than 75% of respondents were using adaptation practices [23]. However, a previous study conducted in the Sekyedumase district of Ghana showed that less than 44% of farmers use adaptation measures due to lack of funds [8].

This study also revealed that farmer-perceived important adaptation practices were different from the actual practices being implemented. Although farmers ranked improved crop varieties (e.g., drought-tolerant and early maturing crops) and irrigation as the most important adaptation strategies, only 14% actually implemented measures in these categories. The majority of respondents (51%) used crop diversification activities (*i.e.*, mixed cropping and crop rotation). Similar findings were reported in Northern Nigeria [5]. Feedback from the group discussions showed that most farmers did not have access to improved crop varieties; hence, they could not implement their most preferred measure. Results of the group discussion showed that farmers are generally aware of the annual recurrent dry spells and droughts. Also, although they view irrigation as the most important solution to these extreme climatic events, they failed to rank it as such. This is because, according to farmers, water resources such as dams and dugouts are very limited in the district. Field observation showed that most of the available water bodies for irrigation are broken down.

Also, in this study, unpredictable weather, high farm input cost, and limited access to timely weather information and water resources were identified as the most critical barriers to adoption. This is likely the case, because in Ghana, the main sources of weather information are television and radio broadcasts. The majority of farmers surveyed did not have electronic gadgets and hence could not readily access weather information. Also, the FGDs revealed that farmers in the Lawra district operate under limited resources due to limited agricultural credit and subsidies. Field observations revealed that the limited number of irrigation facilities (*i.e.*, dams and dugouts) were either broken down or dried out. Similar barriers to adoption have been reported in South Africa [12] and Nigeria [26].

5. Conclusions

With properly tailored policies, smallholder farmers can adjust to climate change and improve their crop production. To do this, climate change policies need to factor in farmers' understanding of the risks they face and potential adaptations to climate change. In this regard, interventions of the Ghanaian government should focus on the development of improved crop varieties and irrigation facilities. More specifically, the development of drought-tolerant crop varieties and the construction of dams and dugouts need to be prioritized in the list of climate change adaptation practices in the Lawra district. Also, there is a need for stakeholders to adhere to proper management and maintenance of existing irrigation facilities.

355

The perception that climate change is also caused by traditional gods and ancestral curses implies that scientists and development experts should consider the cultural and traditional beliefs of farmers when designing adaptation practices. As such, a bottom-up approach must be used to ensure that farmers' beliefs and understanding are a crucial part of the design and dissemination of adaptation practices.

Farmers' access to timely weather information also needs to be prioritized to help farmers in their production decision-making processes (e.g., selection of adaptation options). The Ghana Meteorological Agency and agricultural staff need to be properly trained and resourced to collect, collate, and disseminate accurate weather information timely and widely. Also, the government should boost the capacity of scientists and agricultural staff to develop and promote appropriate and effective technologies to help farmers adapt to climate change. In addition, the prevailing high cost of farm inputs and lack of credit facilities and subsidies require the government to ensure that agricultural loans with flexible terms are made available to farmers to boost their capacity to adapt to the changing climate.

Finally, further research is recommended to assess the feasibility of farm-level adaptation practices to climate change. This will help governments, researchers, non-governmental organizations (NGOs), and farmers to develop and implement adaptation measures that are sustainable, resilient, and reliable.

Acknowledgments: This study was funded by the Kochi University of Technology, Kochi, Japan. The authors gratefully acknowledge the valuable support of the farmers, agricultural officers, and local government staff of the Lawra district of Ghana.

Author Contributions: Francis Ndamani conceived the idea of the study, carried out data collection and data analysis. Tsunemi Watanabe supervised data collection and data analysis. Both authors drafted the manuscript, read and approved the final manuscript.

Conflicts of Interest: The authors declare no conflict of interest.

References

1. USAID. Ghana Climate Change Vulnerability and Adaptation Assessment. Available online: http://www.encapafrica.org/documents/biofor/Climate%20Change%20Assessment_Ghana_%20FINAL.pdf (accessed on 23 April 2015).
2. Jung, G.; Kunstmann, H. High-resolution regional climate modelling for the Volta Basin of West Africa. *J. Geophys. Res.* **2007**, *112*.
3. Laux, P.; Kunstmann, H.; Bárdossy, A. Predicting the regional onset of the rainy season in West Africa. *Int. J. Climatol.* **2008**, *28*, 329–342.
4. Laube, W.; Schraven, B.; Awo, M. Smallholder adaptation to climate change: Dynamics and limits in Northern Ghana. *Clim. Chang.* **2009**, *111*, 753–774.

5. Farauta, B.K.; Egbule, C.L.; Idrisa, Y.L.; Agu, V.C. Farmers' Perceptions of Climate Change and Adaptation Strategies in Northern Nigeria: An Empirical Assessment. Available online: http://www.atpsnet.org/Files/rps15.pdf (accessed on 15 May 2015).

6. Adger, W.N.; Huq, S.; Brown, K.; Conway, D.; Hulme, M. Adaptation to Climate Change in the Developing World. Available online: http://www.glerl.noaa.gov/seagrant/ ClimateChangeWhiteboard/Resources/Uncertainty/climatech/adger03PR.pdf (accessed on 9 May 2015).

7. Intergovernmental Panel on Climate Change (IPCC). *Climate Change 2001: Impacts, Adaptation, and Vulnerability. Intergovernmental Panel on Climate Change*; Cambridge University Press: Cambridge, UK, 2001.

8. Fosu-Mensah, B.Y.; Vlek, P.L.G.; MacCarthy, D.S. Farmers' perceptions and adaptation to climate change: A case study of Sekyeredumase district in Ghana. *Environ. Dev. Sustain.* **2012**, *14*, 495–505.

9. Nhamo, N.; Daniel, M.; Fritz, O.T. Adaptation strategies to climate extremes among smallholder farmers: A case of cropping practices in the Volta Region of Ghana. *Br. J. Appl. Sci. Technol.* **2014**, *4*, 198–213.

10. Okonya, J.S.; Syndikus, K.; Kroschel, J. Farmers' perceptions of and copping strategies to climate change: Evidence from six agro-ecological zones of Uganda. *J. Agric. Sci.* **2013**, *5*.

11. Apata, T.G. Factors Influencing the Perception and Choice of Adaptation Measures to Climate Change among Farmers in Nigeria. Evidence From farm Households in Southwest Nigeria. Available online: http://businessperspectives.org/journals_free/ ee/2011/ee_2011_04_Apata.pdf (accessed on 9 May 2015).

12. Nhemachena, C.; Hassan, R. Micro-Level Analysis of Farmers' Adaptation to Climate Change in Southern Africa. Available online: http://cdm15738.contentdm.oclc. org/utils/getfile/collection/p15738coll2/id/39726/filename/39727.pdf (accessed on 23 May 2015).

13. Kusakari, Y.; Asubonteng, K.O.; Jasaw, G.S.; Dayour, F.; Dzivenu, T.; Lolig, V.; Donkoh, S.A.; Obeng, F.K.; Gandaa, B.; Kranjac-Berisavljevic, G. Farmer-perceived effects of climate change on livelihoods in Wa West District, Upper West Region of Ghana. *J. Disaster Res.* **2014**, *9*, 516–528.

14. Lindsay, P.H.; Norman, D.A. Human Information Processing: An Introduction to Psychology. Available online: http://www.phon.ucl.ac.uk/courses/spsci/audper/ SDT%20Lindsay%20&%20Norman%20App%20B.pdf (accessed on 4 August 2015).

15. Harig, T.; Kaiser, F.G.; Bowler, P.A. Psychological restoration in nature as a positive motivation for ecological behavior. *Environ. Behav.* **2001**, *33*, 590–607.

16. Ministry of Environment, Science and Technology (MEST) of Ghana. Ghana Goes for Green Growth: National Engagement on Climate Change. Available online: http://prod-http-80-800498448.us-east-1.elb.amazonaws.com/w/images/2/ 29/GhanaGreen.pdf (accessed on 4 August 2015).

17. Luni, P.; Maharjan, K.L.; Joshi, N.P. Perceptions and realities of climate change among the Chepong Communities in rural mid-hills of Nepal. *J. Contemp. India Stud. Space Soc.* **2012**, *2*, 35–50.

18. Di Falco, S.; Yesuf, M.; Kohlin, G.; Ringler, C. Estimating the impact of climate change on agriculture in low-income countries: Household level evidence from the Nile Basin, Ethiopia. *Environ. Resour. Econ.* **2012**, *52*, 457–478.

19. Jackson, L.E.; Pascual, U.; Hodgkin, T. Utilizing and conserving agrobiodiversity in agricultural landscapes. *Agric. Ecosyst. Environ.* **2007**, *121*, 196–210.

20. Ndamani, F.; Watanabe, T. Influences of rainfall on crop production and suggestions for adaptation. *Int. J. Agric. Sci.* **2015**, *5*, 367–374.

21. Ghana Statistical Service (GSS). Population by Region, District, Locality of Residence, Age Groups and Sex, 2010. Available online: http://www.statsghana.gov.gh/docfiles/population_by_region_district_locality_of_residence_age_groups_and_sex,_2010.pdf (accessed on 15 May 2015).

22. Uddin, M.N.; Bokelmann, W.; Entsminger, J.S. Factors affecting farmers' adaptation strategies to environmental degradation and climate change effects: A farm level study in Bangladesh. *Climate* **2014**, *2*, 223–241.

23. Devkota, R.P.; Cockfield, G.; Maraseni, T.N. Perceived community-based flood adaptation strategies under climate change in Nepal. *Int. J. Glob. Warm.* **2014**, *6*, 113–124.

24. Mertz, O.; Mbow, C.; Reenberg, A.; Diouf, A. Farmers' perceptions of climate change and agricultural adaptation strategies in rural Sahel. *Environ. Manag.* **2009**, *43*, 804–816.

25. Intergovernmental Panel on Climate Change (IPCC). *Contribution of Working Group II to the Fourth Assessment Report of the Intergovernmental Panel on Climate Change, 2007*; Parry, M.L., Canziani, O.F., Palutik, J.P., Linden, P.J., Hanson, C.E., Eds.; Cambridge University Press: Cambridge, UK, 2007.

26. Kandlinkar, M.; Risbey, J. Agricultural impacts of climate change: If adaptation is the answer, what is the question? *Clim. Chang.* **2000**, *45*, 529–539.

Stressors and Strategies for Managing Urban Water Scarcity: Perspectives from the Field

Vivek Shandas, Rosa Lehman, Kelli L. Larson, Jeremy Bunn and Heejun Chang

Abstract: Largely because water resource planning in the U.S. has been separated from land-use planning, opportunities for explicitly linking planning policies to water availability remain unexamined. The pressing need for better coordination between land-use planning and water management is amplified by changes in the global climate, which will place even greater importance on managing water supplies and demands than in the past. By surveying land and water managers in two urbanizing regions of the western United States—Portland, Oregon and Phoenix Arizona—we assessed the extent to which their perspectives regarding municipal water resource management align or differ. We specifically focus on characterizing how they perceive water scarcity problems (*i.e.*, stressors) and solutions (*i.e.*, strategies). Overall, the results show a general agreement across both regions and professions that long-term drought, population growth, and outdoor water use are the most important stressors to urban water systems. The results of the survey indicated more agreement across cities than across professions with regard to effective strategies, reinforcing the idea that land-use planners and water managers remain divided in their conception of the solutions to urban water management. To conclude, we recommend potential pathways for coordinating the fields of land and water management for urban sustainability.

Reprinted from *Water*. Cite as: Shandas, V.; Lehman, R.; Larson, K.L.; Bunn, J.; Chang, H. Stressors and Strategies for Managing Urban Water Scarcity: Perspectives from the Field. *Water* **2015**, *7*, 6775–6787.

1. Introduction

Despite the breadth and depth of the literatures on water supply and demand management, the operation of these systems have been viewed largely in isolation from land-use planning [1]. Residential development patterns, which in some cases are the direct and intentional outcome of land-use planning initiatives, are a potentially important exogenous influence on water management [2–4]. In fact, largely because water resource planning in the U.S. has traditionally been separated from land-use planning, opportunities to explicitly link planning policies with water availability have gone unexamined [5]. Further, the planning community has neglected the role of water supply, assuming it is readily available for urban (re)development. In short, integrated planning across the land and water sectors is rarely if ever practiced. To better coordinate planning for land and water resources, information about professional perspectives and practices is required. Such information can reveal areas of converging and diverging viewpoints, thereby identifying the potential for collaborations as well as for conflicts as efforts to integrate across sectors continue.

This paper seeks to characterize and compare perceptions among land-use and water-resource planners in two cities of the American West. Using humid Portland, Oregon and arid Phoenix, Arizona as case study samples, we surveyed professionals about various stressors and strategies for their local community water systems. In addition to identifying converging and diverging views, we also examined similarities and differences across the two metropolitan study areas, which may be important in signaling the unique effects of each regional context.

1.1. Pressing Challenges

The pressing need for better coordination between land-use planning and water management is amplified by changes in population and the global climate, which present even greater challenges and uncertainties than in the past. Urban water managers—defined here as those public sector employees responsible for the allocation, administration, and budgetary aspects of water management—in Portland and Phoenix face the complexity of burgeoning growth, with the populations of both urban regions expected to double in coming decades [6,7]. This comes on top of the increasing growth rates of the last decade. Compounding the macro-scale pressures faced in both cities, recent evidence suggests that global climate change will have profound impacts on water resources in the West [8]. The Pacific Northwest (PNW) is experiencing more frequent winter floods and summer warming [9,10], while the American Southwest is currently experiencing a multi-year drought [11].

Long-term climate models project increasing annual temperatures and high precipitation variability (with generally less precipitation in summer) for the PNW [9]. Higher winter temperatures combined with more precipitation will lead to higher snow-line elevations, which in turn will affect the timing of snowmelt and summer flow, ultimately impacting the availability of water resources for the Portland region [12].

With respect to Phoenix, global climate models increasingly conclude the Southwest will likely be warmer and drier in this century than in the last. Regional climate models have recently indicated reductions in surface water runoff in regional watersheds serving metropolitan Phoenix [6]. A report by the National Research Council [13] shows that a warmer and drier future will reduce snowpack, Colorado River flows, and urban water supplies. According to the 5th IPCC assessment report, there is a broad consensus among models that the region will be drier during the 21st Century, and indeed, the transition to a more arid climate is already underway [14]. Although per capita water consumption has decreased since the 1990s in most municipalities in the American west [15–18], the expected combination of reduced water supply and increasing urban populations will likely create an unprecedented state of potential water scarcity.

1.2. Potential Strategies

The process of coordinating water management still involves several policy actions enacted independently and jointly by multiple agents with different goals and objectives, as well as varying levels of influence on outcomes. Water managers have multiple options for adapting to the effects of water supply shortfalls, including taking actions to increase supplies or to decrease demands, both of which can have either short-term or long-term reach. The long-term options, while generally effective in engineering terms, are not always feasible for political and/or financial reasons. Building dams and reservoirs can be cost-prohibitive and requires localities and states to shoulder the burden alone, especially since the U.S. government appears disinterested in funding such projects. The communities designated as hosts often resist large-scale projects. As a result, augmenting supply increasingly does not provide a viable option for the long-term [19]. Rather, U.S. water managers are increasingly forced to use short-term measures that either augment supply or diminish demand [20,21]. These measures include tapping emergency water sources, trucking in water from outside locales, and imposing water-use restrictions on non-essential uses (e.g., lawn irrigation, car washing). We note that such short term measures can prove challenging due to the overall lack of precipitation, and the uncertainties about future rain events.

In the face of new and pressing challenges and a diminished ability to continue adding to the water supply, new solutions must be sought. One area of untapped potential is in collaborating with land-use planners. Earlier studies suggest that the nature of the built environment, landscape treatments, and short-term consumption behavior can also impact residential water use in urban areas [16,22–26]. This is particularly relevant to arid cities, such as Phoenix and Los Angeles, where an estimated 74% of residential water use is for outdoor purposes [27]. Wentz and Gober [22] demonstrated that residential outdoor water use increased with large lots, turf grass (as opposed to drought-tolerant landscaping), and the presence of pools. Similarly, Chang *et al.* [3] found that residential water use was higher in denser older neighborhoods close to downtown than in newer peripheral neighborhoods in the City of Portland, Oregon. House-Peters *et al.* [28] further reported that highly affluent newer neighborhoods had more seasonal water use than denser older low-income neighborhoods in Hillsboro, a suburban city of Portland. Sauri [29] found that demand management policies such as conservation and water pricing may not be sufficient for controlling water consumption levels when low-density urban development continues or income gains occur. In a study of annual residential water use in the Portland metropolitan area, Shandas *et al.,* [23] stressed the importance of investigating behavioral and land-use density barriers that limit the responsiveness of water use to varying climate conditions. Automated timers, for example, generally do not appear to be adjusted in response to weather conditions. The built environment of cities, which is largely controlled by land-use planning, combines with individual and group behaviors to mediate the extent of water consumption [30,31].

1.3. Integration Across Sectors

To address calls for greater integration between water and land managers, many national organizations, professional journals, trade publications, and international conferences are bringing attention to highly inter-connected nature of water and land systems. The National Science Foundation, for example, recently released a call for proposals examining the water–land nexus, and the European Commission is hotly debating the future of this field [32]. Amidst the challenges for integrating water and land systems, little is known about the extent to which the people who are responsible for managing these systems view the same challenges and opportunities. To what degree do land managers agree with water managers in terms of the stressors on the urban water system? What are the similarities and differences between the two groups in terms of strategies for addressing future changes in water supply? These and other questions help to frame a critical issue in the management of water supplies because, arguably, if these two constituencies do not view the problem as the same, then their solutions will also face formidable challenges to implement.

By surveying both land and water managers in two urbanizing regions of the western United States, we sought to assess the extent to which their perspectives align or differ. We specifically focused on their perceptions of the problem (e.g., stressors) and potential solutions (e.g., strategies). While previous studies have looked at single regions [31,33] we sought to examine two regions—Phoenix, Arizona, and Portland, Oregon—that have different land-use policies, climate regimes, and population growth trends. By engaging water managers and land-use planners in these regions, we were able to address two research questions: (1) To what extent do urban water use managers and land-use planners share perspectives regarding the stressors to and strategies for water management both within and between the two locations and professional fields? (2) What factors might help to explain any observed differences and similarities in perspectives about the water management system?

2. Materials and Methods

Portland and Phoenix share the dual challenges of rapid growth and climatic uncertainty. The challenges however occur under different physical geographies, political cultures, and growth-management and land-use policies. These differences allowed us to study perceptions of land-use planners and water-resource managers in two distinct areas, by comparing factors that may exacerbate or mitigate future water shortages, and strategies for ameliorating potential water challenges.

2.1. Survey Design

We used a cross-sectional survey design that was informed by existing literature on the challenges facing the regions [31,34]. An online survey of municipal water managers and land-use planning professionals in the Phoenix and Portland regions constituted the primary vehicle for the empirical assessment, resulting in four samples: Phoenix water managers, Phoenix land planners, Portland water managers, and Portland land planners. The survey was administered in the spring of 2010 to municipal professionals in both regions—32 in metropolitan Phoenix and 25 in the Portland area. By identifying all the municipal land use and water professionals in both regions, we identified nearly all of the potential recipients of the survey. For land planners, our priority was to survey planning directors/managers or the closest equivalent for each city within our study regions. If none could be found, the city/town manager was contacted instead. For water managers, our priority was to target water resource managers/directors or the closest equivalent. While the small number of municipalities in each region did not allow us to conduct statistical tests of inference, they did provide a descriptive assessment of the water scarcity strategies employed in each region, and patterns among survey respondents, comparing the land and water managers and the two study regions.

To address our research questions, we developed survey questions that consisted of two main points—stressors and strategies—with respondents rating a set of options on a scale of one to ten (from 1 = not at all significant to 10 = greatly significant). Twelve stressors and 13 strategies were developed for both potential risk factors and solutions based on a review of literature and discussions with professionals and among our research team. Following from Larson and colleagues [35,36] these items present a variety of natural and anthropogenic risk factors as well as potential voluntary and regulatory-based strategies in both the land and water sectors. For each of the possible factors representing a stressor, respondents were asked, "To what extent are each of the following factors important when considering the future of water supply." For each factor that represented a strategy, respondents were asked, "To what extent are each of the following factors important when considering strategies for addressing shortages in future water supply." Respondents had 12 stressors and 13 strategies to consider, relating to three sources of influence: environment and population, infrastructure and management, and end-user behavior, as shown in Tables 1 and 2. The three categories correspond to the generally accepted challenges facing water resource management, including a growing population and environment, combined with strategies facing the future, such as climate change, technological applications, and human or individual actions. We distinguish two aspects of climate change, namely natural climate variability and human-induced climate change in part because American public opinion continues to be divided about the causes for changes in climate conditions.

Table 1. Stressors listed for consideration in survey with descriptive statistics (Minimum, Maximum, Median, Standard Deviation).

Category of Stressor	Stressor	Min.	Max.	Med.	SD
Environment and Population	Population growth	1	10	7	2.89
	Long-term drought	1	10	7	3.08
	Natural climate variability	1	10	5	2.88
	Human-induced climate changes	1	10	4	2.68
Infrastructure and Management	Inadequate access to water sources	1	10	4	3.02
	Infrastructure to store, treat, deliver water	1	10	4	2.69
	Land-use planning or development	1	10	5	2.46
	Water-resource planning or management	1	10	5	2.66
End-User Behavior	Indoor water uses	1	8	4	2.06
	Outdoor water uses	1	10	6	2.57
	Household or residential water use	1	10	5	2.36
	Industrial or other business water use	1	10	5	2.52

Table 2. Strategies listed for consideration in survey with descriptive statistics (Minimum, Maximum, Median, Standard Deviation).

Category of Strategy	Strategy	Min.	Max.	Med.	SD
Environment and Population	Planning for future climate changes	1	10	5	2.74
	Restricting new building permits	1	10	4	2.56
	Limiting new growth or development	1	10	6	2.97
Infrastructure and Management	More compact or dense communities	1	10	5	3.08
	Retiring agricultural land	1	10	4	2.90
	Acquiring new sources of water	1	10	8	2.73
	Building structures to store water	1	10	4	2.94
	Upgrading water delivery infrastructure	1	10	7	2.53
	Wholesale contracts with other providers	1	10	5	2.93
End-User Behavior	Restrictions or bans to limit water use	1	10	6	2.50
	Increasing the price of water	1	10	7	2.96
	Water conservation education	1	10	8	2.49
	New water efficiency technology	1	10	8	2.53

2.2. Methods of Analysis

Due to the small sample representing diverse municipalities in each region, we developed analytical techniques that emphasize the descriptive and qualitative patterns in the data. These included graphic depictions of the data, information from a series of workshops, and where available, using qualitative responses to flesh out responses to the survey. Specifically, for the purposes of data interpretation, we considered a rating of 7 or above to indicate the respondent considered the strategy or stressor to be important or very important. By looking only at those responses with a rating of 7 or higher we can effectively parse out the differences between the two professions and the two regions using descriptive means.

Across both regions and sectors, 57 surveys were completed and returned for a response rate of 50%—some municipalities had more than one respondent. For Portland, both surveys were returned from 7 towns, either the land or water survey was returned by 10 towns, and 7 towns did not respond at all. For Phoenix, both surveys were returned by 8 towns, one or the other was returned by 17 towns, and 7 towns were not represented in this study. While the response rate was similar across regions, land planners responded at a higher rate. For the water survey, respondents included Public Works Directors, City Engineers, Environmental Managers, and a Town Manager. The respondents from the land-use survey were mostly planners, but also included City Administrators and Community Development Directors from smaller towns. As a whole, the surveys represent a variety of towns across the study regions, ranging from small to large. In accordance with policies governing the ethical treatment of research participants, the survey results are presented anonymously.

3. Results and Discussion

Overall, the results of the surveys show significantly more agreement in identifying the most significant stressors—of which only two stood out—whereas there were many strategies identified as important and a wider range of opinion. There was also somewhat more agreement across cities than across professions, particularly for strategies.

3.1. Stressors

Three stressors emerged as the most widely perceived as important in both cities and by both professions. These were, in ranked order, long term drought, population growth, and outdoor water use (Figure 1). The remaining possible stressors were not viewed as important by the majority of respondents, and there was general consensus that indoor water use is the least important stressor. Phoenix professionals were generally more focused on long-term drought than Portland professionals; 83% of Phoenix water managers identified this as an important contributor to water stress, and 62% of Phoenix land-use planners agreed (Figure 2). Two most widely identified stressors were related to the environment and population, while the third—outdoor water use—was an end-user behavior. Notably, professionals in both locations rated human-induced climate change low—and water professionals rated it very low—as a cause of stress, despite the direct connection between climate change and summer drought in both areas that is projected by current climate models [11,12,37]. While we are not able to discern the reasons for the low rating to human-induced climate change, we note that the term 'human-induced' may reflect the discomfort with the causal connection or attribution of drought events to human activities. These results would be consistent with the earlier point regarding the fact that U.S. public opinion is still divided regarding the causes for climate change.

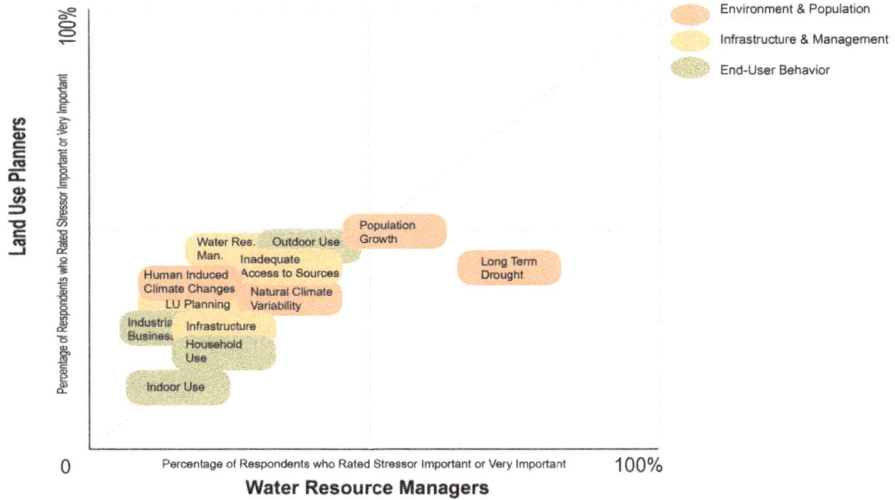

Figure 1. Water Resource Stressors rated at least 7/10 on a scale of importance, by profession.

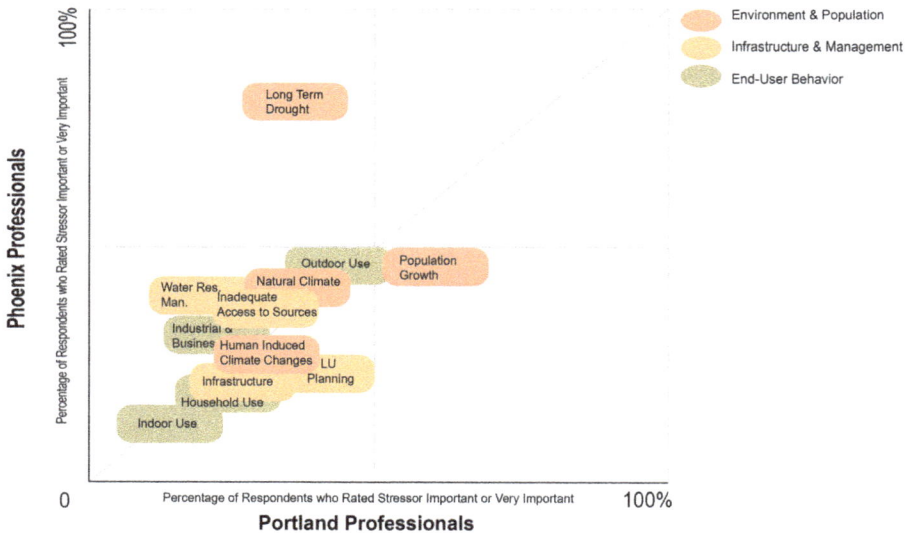

Figure 2. Water Resource Stressors rated at least 7/10 on a scale of importance, by region.

3.2. Strategies

Across both regions and professions, the results suggest greater variation in perspective on the effectiveness of potential strategies than stressors. Between both professions, four strategies were identified as somewhat effective by over 50% of respondents (as compared to only two stressors): water efficiency technology, water conservation education, acquiring new sources of water, and limiting new growth (Figure 3). Although the two biggest stressors are in the environment and population category, namely long-term drought and population growth, the corresponding strategies that directly addressed these stressors were generally not among the highest ranked.

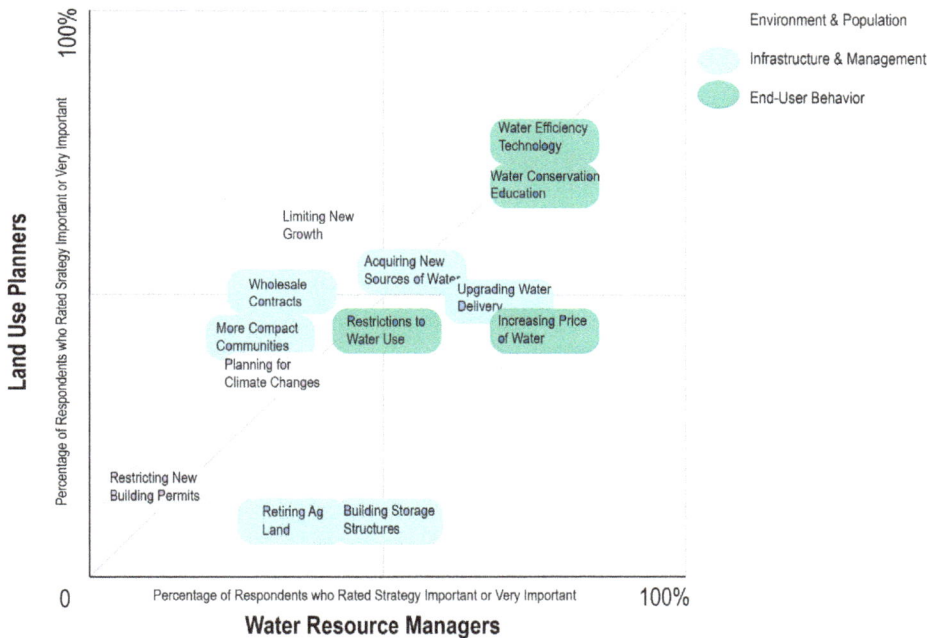

Figure 3. Water Resource Strategies rated at least 7/10 on a scale of importance, by profession.

The strategies relating to end-user behavior generated the most support. Specifically, new water-efficient technology had the most support and least variability of all the strategies in the survey across all categories of respondents. Among water managers, there was consensus that increasing the price of water would be effective, in Portland 70% of water managers ranked it 7/10 or above, and in Phoenix, almost 60% of water managers ranked it 10/10 (Figure 4). Water conservation education was also a highly ranked strategy, although land planners lagged water managers in their enthusiasm for its effectiveness. Outright restrictions to ban or limit water-use was split in among the Portland respondents and were ranked moderately in Phoenix. Although inadequate access to water sources was not rated highly as a cause of stress, there was general agreement in Phoenix that acquiring new sources of water would be a highly effective strategy: half of Phoenix water managers ranked this a 10/10. We note that while many of these strategies were supported, recent research from California suggest that applications of water efficient technology, increasing the prices of water and education [38], may not be highly effective.

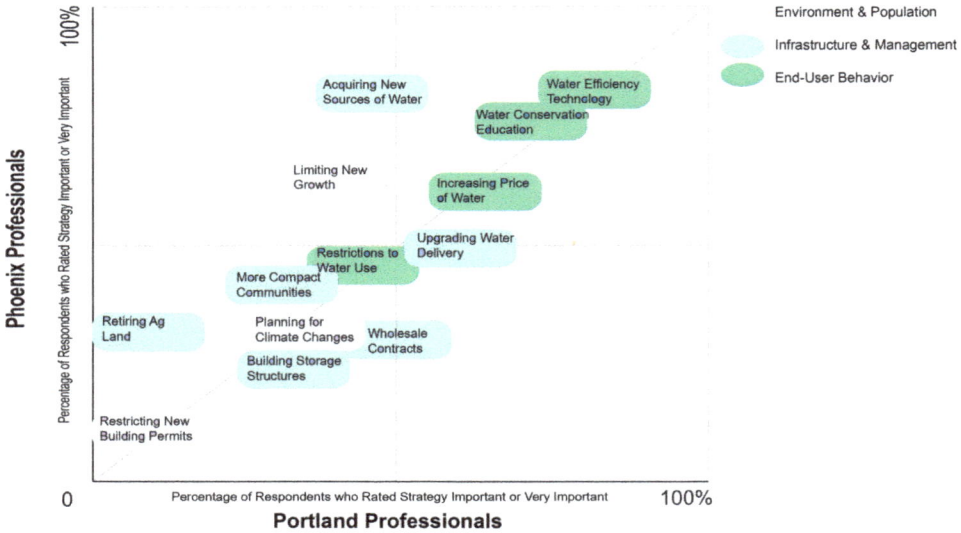

Figure 4. Water Resource Strategies rated at least 7/10 on a scale of importance, by region.

Generally, land-use planners were relatively more supportive than water managers of strategies in their field, such as limiting new growth, and more dense and compact communities. Similarly, water managers were more favorable towards upgrading water delivery systems and increasing the price of water. The main regional difference is that professionals in Phoenix were more likely to support acquiring new sources of water and limiting new growth than those in Portland. Generally, restricting new building permits was not identified but by a few land-use planners and water managers, while limiting new growth was somewhat more highly valued, particularly in Phoenix where a clear majority (75%) of water managers rated it 7/10 or above in effectiveness.

4. Discussion

Despite the sample size of the survey, we observed that across region and profession, survey respondents widely agreed on the sources of stress to urban water systems, while opinions on strategies were much less uniform and a wider array of strategies was identified as effective. The fact that only two stressors were identified by 50% or more of respondents as important while many strategies emerged could suggest that while stress to water resources comes from one or two causes, the solutions are less clear. This is logical because a suite of strategies for each specific region will be necessary, rather than a single silver bullet that fits every region. Alternatively, these results could suggest that the research team more successfully presented options for stressors with which practitioners agreed, while failing to include other important stressors.

While the top two stressors were related to environment and population, the majority of highly ranked strategies were related to end-user behavior. While understandable due to the challenge in the physical and political feasibility of other strategies, research by Lach and colleagues suggest that water managers are highly conservative when attempting to regulate end-user behavior [39]. Accordingly, while a strategy such as acquiring new sources of water might be highly effective, it is not worth consideration if there are no such new sources available or if this strategy would create negative environmental and social outcomes (e.g., on source ecosystems or communities from which water is withdrawn). Further, experiences in California suggest that water agencies are required to balance the cost of water to match what is required to address the demand, and the raising water rates must occur within limits. Local government bodies are much more likely to have the ability to enact policies related to end-use. A more positive finding is that the third most highly rated stressor was outdoor water use, which is highly likely to be impacted by land-use decisions, as described in Section 1.2. Indeed, water provides in both regions have introduced incentives for water conservation through landscape manipulation [40].

Distinctive perspectives were identified along both professional and regional lines. The biggest professional division relates to limiting new growth, which was much more highly ranked by land-use planners, and increasing the price of water, which was much more highly ranked by water managers. This is perhaps due to familiarity level—in that planners use growth measures (e.g., urban growth boundary in Portland) to reduce the physical growth of the urban area—while water manger in Phoenix use instruments such a pricing to mediate demand. The largest regional difference was that Phoenix professionals are much more worried about long-term drought, which is logical as their already arid region was in the midst of a significant drought period at the time of the survey. Professionals in Phoenix are also more optimistic about acquiring new sources of water as a solution.

The practicality of some of the highest ranked solutions is not yet clear. The two top strategies in both regions and professions are education and new technology. While these are important tools, hoped for technological solutions to environmental problems do not always materialize and often create negative, unintended consequences (e.g., dams, canalized water transfer systems, excessive ground water pumping). Conversely, some of the most aggressive strategies were not as highly ranked, perhaps because of perceived political feasibility issues. In particular, raising water prices and limiting growth may not be seen as popular choices by elected officials. Although it is important to remain critical of the practicality and feasibility of any solution, innovative ideas are already being generated, such as tiered water pricing, which was recently implemented in Santa Fe [41] This can also be seen in significant opposition to increasing the price of water in an earlier Phoenix-based survey, which demonstrated that planning professionals opposed those strategies more so than scientists as well as residents [42].

The practitioners' approach to global climate change is of particular importance for the long-term health of the water systems. Human-induced climate change was identified as an important stressor by a minority of water managers and land-use planners. Despite the fact that the two most commonly identified stressors are greatly exacerbated by climate change—long-term drought in Phoenix, and population growth in Portland—climate still is not considered unequivocally as the primary factor in addressing urban water scarcity. The related strategies showed slightly higher interest levels: planning for future climate changes was identified as an effective strategy by a third of Phoenix water and land practitioners, a third of Portland water managers, and almost half of Portland land-use planners.

The lack of recognition of climate change as an important stressor, and only moderate interest in planning for future climate changes as a strategy suggests a serious problem in resource management decision-making in Portland and Phoenix. The professionals most concerned with climate change are land-use planners in the Portland area, suggesting that they are a group to watch and may provide a model. This further suggests that there could be potential opportunities to use land planning to adapt to climate change as it relates to urban water resource management [4,23,24]. The recent experience of California's water management challenges, and the emerging field of 'climate attribution science' may offer many opportunities for further research into the linkages among climate, land use and water recourses planning.

To that end, four plausible directions emerge as a way to address the challenges facing water use in the midst of climate change. First, the gap must be closed between academic and federal concerns over climate change and the lack of concern reported by on the ground practitioners. Meeting mandates to provide water in the short run seems to be central to the administration of urban water resources. Yet, if the frequency of drought and magnitude of droughts are expected to increase in the future as projected by several studies [12,37,40] then engaging practitioners—both land-use planners and water managers—in developing climate sensitive strategies can provide effective guidance to reduce short and long term impacts. Second, a realistic understanding of the scale of the issue is required. Quantification, even as an estimate would help provide a realistic understanding of the improvements in conservation education and technology that would be necessary to offset the large-scale environmental and population challenges that are already emerging. Third, while the communities we surveyed had a high level of consensus regarding the problem (*i.e.*, stressors) and on the top strategies, assessing the effectiveness of those strategies is an area that requires further research and consideration. The consensus itself is a strong foundation for advancing greater coordination between the two fields, and provides an opportunity to build further cross-sector integration.

Finally, the continued rapid population growth that is forecasted for both study areas presents an opportunity to use land-use planning tools to lessen future stresses on urban water systems. The opportunity for linking existing land-use, housing, and transportation models that are common place in urban areas, with water demand models can help in cross-sector communication, particularly early in the land-use planning process in growing cities [43]. Integrating such models with downscaled climate system can also help to inform communities about the potential to greatly impact the stressor and strategies for managing urban water scarcity.

5. Conclusions

We set out to answer two questions. First, to what extent do urban water use managers and land-use planners share perspectives regarding the stressors to and strategies for water management systems both within and between the two locations and professional fields? Overall, survey responses were mostly consistent for the stressors (*i.e.*, problems) facing water resource management, while responses to the effectiveness of strategies were more varied by both region and profession. Professionals expressed the greatest concern about the impact of environmental and population stressors, while strategies related to end-user behavior were found to be the most effective, and there was general consensus that improved water efficiency technology and water conservation education would be effective strategies. Concomitant concerns about long-term drought seems to be pervasive among the respondents from Phoenix, perhaps since the region continues to undergo severe water shortages along with much of the Southwest. The general lack of concern about climate pressures on water resources by Portland respondents is also note worthy, and may suggest a need to further examine why climate concerns may not be central to current water management strategies.

The second research question was: what factors might help to explain any observed differences and similarities in perspectives about the water management system? The general consensus was that environmental and population factors are stressors, but that related strategies are not as effective, likely because these influences are difficult or impossible to control at the level of local government. Conversely, local government bodies may have greater ability and some certainty to influence end-user behavior, which is reflected in these strategies' high ratings. There were categorical divisions in assessing one stressor (long-term drought) and variation in the perceptions of many strategies, notably source acquisition and limiting new growth. The clear consensus that drought is a greater stressor in Phoenix than Portland is likely due to that region's significantly more arid climate with ongoing drought and to the projected impacts of climate change across the southwest. The professional divide between water managers and land planners concerning limiting new growth (favored by land-use planners) and increasing the price of water (favored by water resource managers) might be explained by their differential levels of involvement and knowledge of related programs. Fostering greater interaction between these two groups may be a timely, cost-effective, and prudent means of improving the opportunities for more effective management of this scarce resource.

Taken together, the present paper provides one perspective into the state of water resource management in two geographically diverse urban areas of the U.S. However, the local politics of water management is complex and is often based on accepted behaviors and organizational cultures [38]. The pressing challenges of climate change and precipitation uncertainty pose formidable challenges to traditional approaches. While our approach could not address the reason for specific responses, the summary of responses suggest aligned and divergence regarding the stressors and strategies by region and profession. Now more than ever we need research that articulates how current challenges to water resource management are created, in part, through the social systems we have created and enabled. Our research is one step in that direction, but additional research is needed to assess the extent to which administrative fragmentation of natural resource management is creating water scarcity. The future will require at once the need for collaboration among the diverse groups of interest managing water and land resources, and the sharing of effective strategies across jurisdictions.

Acknowledgments: Financial assistance for this paper comes from the Sector Applications Research Program (SARP), which is part of the Climate Program Office of the U.S. Department of Commerce, National Oceanic and Atmospheric Administration (NOAA) pursuant to NOAA Award No. NA09OAR4310140. In addition, the material is based upon work supported by the National Science Foundation under Grant No. SES-0951366, DMUU: Decision Center for a Desert City II: Urban Climate Adaptation; and Grant No. SES-1462086. Additional financial support was provided by the Institute for Sustainable Solution at Portland State University

(James F. and Marion L. Miller Foundation sustainability grant). The statements, findings, conclusions, and recommendations expressed in this material are those of the research team and do not necessarily reflect the views of NOAA, U.S. Department of Commerce, the National Science Foundation (NSF) or the U.S. Government.

Author Contributions: Vivek Shandas, Kelli L. Larson, and Heejun Chang designed the research, including the development of survey questionnaire. Vivek Shandas and Kelli L. Larson implemented the survey. Rosa Lehman and Jeremy Bunn compiled survey questionnaire and analyzed the data. Vivek Shandas, Rosa Lehman, Kelli L. Larson, Heejun Chang and Jeremy Bunn wrote the manuscript.

Conflicts of Interest: The authors declare no conflict of interest.

References

1. Gober, P.; Larson, K.; Quay, R.; Polsky, C.; Chang, H.; Shandas, V. Why Land Planners and Water Managers Don't Talk to One Another and Why They Should. *Soc. Nat. Resour.* **2013**, *26*, 356–364.

2. Shandas, V.; Parandvash, G.H. Integrating Urban Form and Demographics in Water Demand Management: An Empirical Case Study of Portland Oregon (US). *Environ. Plan. B: Plan. Des.* **2010**, *37*, 112–128.

3. Chang, H.; Parandvash, G.H.; Shandas, V. Spatial Variations of Single Family Residential Water Use in Portland, Oregon. *Urban Geogr.* **2010**, *31*, 953–972.

4. Chang, H.; House-Peters, L. Cities as place for climate mitigation and adaptation: A case study of Portland, Oregon, USA. *J. Korean Geogr. Soc.* **2010**, *45*, 49–74.

5. Macleod, C.J.A.; Blackstock, K.L.; Haygarth, P.M. Mechanisms to improve integrative research at the science-policy interface for sustainable catchment management. *Ecol. Soc.* **2008**, *13*, 48.

6. Gober, P.; Kirkwood, C.W. Climate Change and Water in Southwestern North America Special Feature: Vulnerability assessment of climate-induced water shortage in Phoenix. *Proc. Natl. Acad. Sci. USA* **2010**, *107*, 21289–21294.

7. Hoyer, W.; Chang, H. Development of Future Land Cover Change Scenarios in the Metropolitan Fringe, Oregon, U.S., with Stakeholder Involvement. *Land* **2014**, *3*, 322–341.

8. Georgakakos, A.; Fleming, P.; Dettinger, M.; Peters-Lidard, C.; Terese, T.C.; Richmond, K.; Reckhow, K.; White; Yates, D. Ch. 3: Water Resources. In *Climate Change Impacts in the United States: The Third National Climate Assessment*; Melillo, J.M., Richmond, T.C., Yohe, G.W., Eds.; U.S. Global Change Research Program: Washington, DC, USA, 2014.

9. Mote, P.W.; Salathé, E.P., Jr. Future climate in the Pacific Northwest. *Clim. Chang.* **2010**, *102*, 29–50.

10. Chang, H.; Jung, I.-W.; Steele, M.; Garnett, M. Spatial Patterns of March and September streamflow trends in Pacific Northwest Streams. *Geogr. Anal.* **2012**, *44*, 177–201.

11. MacDonald, G.M. Water, climate change and sustainability in the southwest. *Proc. Natl. Acad. Sci. USA* **2010**, *107*, 21256–21262.

12. Jung, I.-W.; Chang, H. Assessment of future runoff trends under multiple climate change scenarios in the Willamette River Basin, Oregon, USA. *Hydrol. Process.* **2011**, *25*, 258–277.

13. National Research Council. *Colorado River Basin Water Management: Evaluating and Adjusting to Hydroclimatic Variability*; National Academies Press: Washington, DC, USA, 2007; p. 210.

14. Seager, R.; Ting, M.; Li, C.; Naik, N.; Cook, B.; Nakamura, J.; Liu, H. Projections of declining surface-water availability for the southwestern United States. *Nat. Clim. Chang.* **2013**, *3*, 482–486.

15. Gutzler, D.S.; Nims, J.S. Interannual variability of water consumption and summer climate in Albuquerque, New Mexico. *J. Appl. Meteorol.* **2005**, *44*, 1777–1787.

16. Breyer, B.; Chang, H.; Parandvash, H. Land use, temperature and single family residential water use patterns in Portland, Oregon and Phoenix, Arizona. *Appl. Geogr.* **2012**, *35*, 142–151.

17. Chang, H.; Praskievicz, S.; Parandvash, H. Sensitivity of urban water consumption to weather climate variability at multiple temporal scales: The case of Portland. *Int. J. Geospatial Environ. Res.* **2014**, *1*, 7.

18. Mini, C.; Hogue, T.; Pincetl, S. The effectiveness of water conservation measures on summer residential water use in Los Angeles, California. *Landsc. Urban Plan.* **2015**, *94*, 134–146.

19. Platt, R.H. The 2020 Water Supply Study for Metropolitan Boston: The Demise of Diversion. *J. Am. Plan. Assoc.* **1995**, *61*, 185–200.

20. Elliott, C.; Knight, J.; White, G.F. Domestic Water supply: Right or Good? In *Ciba Foundation Symposium on Human Rights in Health*; John Wiley & Sons, Ltd.: Chichester, UK, 1974.

21. Wescoat, J.L., Jr.; White, G.F. *Water for Life: Water Management and Environmental Policy*; Cambridge University Press: Cambridge, UK, 2003.

22. Wentz, E.A.; Gober, P. Factors Influencing Water Consumption for the City of Phoenix, Arizona. *Water Resour. Manag.* **2007**, *21*, 1849–1863.

23. Shandas, V.; Rao, M.; McGrath, M.M. The implications of climate change on residential water use: A micro-scale analysis of Portland (OR). *J. Clim. Water* **2012**, *3*, 225–238.

24. House-Peters, L.; Chang, H. Modeling the Impact of Land Use and Climate Change on Neighborhood-scale Evaporation and Nighttime Cooling: A Surface Energy Balance Approach. *Landsc. Urban Plan.* **2011**, *103*, 139–155.

25. Runfola, D.M.; Polsky, C.; Nicolson, C.; Giner, N.; Pontius, R.G., Jr.; Decatur, A. A growing concern? Examining the influence of lawn size on residential water use in suburban Boston, MA, USA. *Landsc. Urban Plan.* **2013**, *119*, 113–123.

26. Halper, E.B.; Dall'erba, S.; Rosalind, H.B.; Scott, C.A.; Yool, S.R. Effects of irrigated parks on outdoor residential water use in a semi-arid city. *Landsc. Urban Plan.* **2015**, *134*, 210–220.

27. Mayer, P.W.; DeOreo, W.B.; Opitz, E.M.; Kiefer, J.C.; Davis, W.Y.; Dziegielewski, B.; Nelson, J.O. *Residential End Uses of Water. Final Report*; AWWA Research Foundation: Denver, CO, USA, 1999.

28. House-Peters, L.; Pratt, B.; Chang, H. Effects of urban spatial structure, socio-demographics, and climate on residential water consumption in Hillsboro, Oregon. *J. Am. Water Resour. Assoc.* **2010**, *46*, 461–472.

29. Saurí, D. Lights and shadows of urban water demand management: The case of the metropolitan region of Barcelona. *Eur. Plan. Stud.* **2003**, *11*, 229–243.

30. Ouyang, Y.; Wentz, E.A.; Ruddell, B.L.; Harlan, S.L. A Multi-Scale Analysis of Single-Family Residential Water Use in the Phoenix Metropolitan Area. *J. Am. Water Resour. Assoc.* **2013**, *50*, 448–467.

31. Hong, C.Y.; Chang, H. Uncovering the Influence of Household Sociodemographic and Behavioral Characteristics on Summer Water Consumption in the Portland Metropolitan Area. *Int. J. Geospatial Environ. Res.* **2014**, *1*, 2.

32. European Environment Agency. *Water Resources across Europe—Confronting Water Scarcity and Drought*; Technical Report TH-AL-09-002-EN-C; European Environment Agency, Office for Official Publications of the European Communities: Copenhagen, Denmark, 2009; pp. 1725–9711.

33. Wentz, E.A.; Wills, A.J.; Kim, W.K.; Myint, S.W.; Gober, P.; Balling R.C., Jr. Factors Influencing Water Consumption in Multifamily Housing in Tempe, Arizona. *Prof. Geogr.* **2014**, *66*.

34. Larson, K.L.; Polsky, C.; Gober, P.; Chang, H.; Shandas, V. Vulnerability of water systems to the effects of climate change and urbanization: A comparison of Phoenix, Arizona and Portland, Oregon (USA). *Environ. Manag.* **2013**, *52*, 179–195.

35. Larson, K.L. An integrated theoretical approach to understanding the sociocultural basis of multidimensional environmental attitudes. *Soc. Nat. Resour.* **2010**, *23*, 898–907.

36. Larson, K.L.; White, D.; Gober, P.; Harlan, S.; Wutich, A. Divergent perspectives on water resource sustainability in a public-policy-science context. *Environ. Sci. Policy* **2009**, *12*, 1012–1023.

37. Seyranian, V.; Sinatra, G.M.; Polikoff, M.S. Comparing communication strategies for reducing residential water consumption. *J. Environ. Psychol.* **2015**, *41*, 81–90.

38. Vano, J.A.; Nijssen, B.; Lettenmaier, D.P. Seasonal hydrologic responses to climate change in the Pacific Northwest. *Water Resour. Res.* **2015**, *51*, 1959–1976.

39. Lach, D.; Ingram, H.; Rayner, S. Maintaining the status quo: How institutional norms and practices create conservative water organizations. *Texas Law Rev.* **2014**, *83*, 2027.

40. Breyer, B.; Chang, H. Urban water consumption and weather variation in the Portland, Oregon metropolitan area. *Urban Clim.* **2014**, *9*, 1–18.

41. National Public Radio, Santa Fe Cuts Water Consumption by Imposing Tiered Pricing Model. Available online: http://www.npr.org/2015/05/13/406505133/santa-fe-cuts-water-consumption-by-imposing-tiered-pricing-model (accessed on 15 June 2015).

42. Wilder, M.; Scott, C.; Pablos, N.P.; Varady, R.; Garfin, G.M.; McEvoy, J. Adapting Across Boundaries: Climate Change, Social Learning, and Resilience in the U.S-Mexico Border Region. *Ann. Assoc. Am. Geogr.* **2010**, *100*, 917–928.

43. Bouziotas, D.; Rozos, E.; Makropoulos, C. Water and the city: Exploring links between urban growth and water demand management. *J. Hydroinform.* **2015**, *17*, 176–192.

MDPI AG

St. Alban-Anlage 66

4052 Basel, Switzerland

Tel. +41 61 683 77 34

Fax +41 61 302 89 18

http://www.mdpi.com

Water Editorial Office

E-mail: water@mdpi.com

http://www.mdpi.com/journal/water

www.ingramcontent.com/pod-product-compliance
Lightning Source LLC
Chambersburg PA
CBHW051925190326
41458CB00026B/6407